Metal
Ecotoxicology

Metal
Ecotoxicology
Concepts & Applications

Edited by
Michael C. Newman
Alan W. McIntosh

 LEWIS PUBLISHERS

Library of Congress Cataloging-in-Publication Data

Metal ecotoxicology: concepts and applications/edited by Michael C.
  Newman and Alan W. McIntosh
    Materials developed for a special session of the International
Conference on Metals in Soils, Waters, Plants, and Animals, held in
Orlando, Fla., Apr. 30-May 3, 1990.
    Includes bibliographical references and index.
    ISBN 0-87371-411-3
    1. Metals--Environmental aspects--Congresses.   2. Metals-
-Toxicology--Congresses.   3. Pollution--Environmental aspects-
-Congresses.   I. Newman, Michael C.   II. McIntosh, Alan W.
III. International Conference on Metals in Soils, Waters, Plants,
and Animals (1990: Orlando, Fla.)
QH545.M45M46   1991
574.5'222--dc20   91-13943
                                                                  CIP

LEWIS PUBLISHERS, INC.
121 South Main Street, P.O. Drawer 519, Chelsea, Michigan
48118

PRINTED IN THE UNITED STATES OF AMERICA
1 2 3 4 5 6 7 8 9 0

ADVANCES IN TRACE SUBSTANCES RESEARCH

Series Preface

The need to synthesize, critically analyze, and put into perspective the ever-mounting body of information on trace chemicals in the environment provided the impetus for the creation of this series. In addition to examining the fate, behavior, and transport of these substances, the transfer into the food chain and risk assessment to the consumers, including humans, will also be taken into account. It is hoped then that this information will be user-friendly to students, researchers, regulators, and administrators.

The series will have "topical" volumes to address more specific issues as well as volumes with heterogeneous topics for a quicker dissemination. It will have international scope and will cover issues involving natural and anthropogenic sources in both the aquatic and terrestrial ecosystems. To ensure a high quality publication, volume editors and the editorial board will subject each article to peer review.

Thus, **Advances in Trace Substances Research** should provide a forum where experts can discuss contemporary environmental issues dealing with trace chemicals that, hopefully, can lead to solutions resulting in a cleaner and healthier environment.

Domy C. Adriano
Editor-in-Chief

Preface

Our lack of understanding of the fate and effects of metals in the environment has had profound consequences for both human health and ecosystems. Outbreaks of Minamata and Itai-itai disease during the 1950s are stark examples. More recent concerns include those associated with acid precipitation and the presence of organotin complexes contributed by anti-fouling paints. Various adverse environmental and human health effects have occurred, partly because our understanding of metal interactions with abiotic or biotic components of ecosystems is incomplete or inaccurate. It is the purpose of this volume to foster this understanding by presenting current reviews or applications of ecotoxicological concepts, and to do so at all levels of ecological organization.

The topics included in this volume would not be universally accepted as addressing ecotoxicology. Occasionally, the term ecotoxicology has been reserved for work at the community level or higher. Such limitations are not subscribed to herein. Few ecologists would disagree that progress in ecology would have been slowed by exclusion of all but community and systems level research. It seems illogical to assume that growth in this new field of ecology would not be similarly compromised by such a restriction.

The thirteen chapters are arranged in a hierarchical fashion. Chapter 1 and sections of other chapters such as Chapter 8 focus on metals in abiotic components of ecosystems. Chapters 2 through 11 deal primarily with autecology, effects of metals relative to the individual or a single species. The remaining chapters and a large portion of Chapter 10 address metals in marine and freshwater systems in the context of synecology (species associated and interacting as a unit).

Michael C. Newman
Alan W. McIntosh

The Editors

Michael C. Newman is an Associate Research Ecologist at the University of Georgia's Savannah River Ecology Laboratory. He received his Ph.D. (1981) and M.S. degrees in Environmental Sciences from Rutgers University. Earlier he earned M.S. (Zoology) and B.A. (Biological Sciences) degrees from the University of Connecticut.

Dr. Newman's interests include aquatic toxicology and applied aquatic sciences. Factors modifying the accumulation kinetics and toxicokinetics of inorganic contaminants have been the focus of his research during the last decade. Currently, Dr. Newman and coworkers are investigating and modeling population level responses to toxicant-induced stress.

Also active in teaching, he has taught at the University of Connecticut, Rutgers University, the University of California—San Diego, the University of South Carolina, and the University of Georgia.

Alan W. McIntosh received his Ph.D. in Limnology at Michigan State University in 1972. Following several years with the Environmental Health faculty at Purdue University, he joined the Environmental Sciences Department at Rutgers University and served as Director of the Water Resources Research Center there until 1989. In December of that year, he became Director of the Vermont Water Resources and Lake Studies Center and an Associate Professor of Natural Resources at the University of Vermont.

His professional interests include the fate and effects of trace contaminants in fresh waters and the use of stream biota in water quality assessment programs. He currently is active in ongoing research programs on Lake Champlain. He teaches courses in toxic substances and applied limnology at the University of Vermont.

Acknowledgments

This volume contains materials developed for the special session, "The Eco-toxicology of Metals" at the International Conference on Metals in Soils, Waters, Plants and Animals. This Conference was held in Orlando, Florida (U.S.A.), April 30—May 3, 1990. The editors are grateful for the help and advice of the conference director, Dr. Domy C. Adriano (University of Georgia, Savannah River Ecology Laboratory) throughout their efforts. The editors are also very grateful for the excellent and timely reviews of the associated manuscripts. Reviewers included:

Atchison, G. L.	— Iowa State University
Bayne, B. L.	— Plymouth Marine Laboratory
Beyer, W. N.	— USDOI Patuxent Wildlife Res. Center
Bryan, G. W.	— Plymouth Marine Laboratory
Cain, D.	— USDOI Geological Survey
Capuzzo, J. M.	— Woods Hole Oceanographic Institution
Chapman, P. M.	— E.V.S. Consultants
Clements, W. H.	— Colorado State University
Dallinger, R.	— Institut für Zoologie der Universitt Innsbruck
Driscoll, C. T.	— Syracuse University
Gard, T. C.	— University of Georgia
Gulley, D. D.	— University of Wyoming
Guttman, S. I.	— Miami University
Hall, Jr., L. W.	— University of Maryland
Hartwell, S. I.	— University of Maryland
Harrell, Jr., F. E.	— Duke University Medical Center
Hodson, P. V.	— Institut Maurice-Lamontagne
Hopkin, S. P.	— University of Reading
Ireland, M. P.	— University College of Wales
Klerks, P. L.	— University of Toledo
Kramer, V. J.	— University of Alabama
Landrum, P. F.	— NOAA Environmental Research Laboratories
Laughlin, Jr., R. B.	— Fort Pierce, Florida
Leland, H. V.	— USDOI Geological Survey
Lewis, T. E.	— Lockheed Engineering and Sciences, Co.
Little, E. E.	— USDOI National Fisheries Contaminant Research Center
Luoma, S. N.	— USDOI Geological Survey
Mills, G. L.	— University of Georgia
Morgan, A. J.	— University of Wales College of Cardiff
Neff, J. M.	— Arthur D. Little, Inc.
Overnell, J.	— Dunstaffnage Marine Laboratory
Saiki, M. K.	— USDOI Columbia National Fisheries Research Laboratory
Sharp, J. R.	— Southeast Missouri State University
Simkiss, K.	— University of Reading
Specht, W. L.	— Westinghouse Savannah River Co.
Sprague, J. B.	— J. B. Sprague Associates, LTD.
Westerman, A. G.	— Kentucky Department of Environmental Protection
Yan, N. D.	— Dorset Research Centre

CONTENTS

CHAPTER 4
Allometry of Metal Bioaccumulation and Toxicity 91
Michael C. Newman and Mary Gay Heagler

CHAPTER 5
Metal Effects on Fish Behavior—Advances in Determining the
Ecological Significance of Responses 131
Mary G. Henry and Gary J. Atchison

CHAPTER 6
The Developmental Toxicity of Metals and Metalloids in Fish 145
Peddrick Weis and Judith S. Weis

CHAPTER 7
Stochastic Models of Bioaccumulation 171
James H. Matis, Thomas H. Miller, and David M. Allen

CHAPTER 8

Analyzing Toxicity Data Using Statistical Models for Time-to-Death:
An Introduction 207
Philip M. Dixon and Michael C. Newman

CHAPTER 9

Trace Metals in Freshwater Sediments: A Review of the Literature
and an Assessment of Research Needs 243
Alan McIntosh

CHAPTER 10
Samuel N. Luoma and James L. Carter

CHAPTER 11

Genetic Factors and Tolerance Acquisition in Populations Exposed
to Metals and Metalloids 301
Margaret Mulvey and Stephen A. Diamond

CHAPTER 12

Impact of Low Concentrations of Tributyltin (TBT) on Marine
Organisms: A Review 323
Geoffrey W. Bryan and Peter E. Gibbs

CHAPTER 13

Community Responses of Stream Organisms to Heavy Metals:
A Review of Observational and Experimental Approaches

William H. Clements

1

The Influence of Water Chemistry on Trace Metal Bioavailability and Toxicity to Aquatic Organisms

Patrick L. Brezonik, Scott O. King, and Carl E. Mach

Department of Civil and Mineral Engineering, University of Minnesota, Minneapolis, Minnesota 55455

OVERVIEW

Metal ions in natural waters occur as free aquo ions, simple complexes with inorganic ligands, chelates with multidentate organic ligands, and sorbed onto particle surfaces. The organic ligands may be naturally occurring macromolecules with many binding sites or small molecules of natural or anthropogenic origin that form 1:1 or 1:2 chelates. The bioavailability and toxicity of metal ions to aquatic organisms depend strongly on the chemical form in which the metals

1

occur (their "speciation"), and, in turn, speciation depends on solution conditions, especially pH and concentrations of various ligands. Speciation and bioavailability are related quantitatively in terms of the thermodynamic stability of metal-ligand complexes. Chemical principles from which one can predict complex stability can be described in terms of fundamental atomic characteristics of metallic elements, such as, the polarizing power of cations and metal electronegativity. Metal complexing trends, in turn, can be related to metal bioavailability by surface complexation models that describe metal interactions with cell surfaces as a chemical complexation process. Relationships among water chemistry, metal ion speciation, and metal bioaccumulation are illustrated from an ongoing whole-basin acidification experiment at Little Rock Lake in northern Wisconsin.

INTRODUCTION

Quantifying the factors affecting the toxicity and bioaccumulation of trace metals by aquatic organisms is a major goal of aquatic scientists. Overall, these factors can be grouped into four major categories: (1) solution conditions, which affect the chemical form (speciation) of a given metal ion; (2) the nature of the metal ion and trends in complexation chemistry among metals of the periodic table; (3) the nature of the response being measured (acute toxicity, bioaccumulation, various types of chronic effects, etc.); and (4) the nature of the aquatic organisms (e.g., age or life stage, species, position in aquatic food webs). This chapter focuses on the first two categories, in particular, on the effects of variations in chemical composition of natural waters on metal ion speciation and models that predict the chemical behavior and/or bioactivity of metal ions. Chemical speciation plays a controlling role in metal bioaccumulation and toxicity, and chemical models play an important role in predicting trace metal bioaccumulation and/or toxicity to aquatic organisms.

The fact that metal ion bioaccumulation and toxicity are predictable (at least somewhat) from solution conditions and from the aqueous chemistry of metal ions (e.g., complexation trends among the ions) reflects the fact that biological uptake and toxicity are inherently chemical processes: the metal ions react with functional groups (sulfhydryl, amino, carboxyl, hydroxide, oxide) on cell surfaces, membranes, enzymes, etc. In fact, a cell surface interaction model that uses chemical equilibrium principles provides the basis for relating solution chemistry and metal bioaccumulation by aquatic organisms.

EFFECTS OF SOLUTION CONDITIONS ON METAL ION SPECIATION

As summarized in Table 1, the chemical forms in which trace metal ions occur in a given water body depend on numerous physical, chemical, and biological factors. Most of the physical and chemical factors are sufficiently well understood, however, that reasonably accurate predictions can be made concerning

Table 1
Factors Affecting Metal Speciation in Aqueous Solution

I. **Physicochemical Conditions**
   General conditions:
   Ionic strength, temperature, pH, E_H
   Specific ligand binding conditions:
   Concentrations and nature of inorganic ligands: alkalinity (HCO_3^-, CO_3^{2-}), F^-, Cl^-, SO_4^{2-}, HS^- (anoxic water), CN^- (water contaminated with industrial waste)
   Concentrations and nature of organic chelating agents: DOC, aquatic humus, natural organic acids, peptides, polyaminocarboxylates
   Concentrations and nature of particulate matter with surface sites available for metal binding

II. **Time**
   Solutions are not necessarily at thermodynamic equilibrium. Ligand exchange reactions can be quite slow; rates depend on pH, hardness, nature of ligand, and water exchange rates of metals. Redox reactions also can be slow. Some are highly pH dependent [e.g., Fe(II) and Mn(II) oxidation], and photoinduced redox reactions can maintain nonequilibrium conditions.

dissolved metal ion speciation if suitable information is available on solution conditions.[1,2] Numerous computer programs are available (e.g., Westall et al.,[3] Plummer et al.,[4] Nordstrom et al.,[5] Ball et al.[6]) to calculate equilibrium speciation for prescribed conditions of pH and total (analytical) conditions of metals and ligands. For example, Figure 1 illustrates results obtained with MINEQL[3] on variations in the speciation of Cu and Pb versus pH in Little Rock Lake, Wisconsin. This softwater lake has been the subject of a whole-basin acidification experiment,[7] and the chemical speciation of metals over the pH range of the experiment is illustrated in the figure. Prediction of metal levels associated with suspended material, including aquatic microbiota, is more difficult, but, as discussed later in this chapter, advances have been made on this topic in recent years.[8,9]

Among the general physicochemical factors affecting metal ion speciation (Table 1), temperature and ionic strength primarily affect the values of equilibrium constants and activity coefficients. Within the range of ionic strength for freshwaters, ion activity coefficients do not exhibit large variations, but this is an important factor in comparing fresh- and seawater conditions. Similarly, temperature variations in a surface water within a given season (e.g., $\pm 5°C$) are of secondary importance relative to variations in metal speciation, but equilibrium constants for some metal ion reactions, especially those involving solubility of solid phases, can shift significantly between winter and summer conditions. Unfortunately, stability constants for many trace metal complexes have been determined at only one temperature (usually 20° or 25°C), and extrapolation to other temperatures is subject to uncertainty.

Of particular interest to aquatic scientists regarding the effects of complexation on metal availability (bioaccumulation and toxicity) is the presence of natural dissolved organic matter. Humic and fulvic acids (symbolized here as Hum) are widely distributed in surface waters, especially those with visible color. Hum is an important fraction of the dissolved organic matter, even in low-color waters, however, including the open ocean.[10] Other fractions of natural organic matter, including dissolved organic nitrogen compounds (oligo- and polypeptides)[11] and synthetic organic compounds, such as polyaminocarboxylates (e.g., ethylene-tetraacetic acid [EDTA], nitrilotriacetic acid [NTA]), may be important complexing agents for trace metal ions in some waters.

Metal ions complexed with natural macromolecular organic matter or strong synthetic chelating agents generally are considered not to be directly available to aquatic organisms, whereas inorganic complexes generally are. Aluminum fluoride complexes are important exceptions among inorganic complexes of interest in freshwaters.[12] In addition, organometallic forms of several metals (e.g., methylmercury, triorganotin species), which are lipophilic, are bioaccumulated by aquatic organisms much more so than inorganic forms of the metals, and some organometallic forms are toxic to aquatic organisms.

For a variety of reasons, it is difficult to measure stability constants of metals with Hum, and the use of stability constants measured under a given set of solution conditions (so-called "conditional constants") for a different set of conditions (e.g., at a different pH or different set of metal and Hum concentrations) must be done cautiously. Significant advances were made during the past decade in ways to model metal-Hum binding,[13-15] and a sufficient variety of conditional binding constants are now available at least to approximate the metal-binding behavior of natural waters containing Hum.

Finally, it must be noted that metal speciation cannot always be assumed to be an equilibrium condition in natural waters. With regard to complexation reactions, rates of complex formation between free (aquo) metal ions and free

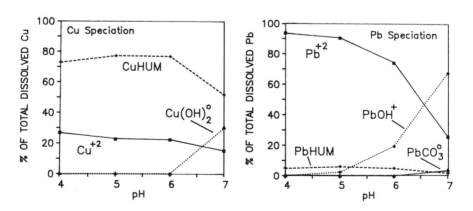

FIGURE 1.   Variation in dissolved chemical forms (speciation) of Cu and Pb versus pH in Little Rock Lake, Wisconsin. Solution conditions are described in the text. Cd and Zn speciation are relatively unchanged over this pH range (see Table 4). HUM represents natural dissolved organic matter.

ligands are very rapid, but nonequilibrium conditions can arise from slow kinetics of ligand exchange. For example, when a strong chelating agent such as EDTA (symbolized as Y^{4-}) is added to a water containing hardness cations (Ca^{2+} and Mg^{2+}) at approximately 10^{-3} M and a trace metal such as Cu^{2+} at approximately 10^{-8} to 10^{-7} M, all the EDTA will react initially with the hardness ions (based on mass action principles), even though equilibrium conditions favor the formation of CuY^{2-}. The ligand exchange reaction: $CaY^{2-} + Cu^{2+} \rightarrow CuY^{2-} + Ca^{2+}$ may be quite slow (minutes to hours), actual rates depending on pH and hardness concentrations.[16]

In general, rates of ligand exchange reactions for metal ions are correlated with water exchange rates of metal ions.[17] The water exchange rate of Cu^{2+} is among the fastest of all metal ions ($\simeq 10^9$/s, which is near the limit of diffusion control), but water exchange rates for Zn^{2+} and Fe^{2+} are about 100 $\times$ slower; Ni^+ water exchange is about $10^{-4} \times$ that of Cu^{2+}. Consequently, equilibration times potentially can be very long (days to months) when strong chelating agents are added to natural waters. The slowness of ligand exchange reactions also has implications regarding the potential toxicity of trace metals added to algal growth media containing strong chelating agents and hardness cations in laboratory studies.

Oxidoreduction reactions also can be slow to reach equilibrium. For example, thermodynamically unfavorable lower oxidation states of iron and manganese can persist in oxygenated waters for days to months under mildly acidic conditions (pH <5 to 6), where rates of Fe^{2+} and Mn^{2+} autoxidation are slow to very slow. Autoxidation rates for both metals increase with the square of hydroxide ion activity over a broad range of pH.[18-22] Photoredox processes also are important in maintaining nonequilibrium conditions for a few metals (Fe, Mn, Cu). For example, Fe(III) complexes with carboxylic acid groups of Hum are subject to a charge-transfer photoredox processes in which Fe is reduced to the nonequilibrium Fe^{2+} state;[23] a similar photoreduction dissolution process produces soluble Fe^{2+} from ferric oxyhydroxides in the presence of organic acids.[24] Insoluble Mn(III,IV) oxides also are photoreduced to soluble Mn^{2+} by marine humic acids[25] and freshwater fulvic acid.[26] As a result, Mn^{2+} concentrations are maintained at levels far above those predicted by equilibrium calculations for surface waters. Finally, nonequilibrium concentrations of Cu^+ are maintained in oxygenated seawater by photoreduction of Cu(II) by superoxide radical anion, O_2^-.[27]

COMPLEXATION TRENDS AMONG METALS: UNDERLYING CHEMICAL FACTORS

The extent of metal complexation for a given set of solution conditions varies widely among metals of interest in natural aquatic systems, and chemists have sought for many years to develop predictive relationships and a rational basis for understanding metal-to-metal patterns of complexation. Three approaches have been used widely in this regard: (1) linear free-energy relationships (LFERs);

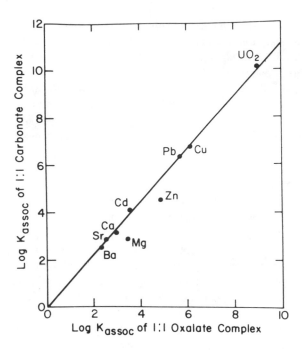

FIGURE 2. Linear free-energy relationship (LFER) showing correlation between stability constants for various metal oxalate complexes and corresponding metal carbonate complexes. Reprinted from Langmuir[82] by permission of the American Chemical Society.

(2) correlations based on metal ion size (most commonly ionic radius or a function of charge:radius ratio); and (3) correlations based on metal ion electronegativity. LFERs are log-log plots of stability constants for a series of metal ions with two different ligands, β_{MiL1} versus β_{MiL2}, or log-log plots of stability constants for two metal ions with a series of ligands, β_{M1Li} versus β_{M2Li} (Figure 2). Such plots are called LFERs because equilibrium constants are log-linearly related to free energies of reaction; in general, $\Delta G^{\circ}_{rx} = -RTln\beta$.[28] Such relationships are useful to estimate the stability constant value for a given complex (e.g., M_1L_j if M_2L_j is known). This approach was recently used to estimate some unknown metal sulfide complexation constants from known values of sulfide solubility products for a variety of metals,[29] but the scarcity of known values limits the accuracy of LFER estimates for these species. Moreover, LFERs do not provide any insight into causes for trends in complexing strength among the various metals.

Correlations involving metal ion radius or charge:radius functions are based on electrostatic factors, and complexes for which such relationships hold (Figure 3) are assumed to involve electrostatic bonds. (Based on Coulomb's law, we would expect stability constants of metals with a given anionic ligand to increase with charge and decrease with radius of the metal ion.) The term z^2/r, called the polarizing power of the cation,[30-32] is widely used in such correlations and often yields good linear relationships. Correlations involving metal electronegativity (E_n) assume a covalent binding model. Covalent bond strength generally increases with similarity in electronegativity of the binding atoms. Increasing values of stability constants for electronegative ligand atoms (such as oxyanions) with increasing metal ion electronegativity (Figure 4) thus support a covalent binding model.

Of course, all chemical bonds are not strictly ionic or covalent, and a continuum exists between these two extremes. Consequently, many metal ions fit correlations with both z^2/r and E_n, and the goodness-of-fit often depends on the nature of the ligand. Lack of fit to a parameter also can be instructive. For example, stability constants for a variety of metal ions with both bicarbonate and sulfate show no dependency on E_n (Figure 4), which suggests these complexes do not involve covalent bonds. In both cases, the complexes are weak ion pair (outer sphere) complexes in which the ligand is bound to the hydrated metal ion by weak electrostatic forces. Even though the slopes of the correlations between stability constants and E_n are zero for HCO_3^- and SO_4^{2-}, the plots still

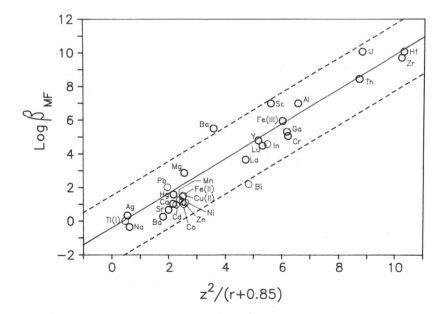

FIGURE 3. Correlation between stability constants for metal fluoride complexes and the polarizing power of the cation, z^2/r; the constant 0.85 is an empirical correction factor. Reprinted from Turner et al.[30] by permission of Pergamon Press.

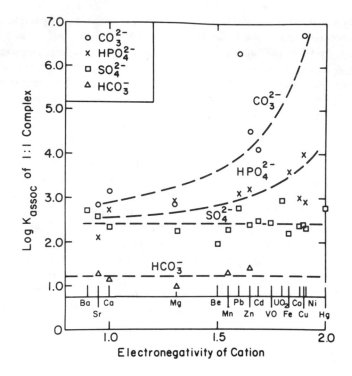

FIGURE 4.   Correlations of stability constants for various metal ion complexes with electronegativity of the metal ion. Reprinted from Langmuir[82] by permission of the American Chemical Society.

are useful in a predictive sense, since they show that the stability constants for bicarbonato and sulfato complexes with a variety of metals all have approximately the same values.

Efforts to obtain a more fundamental (atomic) explanation of the metal ion complexing trends described above for z^2/r and E_n have led to the development of several classification schemes for metals. The Ahrland-Chatt-Davies (ACD) approach[33] and the similar hard-soft acid-base theory[34,35] both divide metal ions into three classes, based on their outer electron orbital configurations (Table 2). Metal ions in class **A** of the ACD system generally correspond to Pearson hard acids and have no electrons in their outer orbitals, i.e., they have an inert gas configuration. Such ions are nonpolarizable (hard) spheres. Class **B** ions (ACD) and Pearson soft acids have 10-12 outer electrons (filled d orbitals) and are highly polarizable.

Definite trends in bonding characteristics and ligand preferences are found between the class **A** and **B** ions (Table 2). In general, A-type cations form weak complexes via electrostatic bonding, and, as a result, strongest complexes are formed with first row ligands (which are smallest and have the highest charge density within a given column). In contrast, **B**-type cations tend to form stronger complexes via covalent binding, and, within a given column, the larger atoms in the 2nd, 3rd, and 4th rows form stronger complexes.

Transition metals occupy a borderline class in both schemes. With incompletely filled d orbitals, they are spherically nonsymmetric (hence polar). Complexation trends within the transition metals depend, however, on crystal field stabilization energy, which is related to the number of d electrons, and they tend not to follow simple patterns applicable to nontransition metals (i.e., correlations with z^2/r or E_n). The Irving-Williams order (Table 2) describes (at least qualitatively) the trend of complexation strengths among the first row transition metals.

Table 2
Metal Classification Schemes and Trends in Ligand Preferences

a. Ahrland-Chatt-Davies Classification of Metal Ions

Class A	Borderline	Class B
Li^+, Na^+, K^+, Be^{2+} Mg^{2+}, Ca^{2+}, Sr^{2+}, Mn^{2+}, Al^{3+}, La^{3+}, Cr^{3+} Inert gas configuration	Fe^{2+}, Co^{2+}, Ni^{2+}, Cu^{2+}, Zn^{2+}, Pb^{2+} 1 to 9 outer shell electrons; not spherically symmetric	Cu^+, Ag^+, Au^+, Tl^+ Hg_2^{2+}, Hg^{2+}, Pd^{2+} Pt^{2+}, Tl^{3+} 10 to 12 outer shell electrons; highly polarizable

b. Pearson Hard and Soft Acids

Hard acids	Borderline	Soft acids
All A cations plus Fe^{3+}, Co^{3+}, Mn^{3+}	All bivalent transition metals plus Zn^{2+}, Pb^{2+}, Bi^{3+}	All B cations minus Zn^{2+}, Pb^{2+}, Bi^{3+}

c. Bonding Characteristics and Ligand Trends for A and B Cations

A-type cations	B-type cations
Strongest complexes with first row ligands; relatively weak electrostatic bonding; entropy changes dominate over enthalpy in energy of reaction $N \gg P > As > Sb$ $O \gg S > Se > Te$ $F \gg Cl > Br > I$	Strongest complexes with heavier ligands in 2nd, 3rd, and 4th rows; mainly strong covalent bonding; enthalpy changes dominate in energy of reaction $N \ll P > As > Sb$ $O \ll S \sim Se \sim Te$ $F \ll Cl < Br < I$

d. Irving-Williams Order

For divalent transition-metal cations, the following well-established sequence of complex stability is based on crystal-field stabilization energies, which depend on the number of d orbital electrons:

$$Mn < Fe < Co < Ni < Cu > Zn$$

The tripartite (**A**-borderline-**B**, hard-borderline-soft) classifications of metal ions have been quantified and also developed in two-dimensional schemes that describe ligand preferences and complexing trends among metals. The degree of **A** character (''**A**-ness'') for a metal ion can be expressed quantitatively in terms of the polarizing power, z^2/r, since **A** ions tend to form electrostatic bonds. Turner et al.[30] divided metals into four classes of increasing degree of hydrolysis, based on z^2/r: I (<2.5); II (2.5 to 7); III (7 to 11); IV (11 to 23). Elements with $z^2/r \geq 23$ are completely hydrolyzed at all pH values in water and exist as oxyanions. Similarly, **B** character (''**B**-ness'') can be quantified by the term $\Delta\beta$, which Turner et al.[30] defined in terms of the difference in metal binding strengths of fluoride and chloride complexes:

$$\Delta\beta = \log\beta_{MF} - \log\beta_{MCl} \tag{1}$$

Since **B**-type cations form weaker fluoride complexes than chloride complexes and the converse is true for **A**-type cations, a large negative value of $\Delta\beta$ is associated with **B**-type cations, and a large positive $\Delta\beta$ is associated with **A**-type cations (Figure 5). Turner et al.[30] divided the range of $\Delta\beta$ into four classes: **A** ($\Delta\beta > 2$), **A'** ($\Delta\beta = 0$ to 2), **B'** ($\Delta\beta = -2$ to 0), and **B** ($\Delta\beta < -2$). Nieboer and Richardson[36] used the product $E_n^2 r$ as a covalent index that groups the **A**, **B**, and borderline ions in a manner similar to $\Delta\beta$ (Figure 5).

Further analysis of binding strengths within metal ion subclasses according to the polarizing power term led to classification of common inorganic ligands into three categories:[30]

> hard: SO_4^{2-}, F^-
> intermediate: OH^-, CO_3^{2-}
> soft: Cl^-

For hard ligands, β_M increases markedly with z^2/r for all cations (both **A** and **B**). For intermediate ligands, β_M increases markedly with z^2/r for **A** and **A'** cations and also increases with increasing **B** character (i.e., with decreasing $\Delta\beta$). Finally, for soft ligands, β_M increases only to a limited extent with z^2/r but it increases markedly with increasing **B** character (decreasing $\Delta\beta$).

The parameters z^2/r and $\Delta\beta$ (or possibly $E_n^2 r$; see McDonald et al.[37]) can be used to develop two-dimensional classification schemes. Figure 6 groups metal ions into four major categories according to their ligand preferences and relative binding strengths. The very weakly complexed cations (Figure 6) include the **A**-type alkali and alkaline earth cations which have low polarizing power and are found primarily as the free aquo ions in both fresh- and seawater. The chloro-dominated cations include weakly polarizing **B** and **B'** cations. Cd and Hg are in the borderline of this grouping insofar as their speciation is chloro-dominated in seawater but not in freshwater. The hydrolysis-dominated cations have rela-

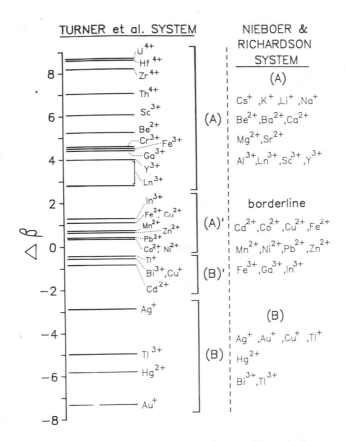

FIGURE 5. Rankings of metal ions according to $\Delta\beta$ of Turner et al.[30] and the covalent index of Nieboer and Richardson.[36] Metal ions are divided into four classes (**A,A',B',B**) in the $\Delta\beta$ ranking and three classes in the covalent index ranking, which is based on metal electronegativity and ionic radius. Redrawn from Turner et al.[30]

tively high values of z^2/r (classes **IIIBB'** and **IVAA'**), and hydroxide and oxo-complexes predominate, especially at high pH. However, under acidic conditions, hydrolysis of some cations in this category is sufficiently weak to allow complexation by hard (e.g., F^-) and intermediate ligands (e.g., CO_3^{2-}). This is especially true for Al with F^-.[38]

The fourth general class of ions in Figure 6 (groups **IIA'** and **IIIA**) is an important residual class whose complexing tendencies are difficult to summarize because they depend strongly on solution conditions (e.g., pH and concentrations of common ligands, such as carbonate and natural organic matter). The usefulness of the bivariate classification scheme to predict metal speciation and bioavailability thus is limited within this category. Included in this group are essential micronutrient elements [Mn(II), Fe(II), Co(II)], a nonessential toxic element (Pb), and essential elements that are toxic to some aquatic organisms at quite low concentrations (Cu, Zn). Tendencies to form complexes vary widely among these metals. Ni and Mn(II) have the weakest tendencies and exist primarily as aquo ions; Cu and Pb have the strongest complexing tendencies and exist primarily as hydroxo, carbonato, or organic complexes under neutral pH conditions. The lanthanides (IIIA) are poorly studied in aquatic systems, but because of their extremely low concentrations are much less important than the other metal ions (IIA') in this category.

In view of the importance of complexation by Hum relative to trace metal bioactivity in aquatic systems, it is interesting to note that Turner et al.[30] found reasonably close correlations between $\log\beta_{Hum}$ and both $\log\beta_{MCO_3}$ and $\log\beta_{MOH}$. In that the binding groups on Hum consist primarily of carboxylate and phenolic groups, correlations with carbonate and hydroxide are not too surprising. As Turner et al.[30] noted, metal complexation with Hum thus is most likely to be important for cations that are significantly complexed by carbonate and hydroxide (described above as "intermediate" ligands).

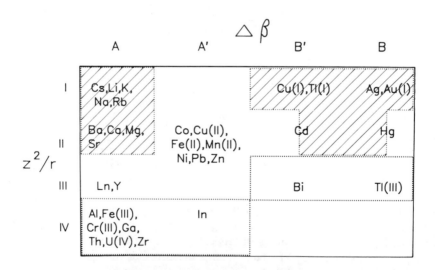

FIGURE 6.   Bivariate metal classification scheme based on ionic and covalent bonding tendencies (cation polarizing power z^2/r and $\Delta\beta$, respectively). The scheme divides metals into four general classes with relatively distinct ligand complexing tendencies: very weakly complexed cations (**IA**, **IIA**); chloro-dominated cations (**IBB'**, **IIBB'**); hydrolysis dominated cations (**IIIBB'**, **IV AA'**); and a residual group (**IIA'**, **IIIA**) of variable speciation. Redrawn from Turner et al.[30]

RELATIONSHIPS BETWEEN ATOMIC PROPERTIES AND METAL BIOACTIVITY

Progress in relating measures of metal bioactivity in aquatic organisms (e.g., bioaccumulation factors and acute toxicity, LC_{50}) to atomic properties or indices thereof (such as softness or extent of **B** character) has been fairly limited, in spite of much interest among scientists in metal contamination of surface waters. Some early examples of such relationships were developed by workers in mammalian toxicology. For example, mouse LD_{50} values of divalent transition metal ions follow the Irving-Williams order.[39-40] The latter authors also related mouse LD_{50} values for a large number of metal ions to the σ parameters of hard and soft acid-base theory:

$$\text{Soft acids: } LD_{50} = 26.1\sigma_p + 0.076; r^2 = 0.85 \tag{2}$$

$$\text{Hard acids: } LD_{50} = -2.85\sigma_k - 5.54; r^2 = 0.99 \tag{3}$$

where σ_k is a hard acid parameter determined from differences in outer orbital and desolvation energies,[41] and σ_p is defined in terms of metal-halide bond energies:

$$\sigma_{p,M} = \frac{(\text{bond energy for MF} - \text{bond energy for MI})}{(\text{bond energy of MF})}$$

$$\tag{4}$$

Thus σ_p is similar to the parameter $\Delta\beta$ defined in Equation (1). σ_p decreases with increasing softness and σ_k decreases with increasing hardness. From the regressions in Equations (2) and (3), it is apparent that metal ion toxicity to mice increases with increasing softness within both the hard and soft acids.[42] Only limited success has been reported in applying metal-softness relationships like those in Equations (2) and (3) to other organisms.[37,43] As the latter authors note, this should not be surprising, since there is likely to be considerable variation among the metals in regard to the chemical characteristic(s) key to their toxic mechanisms.

Kaiser[44] reported success in relating the toxicity of metal ions to various aquatic organisms by empirical regression equations involving several fundamental atomic properties. The equations have the general form:

$$pT = a_0 + a_1\log(AN/\Delta IP) + a_2\Delta E_0 \tag{5}$$

where pT is the negative log of metal ion concentration (M) with a certain toxicity
AN is the atomic number of the metal
ΔIP is the difference in the ion's ionization potential (eV) and the ionization potential of the next lower oxidation state of the element

ΔE_0 is the absolute value of the electrochemical potential between the ion and the first stable reduced state

Coefficients a_0, a_1, and $2a_2$ depend on the group of metals, biota, and type of toxic effect being determined

The terms ΔIP and ΔE_0 are related to outer orbital electronic properties of atoms, and AN is related at least crudely to ionic size. Inclusion of AN in the regressions allows successful predictions for ions having similar ionization and electrochemical potentials but different ionic radii (e.g., Na and Sr). Correlation coefficients as high as 0.96 were reported for toxicity correlations involving various biota and sets of metals, but a given regression equation appears to have little transferability beyond the set of data from which it was generated.

A relatively good predictive relationship has been reported between the log of metal bioaccumulation, logVCF (volume concentration factor), by marine algae and the negative log of the solubility product, $-\log K_{s0}(MOH)$, for 21 different metal hydroxides.[45] In turn, the log concentration of metal ions at which the growth of marine diatoms was reduced by 50% (EC_{50}) was exponentially related to logVCF, and EC_{50} was linearly correlated with $\log K_{s0}(MOH)$. Of course, algae are relatively simple organisms and lack the variety of detoxifying and metal elimination mechanisms found in higher organisms. To this extent, relationships between the aqueous chemistry or atomic properties of metals and their behavior in aquatic microflora should be simpler to derive and more straightforward than relationships involving aquatic fauna.

SURFACE COMPLEXATION MODELS

The interactions of dissolved metals with biological surfaces such as cell membranes (e.g., algal cell walls and gill membranes of higher animals), detritus, and organically coated particles in oceans and lakes can affect the transport, chemistry, bioaccumulation, and toxicity of metals. Biological surfaces are the most important substrate for metal binding in lakes[46] and, in some cases, dissolved metal concentrations are controlled by adsorption to settling biological surfaces.[46-48]

For algae, metal ions must first absorb onto the cell membrane before passing into the cell. The amount of metal bound to the surface thus directly affects the amount taken up. Higher organisms, such as zooplankton and fish, possess several mechanisms for metal uptake (e.g., ingestion of food and passage through membranes) and removal (e.g., defecation and removal of dissolved metals by the liver and kidneys).

The interactions that occur at biological surfaces in natural waters are very complicated. Reactions of metal ions, hardness cations, anions, and organic ligands with the various surface functional groups are numerous and difficult to quantify individually. However, the binding of metal ions to surface functional groups can be modeled effectively by assuming that the process is analogous to metal ion complexation by dissolved ligands. Such models are called surface complexation models. Models similar to the one described below have been used to predict metal binding to algal surfaces[49] and metal binding to gill surfaces of fish.[8,50] Honeyman and Santschi[9] used a surface complexation model to help predict metal residence times in aquatic systems.

Functional groups on the cell membrane behave as weak acids. Their dissociation may be expressed as

$$>ROH_2^+ = >ROH + H^+ \qquad K_{a1}^s \qquad (6)$$

$$>ROH = >RO^- + H^+ \qquad K_{a2}^s \qquad (7)$$

where $>ROH_2^+$, $>ROH$, $>RO^-$ are surface functional groups that are protonated to different extents.

K_{a1}^s and K_{a2}^s are conditional acidity constants that hold only for the experimental conditions under which they are evaluated.

These acidity constants may be expressed as:

$$K_{a1}^s \qquad \{>ROH\}[H^+]/\{>ROH_2^+\} \qquad (8)$$

$$K_{a2}^s \qquad \{>RO^-\}[H^+]/\{>ROH\} \qquad (9)$$

where $\{>ROH_2^+\}$, $\{ROH\}$, and $\{RO^-\}$ are concentrations of the functional groups expressed as mol/g (dry wt)

$[H^+]$ is the hydrogen ion concentration of the surrounding solution (M)

The values of the conditional acidity constants, K_{a1}^s and K_{a2}^s, vary with the extent of surface coverage or surface charge. In order to extend the use of these conditional constants to other conditions they must be corrected for electrostatic interactions at the surface. Therefore, the conditional constants are divided into two factors:

$$K_{a1}^s = K_{a1}^s(\text{int})\exp(- F\Psi_s/RT) \qquad (10)$$

where $K_{a1}^s(\text{int})$, the intrinsic acidity constant, is independent of surface coverage and surface charge (i.e., it is the acidity constant for a hypothetically uncharged surface)

factor $\exp(- F\Psi_s/RT)$ accounts for electrostatic forces involved in adsorption to the surface

F is Faraday's constant

R is the gas constant

Ψ_s is the difference in potential between the surface and the bulk solution and is a function of the surface coverage.

Values of Ψ_s cannot be measured directly and must be extrapolated from measurable data. This is done by use of several theoretical models, the two most common of which are the ideal Nernstian model and constant capacitance model. The former is represented mathematically as

$$\Psi_s = (RT\ln10/F)(pH_{zpc} - pH) \qquad (11)$$

where pH_{zpc}, the pH of zero point of charge of the surface, is determined experimentally by acid-base titration of the surface at different ionic strengths[51]

This model does not work well under natural conditions. The constance capacitance model is a simple double-layer model also known as the Helmholtz model. The surface charge is expressed as

$$\Psi_s = \sigma/\kappa \qquad (12)$$

where σ is the specific surface charge (C/m^2)
κ is the specific capacitance (F/m^2)[52]

The value of σ depends on the concentration of charged species on the surface; Equation (12) thus may be rewritten as

$$\Psi_s = zF\{>R^{z+}\}/\kappa S \qquad (13)$$

where $\{>R^{z+}\}$ is the concentration of surface sites with charge z
S is the specific surface area (m^2/kg)

Substitution of Equation (13) into Equation (10) yields

$$\ln K_a^s = \ln K_a^s(int) - (zF/RT\kappa S)\{>R^{z+}\} \qquad (14)$$

The intrinsic acidity constants can be determined by fitting experimental data to Equation (14). It is important to note that the values of the intrinsic acidity constants depend on which model is used.[52] Also, fit of data to a particular model does not prove that the model is valid.

Binding of metals to biological surfaces is modeled analogously to ligand complexation of dissolved metals. Reactions at the surface can be thought of as a competition between hydrogen ions, metal ions, and hardness cations for the same surface binding sites, a general expression for which is

$$>ROH_x + Me^{2+} = >ROMe^{(2-x)+} + xH^+ \qquad (15)$$

where x is the average number of H^+ ions displaced when a metal ion is adsorbed.

Under ideal conditions, x is an integer (0, 1, or 2), but, under natural conditions, the variety of different sites and reactions results in an average reaction where x is a fractional number, which can be determined by graphical analysis of proton release experiments.[53] The average conditional stability constant for the general equation is

$$K_{Me}^s = \{>ROMe^{(2-x)+}\}[H^+]^x/\{>ROH_x\}[Me^{2+}] \tag{16}$$

where $\{>ROMe^{2-x)+}\}$ and $\{>ROH_x\}$ are the concentration of surface-bound metal ions and hydrogen ions, respectively, expressed in mol/g (dry wt)

Intrinsic stability constants can also be obtained for metal complexation at the surface using methods analogous to the method described above for acidity constants [Equations (10) and (14)]. Equation (16) may be written as

$$lnK_{Me}^s = lnK_{Me}^s(int) - (zF/RT\kappa S)\{>ROMe^{(2-x)+}\} \tag{17}$$

where $K_{Me}^s(int)$ is the intrinsic stability constant and $\{>ROMe^{(2-x)+}\}$ is the concentration of surface-bound metal.

Again, the intrinsic stability constant is determined from measured conditional stability constants at different surface coverages.[52]

The binding of the hardness cations, Ca and Mg, to the surface is modeled exactly as the metal ions above. The interactions of Ca and Mg with the surface are important because they compete for the same binding sites as the metal ions. Many researchers have shown that Ca can ameliorate the toxicity of metal ions.[8,54] Hydrogen ion competition for metal binding sites at low pH also has been reported.[55]

The simple surface complexation model is valid only when all possible surface sites are available in excess. Natural surfaces have been shown to have large distributions of binding sites with different binding energies. Thus a simple one-site model is not satisfactory. Buffle et al.[56] recently developed an approach to interpreting titration curves of metal complexation by using site occupation distribution functions; in addition, several other continuous distribution functions for binding constants are available in the literature (e.g., Turner et al.[13]). Finally, slow kinetics of dissolved metal-ligand complex dissociation at cell surfaces also may play a role in ameliorating the bioactivity (toxicity or bioaccumulation) of metal ions in aquatic systems, but little work has been done on this subject.

UPTAKE MODEL FOR METALS

Studies on the kinetics of metal uptake by aquatic organisms indicate that this is a two-step process consisting of rapid adsorption or binding to the surface, followed by slow, diffusion-controlled transport into the cell interior.[49,57,58] Transport to the interior of the cell may be either by diffusion of the metal ion across the cell membrane or by active transport by a carrier protein (Figure 7). Once in the cell, the metal ions interact with cellular proteins. Some metals, such as Fe and Zn, are required for cell growth, and increasing solution concentrations at low levels may enhance cell growth. At higher (supraoptimal) levels, the metals may exert toxic effects. Other metals, such as Al and Cd, are not known to serve any necessary function in cells and would not be expected to enhance growth even at low levels. Such differences in the use of metals by

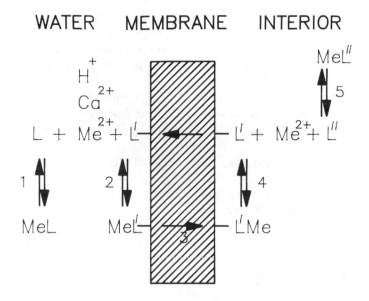

FIGURE 7. Simple schematic diagram of metal uptake through cell membranes. The steps are (1) ligand complexation in solution; (2) competition for surface sites between trace metal ions, hydrogen ion, and hardness cations; (3) transport of metal ions across membrane (usually the rate-limiting step); (4) release of metal from carrier molecule inside cell; and (5) interaction of metal ion with cellular protein.

cells complicate the development of general relationships between aqueous chemistry of metals and their toxicological properties.

APPLICATIONS TO TRACE METAL-BIOTIC INTERACTIONS IN AN ACIDIC LAKE

Acidification of lakes and streams in North America and Europe has altered their trace metal chemistry by: (1) increasing total metal concentrations (Al, Mn, Cd, Pb, Zn[59]; Fe plus the preceding metals except Pb[60]); (2) shifting the speciation of dissolved metals toward free aquo ions[61], the species most toxic to aquatic biota[8]; and (3) reducing particulate metal concentrations in favor of higher dissolved levels.[62] We present trace metal data from the experimental acidification of Little Rock Lake to illustrate these effects.

Little Rock Lake is an uninhabited, forested watershed in Vilas County in northern Wisconsin and has two basins: north (area, 9.8 ha; maximum depth, 10.3 m) and south (area, 8.1 ha; maximum depth, 6.3 m). Typical of acid-sensitive lakes in the region, it is chemically very dilute, low in color, oligotrophic, and has no surface inlets or outlets (i.e., hydrologically, it is a seepage lake). The lake receives 98-100% of its water from precipitation directly onto the lake surface; in some years, a small amount of groundwater flows into the southeast corner of the south basin. Pertinent background conditions were: pH, 6.1; alkalinity, 25 μeq/L; conductivity, 11 μS/cm, and dissolved organic carbon (DOC), $\sim$3 mg/L. Details on the lake and acidification experiment were given by Brezonik et al.[7,63]

Baseline sampling of the lake (1983-1985) demonstrated that the two basins were nearly identical in water chemistry and biota. The basins were divided by a polyvinyl barrier in August 1984, and acidification of the north basin with electrolyte grade H_2SO_4 began after ice-out in April 1985. The north basin is being acidified in three steps to pH 5.6, 5.1, and 4.7 (2 years at each pH), and the south basin is being maintained as a reference. The final acidification step began in 1989. Sample collection and analytical methods for trace methods were described by Mach and Brezonik[62] and Brezonik et al.[60]

Concentrations of "dissolved" Al, Ca, Cd, Fe, Mn, and Zn have increased with decreasing pH in the acidified basin over the period of experimental manipulation (Figure 8). In general, measured changes in dissolved metal ion concentrations agree with results from mesocosm-scale acidification experiments with limnoenclosures in Little Rock Lake and laboratory experiments in which sediments from the lake were extracted at different pH values.[60] In addition, water column increases in Cd and Zn and the lack of increase in Cu at lower pH conform with results of extraction measurements to partition metals in the sediments among major chemical forms.[64] Sedimentary Cd and Zn are associated primarily with the ion-exchangeable fraction and thus are more sensitive to pH changes than Cu, which is associated primarily with the less-labile organic:sulfide fraction in the sediments.

Increased concentrations of the metals shown in Figure 8 have been reported for acidified surface waters in both North America and Europe,[59,65-68] and their increase in the acidified basin of Little Rock Lake is not surprising. Although dissolved Zn in the acidified basin increased with decreasing pH (Figure 8), the observed levels are low compared with acidified surface waters in other regions. For example, Schindler et al.[65] reported Zn levels of approximately 300 μg/L in limnocorrals acidified to pH 5 in Lake 223 of the Experimental Lakes Area (western Ontario). White and Driscoll[68] have observed an annual average Zn concentration of 26 μg/L in Darts Lake (annual average pH 5.1) in the Adirondack region of New York State. The relatively low dissolved Zn level (approximately 6 μg/L) in the acidified basin of Little Rock Lake most likely reflects the hydrologic isolation of the lake and the nature of the local surface strata (noncalcareous glacial till and outwash).

The influence of acidification on dissolved metal speciation in surface waters was reviewed by Campbell and Stokes[55] and Campbell and Tessier.[61] As noted earlier, acidification shifts the speciation of dissolved metals toward free aquo ions. Equilibrium speciation for dissolved trace metals (Al, Cd, Cu, Fe, Mn,

Pb, and Zn) in Little Rock Lake was calculated as described below to illustrate the differences resulting from changing the pH from 6.1 to 4.7. Speciation was calculated with the chemical equilibrium program MINEQL,[3] as modified by Holm et al.[69] to incorporate temperature correction of equilibrium constants. Temperature, pH, alkalinity (HCO^-_3), and filtered (0.4-μm pore-size) concentrations of Ca, Mg, Na, K, Cl, F, SO_4^{2-}, SiO_2, soluble reactive phosphate (SRP), NH_4^+, DOC, Al, Fe, Mn, Cd, Cu, Pb, and Zn from the acidified and reference

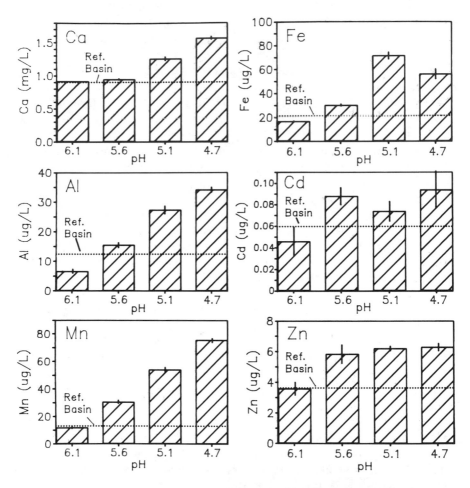

FIGURE 8.   Increase in near-surface "dissolved" Ca, Al, Fe, Mn, Cd, and Zn (0.4-μm pore-size filtered samples) in acidified basin of Little Rock Lake, WI. Bars represent time-averaged metal concentrations during baseline period (pH 6.1) and the three treatment levels (pH 5.6, 5.1, 4.7). Vertical lines at tops of bars show one standard error of the mean; dashed line is mean concentration for reference basin over the study period (1983-1990).

Table 3
Natural Organic Matter Equilibria Used in This Study. LogK (I=0) Used in All Calculations

Reaction	logK	I(M)	pH	logK (I=0)	Reference
$H^+ + L^{2-} = HL^-$	—	—	—	3.0	See footnote f
$2H^+ + L^{2-} = H_2L^\circ$	—	—	—	7.5	See footnote f
$Ca^{2+} + 2L^- = CaL_2^\circ$	3.88[e]	0.02	8	4.2[a]	79
$Mg^{2+} + 2L^- = MgL_2^\circ$	3.74[e]	0.02	8	4.1[a]	79
$Al^{3+} + 3L^- = AlL_3^\circ$	—	—	—	6.2	72
$Fe^{3+} + 3L^- = FeL_3^\circ$	—	—	—	9.4	6
$Mn^{2+} + 2L^- = MnL_2^\circ$	3.9[d]	0.1	5	4.6[a]	80
$Cd^{2+} + 2L^- = CdL_2^\circ$	4.77[e]	0.02	8	5.13[a]	79
$Cu^{2+} + 2L^- = CuL_2^\circ$	6[b]	0.001	6.25	6.1[a]	81
$Pb^{2+} + 2L^- = PbL_2^\circ$	4.4[c]	0.1	6-6.5	5.1[a]	13
$Zn^{2+} + 2L^- = ZnL_2^\circ$	3.7[d]	0.1	5	4.4[a]	8

[a] Adjusted to ionic strength = 0, as in Morel[2]; L^- = carboxylic acid group on a natural organic matter molecule.
[b] Average for fulvic acids from 18 different environments.
[c] Average of four values.
[d] From Figure 8 of cited ref.
[e] Average of four values from two Welsh lakes.
[f] Protolytic equilibria are a consensus of values from Paxeus and Wedborg[75]; Tipping et al.,[76] Lovgren et al.,[77] and Ephraim et al.[78]

basins[60,70] were used as input data. Redox potential (pε) was calculated from pH and dissolved O_2 data (appropriate for an aerobic water column).

Binding constants of metals with natural organic matter were selected from the literature (Table 3). Although the values in Table 3 are conditional constants (and vary with pH), changes over the pH range 4.7 to 6.1 are on the order of 0.5 to 1.0 log unit[71] and do not alter our conclusions. A metal-binding capacity for natural organic matter of 1.5 μmol/mg DOC was assumed, based on literature reviewed by Helmer et al.[72]

Results from chemical equilibrium modeling with MINEQL (Table 4, Figure 1) show that Mn, Cd, and Zn exist predominantly as free aquo ions at both pH 6.1 and 4.7, and little change in speciation occurs over this pH range. Dissolved Cu also shows little change in speciation between pH 6.1 and 4.7, but it exists primarily as complexes with natural organic matter. Only Al and Pb change appreciably in chemical form over the pH range of interest for Little Rock Lake; both show increases of approximately 20% in the free aquo ion.

Equilibrium modeling suggests that levels of dissolved Mn in Little Rock Lake are controlled by the mineral manganite, $MnOOH(s)$,[62] while dissolved Fe levels are controlled by amorphous ferric hydroxide, $[am\text{-}Fe(OH)_3(s)]$. At pH ≥ approximately 5, Al appears to be regulated by a trihydroxide phase slightly more soluble than natural gibbsite but less soluble than microcrystalline gibbsite. Conversely, at pH < approximately 5, dissolved Al in the acidified basin is undersaturated relative to gibbsite. Several explanations have been offered in the literature to explain gibbsite undersaturation at low pH:slow gibbsite dissolution kinetics; Al equilibrium with other minerals like kaolinite or basic Al sulfate minerals; or metal-humate precipitation at low pH. Our studies suggest that gibbsite undersaturation in the acidified basin at pH < approximately 5 is likely due to slow transport of Al from the well-buffered lake sediments to the water column.

Acidification also has changed the distribution of metals between particulate and dissolved forms, as Table 5 shows for the period when the pH difference between the basins was roughly 1.0. Distribution coefficients (K_D) for the metals were calculated as

$$K_D = [M_p/(TSS \times M_d)] \times 10^6 = (L/kg) \tag{18}$$

where M_p and M_d are the particulate and dissolved metal concentrations ($\mu g/L$), respectively

TSS is the total suspended solids concentrations (mg/L).

K_D values for Al, Fe, Mn, and Pb in the acidified basin are 0.5-1.0 log units lower than in the reference basin, demonstrating that metals have reduced affinity for particulate matter at the lower pH of the acidified basin. This may reflect increased competition for surface binding sites by H^+, as predicted by the surface complexation model described in Equations (6) to (17).

As discussed earlier, several variables influence the toxicity of metals to aquatic biota including metal concentration and speciation, type and concentration of available ligands, and type and concentration of dissolved cations (Ca^{2+}, Mg^{2+}, H^+) that compete with trace metals for binding sites on organism surfaces. All these variable were altered by acidifying the north basin of the lake. Trace metal concentrations increased (Figure 8), dissolved organic carbon levels decreased[70], and speciation of some metals shifted toward free aquo ions (Table 4). These changes tend to increase metal bioaccumulation by and toxicity to biota. Conversely, acidification resulted in markedly higher $[H^+]$ and almost a doubling in Ca^{2+} concentration over the pH range 6.1-4.7;[70] these changes tend to ameliorate metal ion toxicity.

Table 4
Calculated Trace Metal Speciation in Little Rock Lake, WI[a]

Metal	% Free Aquo Ion pH 6.1	pH 4.7	Other Dissolved Species[b] (>1%)
Al	<1	18	$AlOH^{2+}$, $Al(OH)_2^+$, $Al(OH)_3^\circ$ $Al(OH)_4^-$, AlHum, AlF^{2+}, AlF_2^+
Mn	91	95	MnHum
Cd	73	75	CdHum, $CdSO_4^\circ$
Cu	22	25	CuHum, $Cu(OH)_2^\circ$
Pb	70	92	PbHum, $PbOH^+$, $PbSO_4^\circ$
Zn	93	94	ZnHum, $ZnSO_4^\circ$

[a] See text for further information.
[b] Hum represents natural organic matter.

Table 5
Volume-Weighted Epilimnetic Concentrations of Particulate and Dissolved Al, Fe, Mn, and Pb in the Acidified and Reference Basins of Little Rock Lake[a]

Metal	Particulate[b] Concentration (μg/L)		Dissolved[c] Concentration (μg/L)		Distribution Coefficient, logK$_D$ (L/kg)	
	Acidified Basin	Reference Basin	Acidified Basin	Reference Basin	Acidified Basin	Reference Basin
Al	2.4	5.4	19	11	5.0	5.5
Fe	7.4	10.3	77	24	4.9	5.4
Mn	0.24	0.74	48	9.1	3.6	4.7
Pb	0.03	0.16	0.07	0.05	5.5	6.3

[a] September, 1987: north basin pH, 5.12; total suspended solids (TSS), 1.3 mg/L; dissolved Ca, 1.2 mg/L. Reference basin pH, 6.13; TSS, 1.6 mg/L; dissolved Ca, 0.8 mg/L.
[b] Particulate matter retained on acid-cleaned Whatman GF/C glass fiber filters digested in 0.1 M HNO$_3$ (Ultrex) for 1 h at 25°C; solubilized Al, Fe, Mn, and Zn measured by graphite furnace AAS.
[c] Filtered through 0.4-μm pore-size Nuclepore polycarbonate membranes.

The largest of the above-mentioned changes has been the increase in H$^+$ caused by experimental acidification. Indeed, H$^+$ activity is approximately 25 times greater in the north basin at pH 4.7 than in the reference basin (pH 6.1). Because metal ions and H$^+$ compete for the same binding sites on organism surfaces, lower pH should result in less surface-bound metal and lower metal burdens in biota when other uptake and elimination mechanisms are not involved. Results of the surface complexation model [Equations (6)-(17)] for conditions similar to those of Little Rock Lake and a surface with carboxylic acid functional groups indeed show that the amount of both bound Ca and Cd decreases when the pH is lowered from 6.1 to 4.0 (Figure 9). The stability constants and surface site concentration for algae were obtained from Crist et al.[58] and Xue et al.[49] Although this is a simplified example, it is useful in illustrating the effect of H$^+$ competition for available surface sites. In addition, the observed increase of Ca in the acidified basin (Figure 8) will also contribute to the decrease in surface-bound Cd.

Measurement of Cd in periphyton from the lake (Figure 10A) agree at least qualitatively with the above prediction; levels from the acidified basin at pH 5.1 were 25% of those in the reference basin. On the other hand, interpretation of pH-related differences for Cd in organisms of other trophic levels is less straightforward. The ranges of Cd body burdens overlap for the midge (*Chaoborus*) and yellow perch (*Perca flavescens*) in the two basins, and Cd concentrations are higher in zooplankton (500-μm mesh size) and amphipods from the acidic basin. These results do not agree with predictions from the simple surface complexation model. Biological factors, such as differing mechanisms of metal uptake and elimination and differences in the life zones inhabited by various trophic levels (e.g., sediments versus the water column), may be more important than simple water chemistry factors, such as H$^+$ competition, in explaining Cd burdens in the vertebrate and invertebrate fauna of the lake. Also, the species composition of the zooplankton community in the north basin changed with acidification,[73] and Yan et al.[74] have shown that different species of zooplankton in a given lake can have larger differences in Cd burdens that those differences observed between basins in Little Rock Lake.

Furthermore, no differences in the Al content of all trophic levels were observed at pH 5.1 in the acidified basin compared with the reference basin (Figure 10B). The lack of response in the higher trophic levels perhaps can be explained by the same biological factors described above for Cd trends. At first, the lack of difference in periphyton Al burdens seems to contradict the surface complexation model, but the similar burdens *may* represent the canceling effects of several opposing trends. As pH declines, H^+ competition for surface sites increases (tending to lower Al accumulation by periphyton). However, at the same time, Al solubility increases, and its speciation changes toward an increasing fraction of Al^{3+}, which should be more strongly complexed than $ALOH^{2+}$ and $AL(OH)_2^+$. Actual dissolved concentrations of Al in the acidified basin increased about twofold in comparison with the reference basin over the pH range in question. Further studies are underway to determine whether these opposing trends provide an adequate explanation for the observed Al burdens in the lake's periphyton.

Equations

1) $RCOOH \rightleftharpoons RCOO^- + H^+$ pK=5.0

2) $RCOO^- + Cd^{+2} \rightleftharpoons RCOO-Cd^+$ pK=−3.8

3) $RCOO^- + Ca^{+2} \rightleftharpoons RCOO-Ca^+$ pK=−3.7

Conditions

$[Ca^{+2}]=2$ mg/L $S_T=0.22$ umol/L

$[Cd^{+2}]=0.05$ ug/L

Species	Percent pH=6.1	pH=4.0
{RCOO⁻}	75	23
{RCOOH}	6	72
{RCOO−Ca⁺}	19	5
{RCOO−Cd⁺}	0.0002	0.00007

FIGURE 9.   Surface complexation model results for simplified Little Rock Lake conditions.

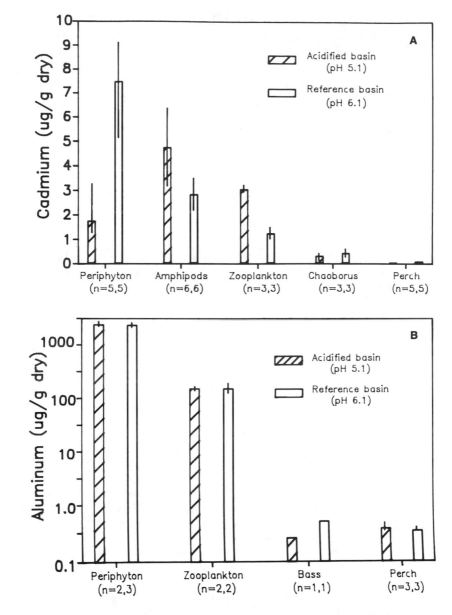

FIGURE 10. Mean (**A**) Cd and (**B**) Al burdens in Little Rock Lake biota. Biotic samples digested with HNO_3/H_2O_2 and analyzed by graphite furnace AAS. Bars define range of concentrations; n, number of analyses for each basin.

Finally, it is clear from Figure 10 that Cd and Al body burdens do not increase at higher trophic levels in the lake, and food chain biomagnification thus is not occurring. In fact, Cd and Al burdens are lowest in perch *(Perca flavescens)*, the highest trophic level sampled. Evidently, their mechanisms for controlling body burdens of nonessential toxic metals are more effective than those of organisms at lower trophic levels.

ACKNOWLEDGMENTS

Studies on Little Rock Lake described here were supported in part by a cooperative agreement with the U.S. Environmental Protection Agency through the Duluth (MN) Environmental Research Laboratory. This paper was not submitted to the EPA for review and no official endorsement should be inferred.

REFERENCES

1. Baes, C. F. and R. E. Mesmer. *The Hydrolysis of Cations. A Critical Review of Hydrolytic Species and Their Stability Constants in Aqueous Solution* (New York: John Wiley & Sons, 1976), pp. 489.
2. Morel, F. M. M. *Principles of Aquatic Chemistry* (New York: Interscience, 1983), pp. 466.
3. Westall, J. C., J. L. Zachary, and F. M. M. Morel. "MINEQL: A Computer Program for the Calculation of Chemical Equilibrium Composition of Aqueous Systems," Tech. Note 18. Dept. Civil. Eng., MIT Cambridge, MA (1976).
4. Plummer, L. N., B. F. Jones, and A. H. Truesdell. "WATEQF, a FORTRAN IV Version of WATEQ, a Computer Program for Calculating Chemical Equilibria of Natural Waters," Water Resour. Invest. Paper 76-13, U.S. Geol. Surv., Reston, VA (1976) (revised 1978, 1984).
5. Nordstrom, D. K., L. N. Plummer, T. M. L. Wigley, T. J. Wolery, J. W. Ball, E. A. Jenne, R. L. Bassett, D. A. Crerar, T. M. Florence, B. Fritz, M. Hoffman, G. R. Holdren, Jr., G. M. Lafon, S. V. Mattigod, R. E. McDuff, F. Morel, M. M. Reddy, G. Sposito, and J. Thrailkill. "Comparison of Computerized Chemical Models for Equilibrium Calculations in Aqueous Systems," in *Chemical Modeling in Aqueous Systems,* E. A. Jenne, Ed., ACS Symp. Ser. 93 (Washington, D.C.: American Chemical Society, 1979), pp. 857-892.
6. Ball, J. W., D. K. Nordstrom, and E. A. Jenne. "Additional and Revised Thermochemical Data and Computer Code for WATEQ2 — A Computerized Chemical Model for Trace and Major Element Speciation and Mineral Equilibria of Natural Waters," U.S. Geol. Surv. WRI 78-116 (1980).
7. Brezonik, P. L., L. A. Baker, J. R. Eaton, T. M. Forst, P. Garrison, T. K. Kratz, J. J. Magnuson, J. E. Perry, W. J. Rose, B. K. Shephard, W. A. Swenson, C. J. Watras, and K. E. Webster. "Experimental Acidification of Little Rock Lake, Wisconsin," *Water Air Soil Pollut.* 31:115-122 (1986).
8. Pagenkopf, G. K. "Gill Surface Interaction Model for Trace-Metal Toxicity to Fishes: Role of Complexation, pH, and Water Hardness," *Environ. Sci. Technol.* 17:342-347 (1983).

9. Honeyman, B. D. and P. H. Santschi. "Metals in Aquatic Systems," *Environ. Sci. Technol.* 22:862-871 (1988).

10. Harvey, G. R. and D. A. Boran. "Geochemistry of Humic Substances in Seawater," in *Humic Substances in Soil, Sediment, and Water,* G. R. Aiken, D. M. McKnight, R. L. Wershaw, and P. McCarthy, Eds. (New York: Interscience, 1985), pp. 233-247.

11. Tuschall, J. R. and P. L. Brezonik. "Characterization of Organic Nitrogen in Natural Waters: Its Molecular Size, Protein Content, and Interactions with Heavy Metals," *Limnol. Oceanogr.* 25:495-504 (1980).

12. Driscoll, C. J., J. P. Baker, J. J. Bisogni, and C. L. Schofield. "Effect of Aluminum Speciation on Fish in Dilute Acidified Waters," *Nature* 284:161-164 (1980).

13. Turner, D. R., M. S. Varney, M. Whitfield, R. F. Mantoura, and J. P. Riley. "Electrochemical Studies of Copper and Lead Complexation by Fulvic Acid. I. Potentiometric Measurements and a Critical Comparison of Metal Binding Models," *Geochim. Cosmochim. Acta* 50:289-297 (1986).

14. Fish, W., D. A. Dzombak, and F. M. M. Morel. "Metal-Humate Interactions. 2. Application and Comparison of Models," *Environ. Sci. Technol.* 20:676-683 (1986).

15. Dzombak, D. A., W. Fish, and F. M. M. Morel. "Metal-Humate Interactions. 1. Discrete Ligand and Continuous Distribution Models," *Environ. Sci. Technol.* 20:669-675 (1986).

16. Hering, J. G. and F. M. M. Morel. "Kinetics of Trace Metal Complexation: Role of Alkaline-earth Metals," *Environ. Sci. Technol.* 20:1469-1478 (1988).

17. Margerum, D. W., G. R. Cayley, D. C. Weatherburn, and G. K. Pagenkopf. "Kinetics and Mechanisms of Complex Formation and Ligand Exchange," in *Coordination Chemistry, Vol. 2,* A. E. Martell, Ed., ACS Monog. 174, (Washington, D.C.: Amer. Chem. Soc., 1978), pp. 1-218.

18. Stumm, W. and G. F. Lee. "Oxygenation of Ferrous Iron," *Ind. Eng. Chem.* 53:143-146 (1961).

19. Sung, W. and J. J. Morgan. "Kinetics and Product of Ferrous Iron Oxygenation in Aqueous Systems," *Environ. Sci. Technol.,* 14:561-568 (1980).

20. Wehrli, B. "Redox Reactions of Metal Ions at Mineral Surfaces," in *Aquatic Chemical Kinetics,* W. Stumm, Ed. (New York: Interscience, 1990), pp. 311-336.

21. Morgan, J. J. "Chemical Equilibria and Kinetic Properties of Manganese in Natural Waters," in *Principles and Applications of Water Chemistry,* S. D. Faust and J. V. Hunter, Eds. (New York: Interscience, 1967), pp. 561-624.

22. Richardson, L. L., C. Aquilar, and K. H. Nealson. "Manganese Oxidation in pH and O_2 Microenvironments Produced by Phytoplankton," *Limnol. Oceanogr.* 33:352-363 (1988).

23. Miles, C. J. and P. L. Brezonik. "Oxygen Consumption by a Photochemical Ferrous-ferric Catalytic Cycle," *Environ. Sci. Technol.* 15:1080-1095 (1981).

24. Waite, T. D. and F. M. M. Morel. "Photoreductive Dissolution of Colloidal Iron Oxide: Effect of Citrate," *J. Colloid Interface Sci.* 102:121-137 (1984).

25. Sunda, W. G., S. A. Huntsman, and G. R. Harvey. "Photoreduction of Manganese Oxides in Seawater and Its Geochemical and Biological Implications," *Nature* 301:234-235 (1983).

26. Waite, T. D., I. C. Wrigley, and R. Szymczak. "Photoassisted Dissolution of a Colloidal Manganese Oxide in the Presence of Fulvic Acid," *Environ. Sci. Technol.* 22:778-785 (1988).

27. Moffett, J. W. and R. G. Zika. "Reaction Kinetics of Hydrogen Peroxide with Copper and Iron in Seawater," *Environ. Sci. Technol.* 21:804-810 (1987).

28. Brezonik, P. L. "Principles of Linear Free Energy and Structure-Activity Relationships and Their Applications to the Fate of Chemicals in Aquatic Systems," in *Aquatic Chemical Kinetics,* W. Stumm, Ed. (New York: Interscience, 1990), pp. 113-143.

29. Elliott, S. "Linear Free Energy Techniques for Estimation of Metal Sulfide Complexation Constants," *Mar. Chem.* 24:203-213 (1988).

30. Turner, D. R., M. Whitfield, and A. G. Dickson. "The Equilibrium Speciation of Dissolved Components in Freshwater and Seawater at 25°C and 1 Atm Pressure," *Geochim. Cosmochim. Acta* 45:855-881 (1981).

31. Millero, F. J. "Thermodynamic Models for the State of Metal Ions in Sea Water," in *The Sea, Vol. 6,* E. D. Goldberg et al., Eds. (New York: Interscience, 1977), pp. 653-693.

32. Phillips, C. S. G. and R. J. P. Williams. *Inorganic Chemistry* (London: Clarendon Press, 1965).

33. Ahrland, S., J. Chatt, and N. R. Davies. "The Relative Affinities of Ligand Atoms for Acceptor Molecules and Ions," *Quart. Rev. Chem. Soc.* 12:265-276 (1958).

34. Pearson, R. G. "Hard and Soft Acids and Bases," *J. Am. Chem. Soc.* 25:3533-3539 (1963).

35. Pearson, R. G., Ed. *Hard and Soft Acids and Bases* (Pennsylvania: Dowden, Hutchinson & Ross, 1973).

36. Nieboer, E. and D. H. S. Richardson. "The Replacement of the Nonedescript term 'Heavy Metals' by a Biologically and Chemically Significant Classification of Metal Ions," *Environ. Pollut.* B1:3-26 (1980).

37. McDonald, D. G., J. P. Reader, and T. R. K. Dalziel. "The Combined Effects of pH and Trace Metals on Fish Ionoregulation," in *Acid Toxicity and Aquatic Animals,"* R. Morris, E. W. Taylor, D. J. A. Brown, and J. A. Brown, Eds., Soc. Exp. Biol. Sem. Ser. 34 (Cambridge: Cambridge Univ. Press, 1989), pp. 221-242.

38. Plankey, B. J., H. H. Patterson, and C. S. Cronan. "Kinetics of Aluminum Fluoride Complexation in Acidic Waters," *Environ. Sci. Technol.* 20:160-165 (1986).

39. Shaw, W. H. R. "Cation Toxicity and the Stability of Transition Metal Complexes," *Nature* 192:754-755 (1961).

40. Jones, M. M. and W. K. Vaughn. "HSAB Theory and Acute Metal Ion Toxicity and Detoxification Processes," *J. Inorg. Nucl. Chem.* 40:2081-88 (1978).

41. Ahrland, S. "Thermodynamics of Complex Formation between Hard and Soft Acceptors and Donors," *Structure Bonding* 5:118-149 (1968).

42. Williams, M. W. and J. E. Turner. "Comments on Softness Parameters and Metal Ion Toxicity," *J. Inorg. Nucl. Chem.* 43:1689-91 (1981).

43. Turner, J. E., M. W. Williams, K. B. Jacobson, and B. E. Hingerty. "Correlations of Acute Toxicity of Metal Ions and the Covalent/Ionic Character of their Bonds," in *Quantitative Structure Activity Relationships in Toxicology and Xenobiochemistry,* M. Tichy, Ed. (New York: Elsevier, 1985), pp. 1-8.

44. Kaiser, K. L. E. "Correlation and Prediction of Metal Toxicity to Aquatic Biota," *Can. J. Fish. Aquat. Sci.* 37:211-218 (1980).

45. Fisher, N. S. "On the Reactivity of Metals for Marine Phytoplankton," *Limnol. Oceanogr.* 31:443-449 (1986).

46. Sigg, L. "Surface Chemical Aspects and The Distribution and Fate of Metal Ions in Lakes," in *Aquatic Surface Chemistry,* W. Stumm, Ed. (New York: John Wiley & Sons, 1987), pp. 319-349.

47. Sigg, L. "Metal Transfer Mechanisms in Lakes; the Role of Settling Particles," in *Chemical Processes in Lakes,* W. Stumm, Ed. (New York: Interscience, 1985), pp. 283-307.

48. Whitfield, M. and D. R. Turner. "The Role of Particles in Regulating the Composition of Seawater," in *Aquatic Surface Chemistry,* W. Stumm, Ed. (New York: John Wiley & Sons, 1987), pp. 457-494.

49. Xue, H., W. Stumm, and L. Sigg. "The Binding of Heavy Metals to Algal Surfaces," *Water Res.* 22:917-926 (1988).

50. Neville, C. M. and P. G. C. Campbell. "Possible Mechanisms of Aluminum Toxicity in a Dilute, Acidic Environment to Fingerlings and Older Life Stages of Salmonids," *Water Air Soil Pollut.* 42:311-327 (1988).

51. Stumm, W. and J. J. Morgan. *Aquatic Chemistry,* 2nd ed. (New York: John Wiley & Sons, 1981).

52. Schindler, P. W. and W. Stumm. "The Surface Chemistry of Oxides, Hydroxides and Oxide Minerals," in *Aquatic Surface Chemistry,* W. Stumm, Ed. (New York: John Wiley & Sons, 1987), pp. 83-110.

53. Balistrieri, L. S. and J. W. Murray. "Metal-Solid Interactions in the Marine Environment: Estimating Apparent Equilibrium Binding Constants," *Geochim. Cosmochim. Acta* 47:1091-1098 (1983).

54. Wright, D. A. and J. W. Frain. "The Effect of Calcium on Cadmium Toxicity in the Freshwater Amphipod, *Gammarus pulex,* (L.)," *Arch. Environ. Contam. Toxicol.* 10:321-328 (1981).

55. Campbell, P. C. G. and P. Stokes. "Acidification and Toxicity of Metals to Aquatic Biota," *Can. J. Fish. Aquat. Sci.* 42:2034-2049 (1985).

56. Buffle, J., R. S. Altman, M. Filella, and A. Tessier. "Complexation by Natural Heterogeneous Compounds: Site Occupation Distribution Functions, A Normalized Description of Metal Complexation," *Geochim. Cosmochim. Acta* 54:1535-1553 (1990).

57. Davies, A. G. "Pollution Studies with Marine Plankton. Part 2. Heavy Metals," *Adv. Mar. Biol.* 15:381-508 (1978).

58. Crist, R. H., K. Oberholser, D. Schwartz, J. Marzoff, D. Ryder, and D. R. Crist. "Interactions of Metals and Protons with Algae," *Environ. Sci. Technol.* 22:755-760 (1988).

59. Campbell, P. G. C., P. M. Stokes, and J. N. Galloway. "Acid-Deposition: Effects on Geochemical Cycling and Bioavailability of Trace Elements," U.S. National Academy of Sciences, Royal Society of Canada, Academia de la Investigacion Cientifica of Mexico, Tri-Academy Committee of Acid Deposition, Sub-Group on Metals (Washington D.C.: National Academy of Sciences, 1985).

60. Brezonik, P. L., C. E. Mach, G. Downing, N. Richardson, and M. Brigham. "Effects of Acidification on Minor and Trace Metal Chemistry in Little Rock Lake, Wisconsin," *Environ. Toxicol. Chem.* 9:871-885 (1990).

61. Campbell, P. C. G. and A. Tessier. "Metal Speciation in Natural Waters: Influence of Environmental Acidification," in *Sources and Fates of Aquatic Pollutants,* R. A. Hites and S. J. Eisenreich, Eds., Adv. Chem. Ser. 216, (Washington, D.C.: American Chemical Society, 1987), pp. 185-208.

62. Mach, C. E. and P. L. Brezonik. "Trace Metal Research at Little Rock Lake, WI: Background Data, Enclosure Experiments, and the First Three Years of Acidification," *Sci. Total Environ.* 87/88:269-285 (1989).

63. Brezonik, P. L., L. A. Baker, N. Detenbeck, J. R. Eaton, T. M. Frost, P. J. Garrison, M. D. Johnson, T. K. Kratz, J. J. Magnuson, J. H. McCormick, J. E. Perry, W. J. Rose, B. K. Shephard, W. A. Swenson, C. J. Watras, and K. E. Webster. "Experimental Acidification of Little Rock Lake, Wisconsin: Baseline Studies and Predictions of Lake Responses to Acidification," Spec. Rep. No. 7, Water Resour. Res. Center, St. Paul, MN: University of Minnesota, 1985).

64. Downing, G. M. "Trace Metal Speciation and Response to Acidification in Sediments from Little Rock Lake, WI," M.S. Thesis, Univ. of Minnesota, Minneapolis, MN (1986).

65. Schindler, D. W., R. H. Hesslein, R. Wagemann, and W. S. Broecker. "Effects of Acidification on Mobilization of Heavy Metals and Radionuclides from the Sediments of a Freshwater Lake," *Can. J. Fish Aquat. Sci.* 37:373-377 (1980).

66. Schindler, D. W., R. Wagemenn, R. B. Cook, T. Ruszczynski, and J. Prokopowich. "Experimental Acidification of Lake 223, Experimental Lakes Area: Background Data and the First Three Years of Acidification," *Can. J. Fish. Aquat. Sci.* 37:342-354 (1980).

67. White, J. R. and C. T. Driscoll. "Manganese Cycling in an Acidic Adirondack Lake," *Biogeochemistry* 3:87-103 (1987).

68. White, J. R. and C. T. Driscoll. "Zinc Cycling in an Acidic Adirondack Lake," *Environ. Sci. Technol.* 21:211-216 (1987).

69. Holm, T. R., S. J. Eisenreich, H. L. Rosenberg, and N. L. Holm. "Groundwater Geochemistry of Short-term Aquifer Thermal Energy Storage Test Cycles," *Water Resour. Res.* 23:1005-1019 (1987).

70. Weir, E. P. "Acid Neutralization Processes in Little Rock Lake, WI: Laboratory and Whole-Lake Observations," M.S. Thesis, Univ. of Minnesota, Minneapolis, MN (1989).

71. Pott, D. B., J. J. Alberts, and A. W. Elzerman. "The Influence of pH on the Binding Capacity and Conditional Stability Constants of Aluminum and Naturally-Occurring Organic Matter," *Chem. Geol.* 48:293-304 (1985).

72. Helmer, E. H., N. R. Urban, and S. J. Eisenreich. "Aluminum Geochemistry in Peatland Waters," *Biogeochemistry* 9:247-276 (1990).

73. Frost, T. and P. K. Montz. "Early Zooplankton Response to Experimental Acidification in Little Rock Lake, Wisconsin, U.S.A.," *Verh. Int. Ver. Limnol.* 23:2279-85 (1988).

74. Yan, N. D., G. L. Mackie and P. L. Dillon. "Cadmium Concentrations of Crustacean Zooplankton of Acidified and Nonacidified Canadian Shield Lakes," *Environ. Sci. Technol.* 24:1367-72 (1990).

75. Paxeus, N. and M. Wedborg. "Acid-Base Properties of Aquatic Fulvic Acid," *Anal. Chim. Acta,* 169:87-98 (1985).

76. Tipping, E., C. A. Backes, and M. A. Hurley. "The Complexation of Protons, Aluminum and Calcium by Aquatic Humic Substances: A Model Incorporating Binding-Site Heterogeneity and Macroionic Effects," *Water Res.* 22:597-611 (1988).

77. Lovgren, L., T. Hedlung, L. O. Öhman, and S. Sjoberg. "Equilibrium Approaches to Natural Water Systems. 6. Acid-base Properties of a Concentrated Bog Water and Its Complexation Reactions with Aluminum (III)," *Water Res.* 21:1401-1407 (1987).

78. Ephraim, J., S. Alegret, A. Mathuthu, M. Bicking, R. L. Malcolm, and J. A. Marinsky. "A Unified Physico-chemical Description to the Protonation and Metal Ion Complexation Equilibria of Natural Organic Acids (Humic and Fulvic Acids). 2. Influence of Polyelectrolyte Properties and Functional Group Heterogeneity on the Protonation Equilibria of Fulvic Acid," *Environ. Sci. Technol.* 30:354-366 (1986).

79. Mantoura, R. F. C., A. Dickson, and J. P. Riley. "The Complexation of Metals with Humic Materials in Natural Waters," *Estuar. Coastal Mar. Sci.* 6:387-408 (1978).

80. Buffle, J. "Natural Organic Matter and Metal-Organic Interactions in Natural Waters," in *Metal Ions in Biological Systems, Vol. 18,* H. Sigel, Ed. (New York: Dekker, 1984), pp. 165-221.

81. McKnight, D. M., G. L. Feder, E. M. Thurman, R. L. Wershaw, and J. C. Westall. "Complexation of Copper by Aquatic Humic Substances from Different Environments," *Sci. Total Environ.* 28:65-76 (1983).

82. Langmuir, D. "Techniques of Estimating Thermodynamic Properties for Some Aqueous Complexes of Geochemical Interest," in *Chemical Modeling in Aquatic Systems,* E. A. Jenne, Ed., ACS Symp. Ser. 93 (Washington, D.C.: American Chemical Society, 1979), pp. 353-387.

2

Evaluating the Effectiveness of Metal Pollution Controls in a Smelter by Using Metallothionein and Other Biochemical Responses in Fish

John F. Klaverkamp, Michael D. Dutton, Henry S. Majewski, Robert V. Hunt, and Laurie J. Wesson

Department of Fisheries and Oceans, Freshwater Institute, 501 University Crescent, Winnipeg, Manitoba R3T 2N6

OVERVIEW

White suckers *(Catostomus commersoni)* and sediments from lakes affected by atmospheric deposition of Cu, Cd, and Zn emitted by a base metal smelter were assessed before and after the installation of electrostatic precipitators in 1982. In 1981, suckers from a lake (Hamell) close (5.8 km) to the smelter had higher concentrations of hepatic and renal "metallothionein" (MTN) and were more resistant to Cd toxicity than those caught from a lake (Thompson) approximately 20 km away. Sediment core profiles of metal concentrations in these two lakes demonstrated the influence of a tall stack (251 m), which was installed in 1974, and the reduced quantities of metals deposited on lakes close to the smelter. In 1986, MTN and other biochemical indicators of metal exposure and stress in suckers and sediment core profiles of metal concentrations from six other lakes in the vicinity (9.8 to 31.0 km) of the smelter were analyzed. Suckers could not be caught in Hamell Lake, nearest to the smelter, in 1986. Although metal concentrations in surficial sediments from these six lakes decreased with increasing distance from the smelter, there were no pronounced decreases in these concentrations that could be attributed to the 1982 installation of electro-static precipitators. Hepatic and renal MTN concentrations were not less in 1986 samples than those collected in 1981. MTN concentrations in kidney of 1986 samples were inversely related to the distance of the lake to the smelter. MTN, estimated by a ^{203}Hg displacement method in 1986, demonstrated a complete saturation of metal-binding sites in kidneys from fish in lakes close to the smelter. The physiological significance of this saturation was assessed by measuring ascorbic acid and acid-soluble thiols in kidney and electrolytes in plasma. Elec-trolyte differences in sucker plasma corresponded with elevated tissue metal burdens. Decreased plasma Na and K were observed in suckers from a lake (Nesootao) over 12 km from the smelter, and similar disturbances were observed in suckers caught as far as 25 km away (Neso Lake). Suckers from these two lakes had the two highest concentrations of total metals and MTN in their livers. Acid-soluble thiol concentrations in posterior kidney were elevated in Nesootao Lake white suckers, indicating an elevation in MTN, and possibly glutathione, in response to metal exposure. Renal MTN concentrations were highest in these fish. Suckers from this lake had significantly elevated Cd concentrations in liver and kidney. Ascorbic acid concentrations in posterior kidney increased with decreasing distance from the smelter. Increases in ascorbic acid were related more to the degree of sediment metal than to the degree of tissue metal loading in these fish. In comparing the biochemical and sediment results obtained in 1981 and 1986, the installation of electrostatic precipitators in 1982 appears to have had little or no effect in reducing metal pollution in nearby lakes.

INTRODUCTION

Since 1930, the metal smelter in Flin Flon, Manitoba, Canada has been in operation and discharging sulfur dioxide, Cd, Cu, Zn, and other elements to the atmosphere.[1] The smelter is located in a remote area 600 km northwest of Winnipeg, Manitoba and is near the southwest boundary of the Precambrian Shield. This area is in a geological transition zone, with most rock being Pre-cambrian in origin; however, outliers of dolomite and Pleistocene glacial debris are also observed.[2] Additional details on the history of development, smelter product and waste processes, geology and climate, and contaminant discharges have been described.[1]

During the past two decades, operations have been modified to decrease contaminant deposition within the immediate area of the town and reduce par-ticulate emissions to the atmosphere. In 1974, a 251-m stack was installed to replace two older stacks of 58 and 69 m in height. The tall stack resulted in less deposition of sulfuric acid and metals in the town, but it spread these contaminants over a wider geographical area. In the late 1970s, the smelter was discharging 2,480 t of Zn, 135 t of Cu, and 65 t of Cd to the atmosphere per year.[3] In 1982, electrostatic precipitators were installed in the stack, resulting in reductions of particulate emission[4] by 85% and 20-fold reductions in Zn.[5]

Aquatic ecosystems in the vicinity of the smelter have been analyzed throughout this period to monitor the extent of contamination, nature of adverse effects, and effectiveness of pollution controls. Analyses of water samples collected from 31 lakes in 1973 and 1974 demonstrated that As, Cu, and sulfate concentrations were much higher in lakes near the smelter.[6] These investigators also collected fish from eight lakes close to the smelter and concluded that these fish "were more tolerant of these high zinc concentrations than would be expected on the basis of the responses of fish and other aquatic organisms to similar concentrations of zinc in some laboratory toxicity tests " From studies conducted in 1976 and 1977, however, McFarlane and Franzin[7] concluded that fish from Hamell Lake, located about 5 km from the smelter, were experiencing reduced spawning success, reduced larval and egg survival, smaller egg size, and reduced longevity. From rain and snow samples collected during these 2 years, the smelter was found to be the source of elevated As, Cd, Cu, Pb, and Zn deposited over an area of 250,000 km^2 surrounding the smelter.[8] These authors also concluded that atmospheric emissions of Cd, Cu, and Zn were transported distances of up to 48, 264, and 284 km, respectively. Accumulation of metals by fish from lakes in the area was related to distance from the smelter and Ca concentrations in water.[7] More recent estimates[9] of atmospheric transport using enrichment of surficial metals in peatlands show that deposition is negligible beyond 100 km for Cu and 77 km for Zn. The predominant wind direction is from the northwest.[8,9]

A recent study[10] provided more extensive analyses of metal accumulation by fish in affected lakes and an evaluation of pollution control measures (electrostatic precipitators) installed in 1982. This study found that (1) limnological parameters of lake size, alkalinity, dissolved inorganic and organic carbon (DIC and DOC, respectively), conductivity, and Ca were not strongly correlated with metal concentrations in the bottom-feeding white sucker *(Catostomus commersoni)* and the top predator, northern pike *(Esox lucius)*; (2) metal concentrations in surface sediments were poor predictors of metal concentrations in these fish; (3) concentrations of Cd and Se in liver of northern pike and of Cd, Cu, Se, and Zn in liver of white sucker were highest in fish from the most contaminated lakes; (4) the tall stack was probably responsible for decreased concentrations of some metals in livers from these fish and in surficial sediments from lakes close to the smelter; and (5) the area receiving significant metal deposition is considerably smaller ($\leq$ 10,000 km^2) than previous estimates. Sections of sediment cores indicated that deposition of Cd, Cu, and Zn on lakes did not extend beyond 68 km northwest of the smelter and that the eastern limit of the deposition zone was about 45 km. This apparent decrease in affected geographical area may be due, in part, to reductions in particulate emissions.

To evaluate the relevance of these decreased metal loadings to fish from affected lakes, knowledge of the adverse effects produced by the metals in natural fish populations is required. The use of biochemical indicators of contaminant-induced stress in natural fish populations is a rapidly developing area in ecotoxicology.[11-13] These indicators, for example, have recently been used to demonstrate that Atlantic salmon from acidified Nova Scotia rivers have impaired osmoregulatory abilities, altered thyroid hormones, depleted acid-soluble-thiols (AST), and ascorbic acid (AsA) reserves, elevated liver glycogen, and decreased Ca and P in bone.[14,15]

One of the most promising biochemical indicators for exposure to Cd, Cu, and Zn is the family of metal-binding proteins, the metallothioneins.[16-18] Synthesis of these cysteine-rich proteins is induced by exposure to metals in groups IB and IIB of the periodic table.[19-21] Metallothioneins (MTN) are thought to function in the transport, storage, and detoxification of these metals.[22-25]

The purposes of this chapter are twofold. First, comparisons are made of metal concentrations in sections of sediment cores taken from lakes receiving atmospheric deposition of metals before and after the installation of electrostatic precipitators in 1982. Second, the availability of these metals to and responses by the bottom-feeding white sucker are evaluated by analyzing for MTN and other biochemical indicators of metal exposure and toxicity.

MATERIALS AND METHODS
Study Lakes and Water Sampling

Figure 1 presents a map of the study area, and the surface areas, maximum depths, and distances from the smelter of lakes. Chemical parameters of the study lakes are presented in Table 1. The study area is near the southwest boundary of the Precambrian Shield. Surface water samples were collected in acid-washed polyethylene bottles. Analyses for pH, SO_4^{2-}, Ca^{2+}, Mg^{2+}, Na^+, Cl^-, dissolved organic carbon (DOC), dissolved inorganic carbon (DIC), and conductivity were conducted by the methods of Stainton et al.[26] Alkalinity was calculated by the method of Schindler et al.[27] as modified by Harrison and Klaverkamp:[10] Alkalinity (mEq/L) = $[Na^+]$ + $[K^+]$ + $[Ca^{2+}]$ + $[Mg^{2+}]$ − $[Cl^-]$ − $[SO_4^{2-}]$.

Sediment Sampling and Metal Analysis

Sediment cores were collected as described by Harrison and Klaverkamp[10] by Scuba at maximum depth in each lake, with the exception of Twin Lake, where cores were collected in 1986 at the maximum depth (8 m) of the smaller (east) basin of the lake. In 1981, only one core was obtained from the deepest parts of Hamell and Thompson Lakes. Each core was sectioned into 0.5-cm segments for 10 cm immediate below the sediment-water interface and into 3.0-cm segments from 20 cm and deeper. In 1986, four cores were obtained from each of the six lakes. From each core, ten sections of 1-cm thickness were removed

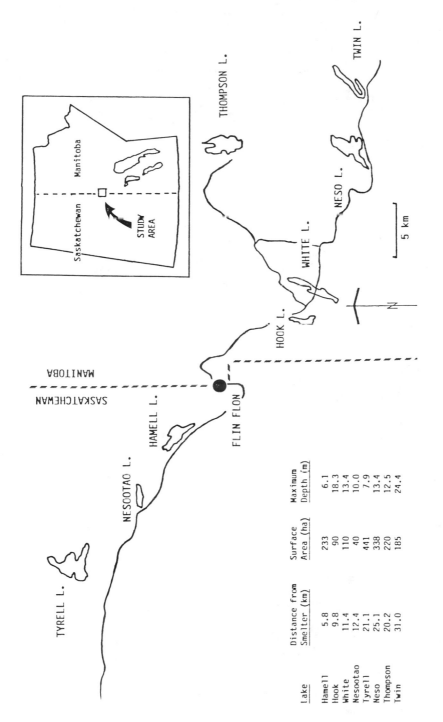

FIGURE 1. Map of the study area and physical parameters of the lakes.

Lake	Distance from Smelter (km)	Surface Area (ha)	Maximum Depth (m)
Hamell	5.8	233	6.1
Hook	9.8	90	18.3
White	11.4	110	13.4
Nesootao	12.4	40	10.0
Tyrell	21.1	441	7.9
Neso	25.1	338	13.4
Thompson	20.2	220	12.5
Twin	31.0	185	24.4

Table 1
Chemical Parameters of Study Lakes

	1981		1986					
	Hamell	Thompson	Hook	White	Nesootao	Tyrell	Neso	Twin
Sodium (mg/L)	1.45	1.91	2.12	2.00	1.59	1.49	1.37	1.21
Potassium (mg/L)	1.37	1.11	1.25	1.05	1.09	0.97	0.71	0.69
Calcium (mg/L)	14.7	15.8	35.6	16.3	8.7	5.6	8.2	11.5
Magnesium (mg/L)	2.87	4.22	7.53	5.66	2.04	1.92	2.43	3.43
Chloride (mg/L)	2.8	0.4	2.6	1.3	1.0	0.7	0.7	0.5
Sulfate (mg/L)	14.60	8.60	12.40	9.13	10.00	5.77	3.27	3.84
Conductivity (μS/cm)	100	110	229	126	70	50	65	83
pH	7.5	7.6	8.3	8.0	7.5	7.4	7.6	7.8
DIC (as HCO_3) (mg/L)	46.4	59.8	119.6	44.05	11.6	10.4	7.3	28.1
DOC (as C) (mg/L)	10.0	12.4	10.0	14.6	15.5	15.1	14.5	12.6
Alkalinity (mEq/L)	687	1060	2194	1168	464	389	600	832

immediately below the sediment-water interface. Below 10 cm, 3-cm sections were removed to a maximum depth of 25 cm.

Sediment core sections were dried to constant weight, powdered, and digested by a nitric, perchloric, and hydrofluoric acid digestion procedure.[28] Cd, Cu, and Zn were analyzed by atomic absorption on a Varian AA-5 using an air-acetylene flame. Only Cd reached detection limits of the method (1.0 μg/g) in the deepest core sections.

Fish Sampling and Metal Analysis

In 1981, adult white suckers were captured from Hamell or Thompson Lakes by gill nets in the evenings before a toxicity test (see below) and were held in pens of nylon netting and wood frame (2.0 m × 1.5 m × 1.5 m). Fish were removed from the gill nets at 20- to 30-min intervals. The next morning five suckers were removed from the holding pen for the toxicity test conducted over the course of that day. Eight additional suckers were removed from the holding pen, anesthetized, and dissected to remove liver and kidney for analyses of metals and MTN.

In 1986, adult white suckers were also captured using gill nets. Fish were removed from the nets approximately every 30 min. If more fish were collected than could be processed within 10 min, the fish were transferred to a nylon-net holding pen. Fish were anesthetized with buffered tricaine methanesulfonate (MS-222). Livers and kidneys were removed, immediately frozen in liquid nitrogen, and stored on dry ice in the field. Samples were stored at −120°C upon return to the laboratory.

Liver samples (approximately 0.5 g wet wt) were digested in batches with nitric and sulfuric acids, and hydrogen peroxide.[29] Samples were made up to 25 mL with distilled, deionized water. Cd, Cu, and Zn were measured by atomic absorption on a Varian AA-5 with background correction, using an air-acetylene flame.

For comparisons of (wet wt) tissue metal determinations made in this study with dry weight determinations from other studies, we assume that dry wt = 0.25 (wet wt).

Toxicity Tests (1981)

Toxicity tests were conducted on July 13, 14, and 16 in Thompson Lake, and on July 26 and 29 in Hamell Lake. On the morning of a test, five circular fiberglass tubs of 1000-L capacity received 750 L of lake water. Tubs were placed into shallow areas within the lake close to the shoreline. Lake water circulated around the partially submerged tubs to maintain a reasonably constant temperature over the course of a test. In general, a temperature increase of 1°C occurred during a test. The maximum temperature observed was a 2°C increase to 22°C over the course of a day. Individual suckers, captured the night before, were placed into each tub at the start of a test.

For each test, one tub served as a control and four others received high concentrations of Cd. Preliminary experiments demonstrated that concentrations from 10 mg Cd/L to 100 mg Cd/L produced death in 13 h or less. Rates of respiration in control suckers remained constant at 25 ± 2 opercular beats per minute over the 13-h duration of experiments. There were no apparent signs of stress in these fish. Exposure to Cd produced a pronounced increase in respiration rate, up to 100 beats per minute in the early hours of a test, but, in the later stages, there was a dramatic decrease in rate. Cadmium-exposed fish swam erratically at the water-air surface and completely lost equilibrium. Criterion of death was cessation of respiration for 5 min.

When a test was completed, length, weight, and sex data were obtained; pectoral fins and scales were removed for aging.[29] Eight additional suckers were removed from the holding pens in each lake. These fish were anesthetized and dissected to remove liver and kidney for metal and MTN analyses using gel filtration.

MTN Analysis by Gel Filtration Chromatography

In 1981, livers from two fish were pooled and homogenized in 0.9% (w/v) saline 1:1 using a Polytron at highest speed for 30 sec. The same procedure was used for kidneys. Four gel filtration analyses of each organ were conducted on the eight fish from each of Thompson and Hamell Lakes.

A portion of the homogenate was analyzed for total concentrations of Cd, Cu, and Zn in the organ, and the remainder was centrifuged at 27,000 g for 20 min at 3°C. The supernatant was heated at 70°C for 4 min and centrifuged again at 27,000 g for 20 min at 3°C. This supernatant was applied to a gel filtration column (1.6 × 100 cm) containing Sephadex G-75 and was eluted with ammonium formate buffer (0.09 M), pH 8.0, at a flow rate of 0.4 mL/min. The column was calibrated using a kit of low-molecular weight proteins (Pharmacia Fine Chemicals). Fractions of 7.5 mL were collected and analyzed by flame atomic absorption spectrophotometry for Cd, Cu, and Zn without digestion.[23]

MTN in these chromatographed cytosols was determined by coelution of Cd, Cu, and Zn in chromatographic fractions corresponding to the elution volume of MTN from column calibration. MTN concentration in the cytosols (nmol MTN/g) was calculated under the assumption that MTN binds 7 g-atom of Zn or Cd per mol[25] and 12 g-atom Cu per mol.[19]

In 1986, pooled samples of liver and kidney of suckers from each lake were homogenized using a Polytron in two volumes of 10 nM phosphate buffer, pH 7.6, with 5 nM 2-mercaptoethanol, 0.02% (w/v) sodium azide, and 0.15 M KCl. Phenylmethylsulfonylfluoride (PMSF) was added to the mixture (0.1 M) as a protease inhibitor. Homogenates were centrifuged at 30,000 g for 30 min at 4°C. The surface lipid layer was removed from the surface of the supernatant.

The final supernatant (cytosol) was further fractionated by gel filtration chromatography. The cytosol was filtered through a 0.45-μm cellulose acetate filter and chromatographed on a 1.9 × 100 cm column of Sephadex G-75 at a flow rate of 0.5 mL/min. Fractions of 10 mL were collected and analyzed for Cd, Cu, and Zn as described above. The chromatography column was calibrated with Blue Dextran 2000, bovine serum albumin, ovalbumin, chymotrypsinogen A, ribonuclease A, and MTN (Sigma rabbit liver MTN-II). MTN in chromatographed cytosols was determined as described above.

MTN Analysis by Mercury Saturation

In 1986 only, MTN was also estimated using a Hg-saturation assay.[30] Briefly, the tissues were homogenized in four volumes of 0.9% (w/v) NaCl and heat-treated at 100°C for 10 min in 1.5-mL polypropylene microcentrifuge tubes. The heat-treated homogenates were cooled on ice for 5 min and centrifuged for 10 min at 10,000 g at room temperature in a benchtop microcentrifuge (Eppendorf 5412). The resulting supernatants were stored at −120°C until analyzed.

The assay was performed on four serial dilutions of each sample. Four 1.5-mL microcentrifuge tubes were prepared for each dilution series. NaCl (200 μL of 0.9%) (w/v) were pipetted into the second and third tube, and 100 μL were added to the fourth tube. Heat-treated supernatant (300 μL) was added to the first tube; 100 μL were withdrawn and transferred to the second tube and mixed by several draws of the pipetter. The procedure was repeated for the third and fourth tubes to provide 200 μL of 1:1, 1:3, 1:9, and 1:18 dilutions of the sample in the centrifuge tubes, respectively.

Mercury was used to saturate the metal-binding sites of MTN. To the tubes, 200 μL of ^{203}Hg-labeled $HgCl_2$ (containing 10 μg Hg and 10,000 cpm for liver analyses; 5 μg Hg and 5000 cpm for kidney analyses) in 10% (w/v) trichloroacetic acid (TCA) were added, mixed by vortex, and incubated for 10 min at room temperature. To end the saturation step, 400 μL of 50% (w/w) hens' egg white in 0.9% (w/v) NaCl was added to the tubes. Egg white denatures on contact with the acid and is removed from solution with non-MTN-bound Hg by centrifugation (10,000 g for 3 min). Total activity samples and egg white blanks were assayed with each set of unknowns to provide the specific activity used in the assay and to ensure that the added egg white was sufficient to remove all the unbound Hg in the assay tubes, respectively. The molar binding capacity of the MTN in each sample was determined using the following equation:

$$\text{nmol Hg bound/g tissue} = (\text{cpm[sample]} - \text{cpm[blank]})/$$

$$(\text{cpm[total]}) \times (10 \ \mu\text{g Hg/sample})/(0.2006 \ \mu\text{g Hg} \cdot \text{nmol}^{-1})/$$

$$(0.2 \ \text{mL/sample}) \times \text{tissue dilution}$$

where tissue dilution refers to g of supernatant per g of tissue.
The nmol Hg bound per gram of tissue can be converted to nmol MTN/g, based on the assumption that MTN binds 7 g-atom/mol.[25]

Biochemical Analysis (1986)

Plasma osmolality was measured by freezing point depression (Precision Osmette Model 5004). Na and K were measured by flame photometry (IL943) and chloride was determined coulometrically with a Corning Model 925 chloride titrator. Plasma Ca^{2+} concentrations were measured using a Sigma test kit (No. 586) as were plasma glucose (Sigma test kit No. 510) and protein (Sigma test kit No. P5656).

Acid-soluble thiol concentrations in liver and kidney were determined using the 5,5'-dithiobis(2-nitrobenzoic acid) (DTNB) method of Moron et al.[31] Tissue ascorbic acid concentrations were analyzed using Folin-Ciocalteu phenol reagent in acidic pH.[32] Tissue protein was measured using a modification of the micro-Lowry method.[33]

Statistical Analysis (1986)

Analyses of variance were computed using the GLM procedure of the Statistical Analysis System.[34] A two-way factorial arrangement using group and sex was applied to test for significant differences in fish from the terminal sampling. Differences between group means were examined using Tukey's Studentized Range test. White suckers from sampling sites in the vicinity of the smelter were compared with reference fish from the most distant lake (Twin Lake). Statistically significant differences between groups have been indicated at $p < 0.05$ and $p < 0.10$. Multiple regression analysis was used to develop predictive equations for metal concentrations in liver and kidney of fish, as well as to determine the relationship between surficial sediment metals and limnological variables. Pearson product-moment correlation coefficients were used to establish significant ($p < 0.1$) relationships between measured biochemical parameters and metal levels in either surficial sediment or fish tissue.

RESULTS

Because of previous research on suckers in Hamell and Thompson Lakes, these lakes were chosen for studies in 1981 on sediment core profiles of Cd, Cu, and Zn and on MTN content in liver and kidney. In 1986, suckers could not be captured in Hamell Lake, however, despite setting nets in areas where they had been caught in previous years. Criteria for lake selection in 1986 were ease of access for boats and sampling gear, and increasing distances from the smelter in both easterly and westerly directions (Figure 1). From a previous survey,[10] the lakes selected in 1986 were known to be in the geographical area receiving atmospheric deposition of metals emitted by the smelter.

The lakes ranged in size from 40 ha (Nesootao) to 441 ha (Tyrell). Maximum depths ranged from 6.0 m to 24.5 m (Figure 1). Sulfate concentrations decreased with increased distance from the smelter ($r = -0.94$), but pH was at 7.4 or above for all the lakes (Table 1). K and Cl were also inversely related, $r = -0.92$ and -0.82, respectively, to distance from the smelter. Alkalinity, dissolved inorganic carbon (DIC), and Ca showed no statistically significant relationships ($r = -0.33$, -0.46, and -0.47, respectively) to distance from the smelter.

Profiles of metal concentrations in individual sediment cores obtained in 1981 for Hamell and Thompson Lakes are presented in Figure 2. Although estimates of dates were not obtained for these samples, the top 17 to 20 cm of sediment illustrates anthropogenic increases in metal concentrations over natural background found at 23 cm and deeper. The reductions in metal concentrations from about 5 cm to the sediment-water interface (0 cm) probably represent decreased metal deposition onto lakes close to the smelter due to installation of the taller stack. It is reasonable, therefore, to assume that 17 to 20 cm depth in cores represents the onset of smelter operations in 1930 and that decreases in metal concentrations beginning at 4 to 5 cm in Hamell Lake represent the installation of the "super" stack in 1974 (see Discussion).

Concentrations of metals in "surficial" (Table 2) and "peak" segments of sediment cores are well above "background," indicating that all lakes were receiving atmospheric deposition of metals. With the exception of Cu in sediments from Thompson Lake, ratios of surficial or peak to background metal concentrations are higher in Hamell, Hook, White, and Nesootao Lakes (Table 3), which are closer (>13 km) to the smelter than Thompson, Tyrell, Neso, and Twin Lakes (>20 km). Because the peak for 1986 is 0.25 cm deeper than those sections used for peak in 1981, they probably represent approximately the same time and can be compared to surficial segments in order to determine the influence of the tall stack and electrostatic precipitators. The surficial sediment sections of Hamell Lake contained 46% of Cu, 63% of Zn, and 47% of Cd found in the peak segments. This demonstrates the effect of the tall stack in reducing metal deposition in areas close to the smelter. Consistent with this are the slight increases, 103% for Cu and 114% for Zn, of surficial Cu and Zn concentrations over those observed in peak segments in Thompson Lake. In the 1986 samples, with the exception of Zn in sediments of White Lake, concentrations of Cd, Cu, and Zn in surficial sediments were less than those in peak sections. These decreases, which reflect the additional influence of electrostatic precipitators, are small, especially in comparison to decreases in total Cd, Cu, and Zn observed in Tyrell Lake where surficial sections contained 80, 84, and 84% of their respective concentrations in peak segments.

Comparisons of metal concentrations in livers and kidneys from suckers obtained in 1981 (Table 4) to those captured in 1986 (Table 5) did not show pronounced decreases over this time. In 1981, with the exception of Cu in liver, fish from Hamell Lake had higher concentrations of metals in their livers and kidneys than those observed in Thompson Lake fish. In kidneys, Cd, Cu, and Zn concentrations were approximately 2.7, 2.3, and 2.5, respectively, times higher in suckers from Hamell Lake. With the exception of fish from Neso Lake, mean Cu concentrations in liver were lower in 1986 than those observed in 1981, and renal Cu concentrations were also lower in the 1986 samples.

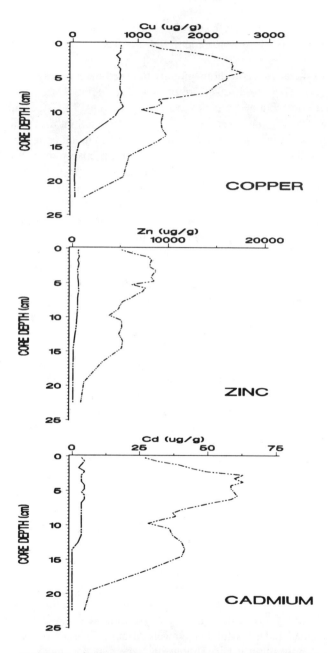

FIGURE 2.   Concentrations (μg/g, dry weight) of Cu (top panel), Zn (middle panel), and Cd (bottom panel) in cores of sediments obtained from the deepest parts of Thompson and Hamell Lakes. The lines showing highest concentrations for each metal represent sediment from Hamell Lake.

Table 2
Concentrations (µg/g, Dry Weight) of Cu, Zn, and Cd in Sediment Cores

	Surficial[b]			Peak[c]			Background[d]		
	Cu	Zn	Cd	Cu	Zn	Cd	Cu	Zn	Cd
1981									
Hamell	1,180	5,140	27	2,580	8,230	58	87	150	1
Thompson	740	630	3	720	550	4	27	99	<1
1986[a]									
Hook	793 ± 110	9,820 ± 929	65 ± 3	1,245 ± 67	23,900 ± 4,140	92 ± 5	65 ± 3	92 ± 5	1 ± 0.5
White	258 ± 12	4,497 ± 486	8 ± 1	288 ± 21	3,730 ± 822	8 ± 1	55 ± 14	123 ± 32	0.6 ± 0.1
Nesootao	413 ± 34	2,430 ± 223	14 ± 1	433 ± 8	2,645 ± 74	15 ± 0.5	42.2 ± 2	79 ± 5	1 ± 0.5
Tyrell	130 ± 4	900 ± 27	4 ± 0	153 ± 5	1,078 ± 66	5 ± 0	35 ± 2	96 ± 8	1 ± 0.5
Neso	105 ± 3	688 ± 98	2.3 ± 0.3	115 ± 3	783 ± 16	2.3 ± 0.3	30 ± 1	110 ± 4	1 ± 0.5
Twin	130 ± 4	688 ± 98	3.3 ± 0.3	145 ± 3	853 ± 74	4.0 ± 0.4	66 ± 1	94 ± 3	1 ± 0.5

a Data are expressed as X ± SEM, n = 4.
b Surficial were concentrations in the top 0.5 cm in 1981 and the top 1.0 cm in 1986.
c Peak were concentrations at 4.25 cm in 1981 and at 4.5 cm in 1986.
d Background were concentrations at 23 cm or deeper.

Table 3
Ratios of Concentrations of Cu, Zn, and Cd in Surficial and Peak Segments to Background Concentrations

Lake	DFS[a]	Cu			Zn			Cd		
		S[b]	P[c]	S/P(%)	S	P	S/P(%)	S	P	S/P(%)
Hamell	5.8	13.6	29.7	45.8	34.3	54.9	62.5	27	58	46.6
Hook	9.8	12.2	19.2	63.5	106.7	259.8	41.1	65	92	70.7
White	11.4	4.7	5.2	90.4	36.6	30.3	120.8	13.3	13.3	100
Nesootao	12.4	9.8	10.3	95	30.8	33.5	91.9	14	15	93.3
Thompson	20.2	27.4	26.7	102.6	6.4	5.6	114.3	3	4	75
Tyrell	21.1	3.7	4.4	84.0	9.4	11.2	83.9	4	5	90
Neso	25.1	3.5	3.8	92.1	6.2	7.1	87.3	2.3	2.3	100
Twin	31.0	2.0	2.2	90.9	7.3	9.1	90.2	3.3	4.0	92.5

a DFS, distance (km) from smelter.
b S, surficial (see Table 2).
c P, peak (see Table 2).

Table 4
Concentrations of Cu, Zn, and Cd in Livers and Kidneys of White Suckers Caught in 1981

Lake	DFS	n	Length(mm)	Wt(g)	M	F	Liver			Kidney		
							[Cu]	[Zn]	[Cd]	[Cu]	[Zn]	[Cd]
Hamell	5.8	8	447 ± 31	1314 ± 325	6	2	386 ± 94	593 ± 134	12 ± 12	143 ± 68	721 ± 128	41 ± 35
Thompson	20.2	8	387 ± 26	851 ± 140	3	5	508 ± 191	437 ± 71	4 ± 2	63 ± 11	288 ± 11	15 ± 15

a n = 4; data expressed as X ± SD.

Table 5
Characteristics of White Suckers Caught in 1986 and Metal Concentrations in Their Livers ([Me] Liver) and Kidneys ([Me] Kidney)[a]

Lake	DFS	n	Length(mm)	Wt(g)	M	F	Age	Liver [Cu]	Liver [Zn]	Liver [Cd]	Kidney [Cu]	Kidney [Zn]	Kidney [Cd]
Hook	9.8	10	396 ± 23	873 ± 150	8	2	10.2 ± 1.3	287 ± 154	464 ± 131	8.7 ± 3.6	—	NA[b]	—
White	11.4	8	393 ± 21	799 ± 123	3	5	9.9 ± 1.9	343 ± 162	504 ± 77	6.5 ± 2.9	—	NA	—
Nesootao	12.4	8	386 ± 29	853 ± 193	1	7	7.9 ± 1.1	346 ± 97	598 ± 162	16.9 ± 9.4	41.6 ± 13	568 ± 157	33.1 ± 9.4
Tyrell	21.1	9	378 ± 30	881 ± 167	4	5	10.1 ± 1.7	144 ± 104	351 ± 34	3.2 ± 1.7	31.7 ± 6.0	380 ± 58	4.9 ± 4.9
Neso	25.1	11	389 ± 26	807 ± 169	7	4	9.3 ± 2.7	397 ± 148	495 ± 79	9.0 ± 5.	35.1 ± 11.4	359 ± 57	19.0 ± 16.0
Twin	31.0	11	419 ± 34	1045 ± 266	5	6	8.5 ± 3.2	300 ± 150	465 ± 114	4.7 ± 3.2	33.1 ± 17.3	438 ± 222	12.7 ± 11.5

[a] Metal concentration data are presented as X ± SD.
[b] NA, not analyzed.

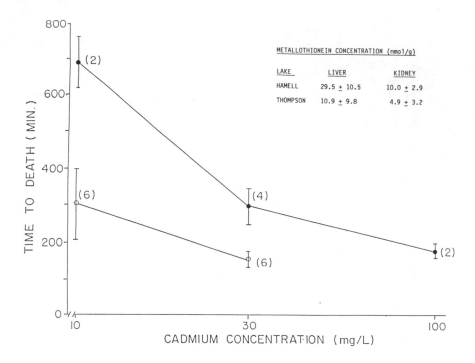

FIGURE 3.   Mortality curves of times to death by white suckers exposed to Cd and concentrations of MTN in liver and kidney of these fish. Numbers of fish are in parentheses. Data are expressed as X ± SD.

Estimates of MTN concentrations in liver and kidney from fish captured in 1981 using gel filtration and the metal summation approach are presented in Figure 3. Average concentrations were 2.7- and 2.0-fold higher in liver and kidney, respectively, of suckers from Hamell Lake.

The higher MTN concentrations found in Hamell Lake white suckers correspond to greater resistance to acute lethality produced by high concentrations of Cd (Figure 3). Suckers from Hamell Lake survived 2.3 and 1.9 times longer at 10 mg Cd/L and 30 mg Cd/L, respectively. The ten Hamell Lake suckers used in these toxicity tests weighed 1132 g ± 186 and were 433 mm ± 23 in length. The fifteen Thompson Lake suckers were lighter at 763 g ± 131 and shorter at 382 mm ± 24. This group of fish was 6.9 years ± 2.1 years of age and consisted of 9 males and 6 females, whereas Hamell Lake suckers were 10.0 years ± 0.9 in age and were males only.

In samples obtained in 1986, additional results were obtained on the subcellular distribution of these metals (Table 6). The mean percentages in cytosol of total Cd, Cu, and Zn in liver were 36, 51, and 45, respectively. For the kidney, these percentages were 56, 98, and 28, respectively. The percentage in cytosol of total Cu in liver increased ($r = 0.85$; $p = 0.03$) with increasing distance from the smelter. Although the results on metals in kidney cytosol are limited to four lakes, there was an inverse relationship between percentage in cytosol of Cd and Zn with distance from the smelter. Percentages of cytosolic metals bound to MTN were highest for Cu, 93% in liver and 71% in kidney, and lowest for Zn, 44% in liver and 37% in kidney. There were no significant relationships between percentage of cytosolic metals bound to MTN and distance from the smelter (the largest correlation coefficient was for Zn in kidney, $r = -0.61; p = 0.27$).

Two methods, metal summation in gel filtration fractions containing MTN and [203]Hg displacement, were used to estimate concentrations of this metal-binding protein in 1986 samples (Table 7). Comparing results obtained by metal summation to those observed in 1981 (Figure 3) demonstrates no differences between years. Concentration ranges of MTN in liver of 11.6 nmol/g in fish from Tyrell Lake to 41.1 nmol/g in Neso Lake suckers were observed. Similar comparisons can be made for concentrations found in kidney. These concentrations, however, were inversely correlated ($r = -0.86; p = 0.06$) to distance from the smelter. Estimates of MTN in 1986 liver samples using the [203]Hg displacement method were higher than those obtained using metal summation. The saturation index (percentage of total metal-binding sites occupied by metals) was less than 100 in livers of fish from all six lakes and was directly correlated ($r = 0.73; p = 0.1$) to distance from smelter. In 1986 kidney samples, MTN estimates using [203]Hg were about the same as those obtained using metal summation. Saturation indices in kidney, however, were 100 or greater, indicating complete saturation of metal-binding sites in kidney of fish from four of the five lakes.

An evaluation of whether saturation of MTN metal-binding sites in kidney corresponded to symptoms of metal toxicity was made by analyzing for other biochemical responses. Results from Twin Lake were used as reference values because this lake was farthest from the smelter. Blood electrolyte data (Figure 4) and concentrations of ascorbic acid (AsA) and acid-soluble thiols (AST) in kidney (Figure 5) demonstrated differences which may be due to metal deposition or accumulation. Suckers from Neso Lake had significantly lower plasma osmolality (Figure 4a) and Cl (Figure 4b). Fish from Neso and Nesootao Lakes had significantly lower plasma Na (Figure 4c) and K (Figure 4d). Suckers from Nesootao and Neso had the highest and second highest concentrations of total metals in their livers: 961 ± 86 nmol/g and 901 ± 61 nmol/g, respectively (Table 5). These two populations had the highest and second highest concentrations of Cd in their livers and posterior kidneys (Table 5). Hepatic MTN concentrations were also highest in fish from these two lakes (Table 7). Coincident with these observations are elevated concentrations of AST (Figure 5b) and MTN (Table 7) in kidneys of suckers from Nesootao Lake. Concentrations of AsA (Figure 5a) in posterior kidney were highest in Hook Lake and directly correlated ($r = 0.90; p = 0.08$) with metal concentrations in sediments, whereas ASTs were positively correlated with kidney Cu ($r = 0.95; p = 0.05$) and Cd

Table 6
Percentages of Concentrations of Cu, Zn, and Cd in the Cytosolic Pool (% Cyto) and of Cytosolic Pool in the MTN Fraction (% MTN) of the 1986 Samples

	Liver						Kidney					
	Cu		Zn		Cd		Cu		Zn		Cd	
Lake	%Cyto[a]	%MTN[b]	%Cyto	%MTN	%Cyto	%MTN	%Cyto	%MTN	%Cyto	%MTN	%Cyto	%MTN
Hook	20	91	37	49	47	80	NA[c]	64	NA	38	NA	58
White	44	90	46	45	32	58	NA	NA	NA	NA	NA	NA
Nesootao	60	95	50	57	22	80	100	92	31	47	59	55
Tyrell	51	91	38	20	35	36	100	67	36	33	71	46
Neso	67	95	57	51	31	69	100	71	29	33	59	58
Twin	62	94	40	39	48	62	67	70	15	36	35	55
X ± SD	51 ± 17	93 ± 2	45 ± 8	44 ± 13	36 ± 10	64 ± 16	92 ± 17	71 ± 7	28 ± 9	37 ± 6	56 ± 15	54 ± 5

a %Cyto, percentage of total in cytosolic pool.
b %MTN, percentage of cytosolic in MTN fractions.
c NA, not analyzed.

Table 7
Estimates of MTN Concentrations (nM/g) and Saturation Indices in Liver and Kidney from White Suckers Captured in 1986[a]

	Liver				Kidney		
Lake	DFS	[Cu + Zn + Cd]	^{203}Hg[b]	Saturation Index	[Cu + Zn + Cd]	^{203}Hg	Saturation Index
Hook	0.8	16.4	66.7 ± 27.5	24.6	11.8	11.8 ± 3.5	100
White	11.4	27.1	65.2 ± 21.0	41.6	NA[c]	6.4 ± 2.6	—
Nesootao	12.4	38.9	96.7 ± 34.1	40.2	17.1	14.9 ± 6.4	114.8
Tyrell	21.1	11.6	44.3 ± 33.2	26.2	9.2	8.0 ± 3.6	115.0
Neso	25.1	41.1	81.1 ± 24.6	50.7	7.5	6.9 ± 3.5	108.7
Twin	31.0	24.7	32.4 ± 14.1	76.2	5.5	7.0 ± 3.2	78.6

a n, same as Table 5.
b ^{203}Hg data are expressed as $\bar{X}$ ± SD.
c NA, not analyzed.

$(r = 0.90; p = 0.10)$ as well as liver Zn $(r = 0.77; p = 0.07)$ and Cd $(r = 0.91; p = 0.01)$. No significant interlake differences in either AsA or AST were observed in sucker livers.

DISCUSSION

Biological, chemical, and physical factors, individually and in combination, are known to affect metal accumulation in natural freshwater fish populations. These factors can include lake water and sediment chemical parameters, especially Ca, pH, organic acids, and sulfur; chemical characteristics of the metal, such as lipid solubility and speciation constants; physical parameters, such as temperature; and biological considerations, such as essentiality and nonessentiality for survival and reproduction (see this volume, Chapters 1, 4, 8, and 10). Although numbers of lakes and biological species reported in this study were

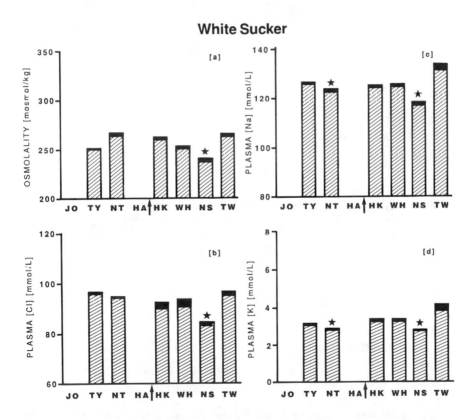

FIGURE 4. Plasma osmolality (a), CL (b), Na (c), and K (d) in white suckers from six lakes situated near a base metal smelter. Each bar represents the mean (hatched) and SEM (solid). Values which are significantly different from those in the reference lake (Twin) are makred with a star ($p \leq 0.05$). Each lake is shown in its relative position to the smelter (↑).

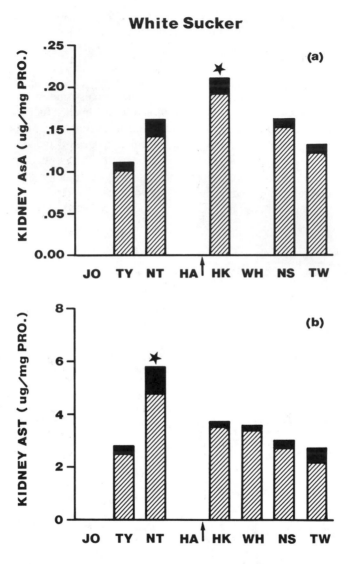

FIGURE 5. Posterior kidney ascorbic acid (a) and acid-soluble thiol (b) concentrations in white suckers from lakes situated near a base metal smelter. See Figure 4 for further explanations.

limited, there were no strong relationships between individual controlling factors, such as Ca or DOC in water and metal accumulation in white suckers. Research in these lakes on other fish species and on relationships of metal-binding affinities in sediment to bioavailability is in progress.[15,35]

Of the abiotic variables used to indicate metal pollution in the study lakes, depth profiles of metals in sediment cores reflect trends in atmospheric metal deposition from the smelter (Figure 2). Background concentrations of Cd, Cu, and Zn in the deepest sections of the cores (Table 2) are similar to background levels reported by Johnson et al.[36] in four Precambrian Shield lakes near Sault Ste. Marie, Ontario. The magnitude of metal deposition onto the Flin Flon lakes is revealed in the surficial sediments, in which Cd, Cu, and Zn concentrations were 2-20 times greater than the surficial concentrations present in Ontario lakes not receiving smelter fallout.[36-37] Ratios of surficial-to-background metal concentrations (present-to-background ratios, or P:B ratios) in the sediments of the Flin Flon lakes decreased with distance from the smelter (Table 3), with the largest proportion of metal deposition being restricted to a radius of 20 km around the smelter. This interpretation is consistent with that of Harrison and Klaverkamp[10] who concluded that the radius of metal deposition around the Flin Flon smelter was much smaller (<40 km) than earlier estimates.[8]

The highest metal concentrations within the sediment cores occurred between 2 and 5 cm below the sediment-water interface (Figure 2, Table 2). Although no attempts were made to establish the geochronology of the sediment cores from the study lakes, the profiles of metal concentrations in the cores were used to infer date estimates using assumed sedimentation rates for the lakes. Harrison and Klaverkamp[10] used sedimentation rate estimates of 2.0 mm/year, similar to those used for lakes in southern Ontario,[36,38] to establish a tentative geochronology of sediments from lakes near Flin Flon. This estimate would place 1930 (the year the smelter began operation) at approximately 11.2 cm depth in the sediment, while 1974 (the year the 251-m stack was commissioned) would be approximately 2.4 cm below the sediment-water interface. If actual sedimentation rates were greater than 2.0 mm/year, these dates would correspond to deeper sediment sections.

The metal profiles of Thompson Lake sediments generally support the assignment of a 1930 date to a depth of approximately 11 cm. Downward mobilization of metals, possibly enhanced by bioturbation, could account for the broadening of the metal-enriched zone toward 14 cm depth. In Hamell Lake sediments, presumably greater metal loadings resulted in deeper mobilization of metals, making it difficult to assign the year 1930 to a specific core section in that lake. Under the assumption that the highest sediment metal concentrations correspond to the period up to 1974, ratios of surficial (present) metal concentrations to peak metal concentrations should reveal the extent of reductions in the atmospheric release of metals from the smelter. These ratios (Table 3) indicate that major reductions in metal deposition have occurred only in the two lakes within 10 km of the smelter (Hook and Hamell Lakes). In the remaining six study lakes, reductions were modest, at best, with none of the lakes realizing more than a 25% decrease in recent metal deposition.

Electrostatic precipitators, which remove particulates from flue gases,[39] began operation at the Flin Flon smelter in 1982. Although snow sampling has demonstrated as As and sulfate deposition has been reduced since precipitators were installed,[5] sediment metal profiles from 1986 do not indicate that this reduction was substantive. Rather, the reductions in sediment metal concentrations appear to be mostly attributable to the onset of use of the 251-m stack in 1974.

The acidification of lakes which has occurred around the smelter at Sudbury, Ontario[40] was not observed in lakes around the Flin Flon smelter, even though sulfate concentrations in the Flin Flon lakes indicate that SO_2 deposition is high. The absence of an acidification problem near Flin Flon is due, in part, to the location of the smelter in a geological transitional zone between the Precambrian Shield and the sedimentary region of the North American prairies. Outcroppings of calcareous glacial till are also widespread in the area and may provide additional buffering capacity to lake catchments.

Many of the ores processed by the Flin Flon smelter contain low concentrations of alkali metals (e.g., Na and K) and alkaline earth metals (e.g., Ca and Mg).[41,42] The alkali metals have a relatively high affinity for SO_2 (Slack and Hollinden, 1975), which could explain the correlation between K^+ and SO_4^{2-} concentrations in the study lakes and indicate codeposition of these ions. The negative correlation between Cl^- and distance from the smelter has been observed previously near Flin Flon[10] and Sudbury[43] and probably reflects the presence of Cl and alkali metals in the stack emissions.

The major metals, Cu and Zn, in the atmospheric discharges of the smelter are essential trace elements, which are homeostatically regulated in vertebrates.[44,45] Liver and kidney are generally used for monitoring these metals in fish, because a considerable proportion of the Cu and Zn in the body is contained in these two organs. Liver has been described as the most metabolically active metal-containing compartment in vertebrates.[46] Elevated tissue concentrations of these essential metals would indicate that homeostasis has been altered and metal-induced stress may be occurring. These elevated concentrations result from the combination of increased metal uptake and overburdening of excretory mechanisms.[47]

The Hamell Lake suckers probably represented the most severely exposed population sampled, while the other populations would be found on an exposure gradient between that of Hamell Lake and those of suckers from unexposed populations. Metal concentrations reported in suckers not exposed to high levels of Cd, Cu, or Zn were approximately 0.7 nmol Cd/g, 197 nmol Cu/g, and 418 nmol Zn/g in liver, and about 1.9, 30 and 407 nmol/g, respectively, in kidney.[48] In Hamell Lake suckers captured in 1981, all three metals in both liver and kidney were considerably elevated (Table 4) relative to these "normal" concentrations. Because the suckers collected in 1981 from this lake were older, with a predominance of males, and because suckers could not be caught from this lake in 1986, they may have become extinct during this interim. In Thompson Lake suckers sampled in 1981, Cd and Cu in liver and kidney were also elevated (Table 4) relative to concentrations reported by Bendell-Young and Harvey.[48]

Copper concentrations in liver of 1986 samples (Table 5) were reduced slightly relative to those observed (Table 4) in 1981. Concentrations of Cu in kidney were considerably reduced, with values approaching those reported by Bendell-Young and Harvey.[48] No reductions in Cd or Zn were evident in either tissue. It is unclear whether the reductions in Cu were due to the installation of electrostatic precipitators, because no pronounced effect of the precipitators was evident in the sediment metal profiles. It is more likely that the reduced concentrations of Cu seen in suckers in 1986 may be a general trend due to longer-term reductions in metal deposition to the study lakes since the 251-m stack began operation. The slow reduction of the Cu concentrations in suckers probably indicates that the fish are still stressed by high levels of metals in their environment.[7] Certainly, the concentrations of Cd, Cu, and Zn in surficial sediments are still elevated relative to concentrations observed in lakes not receiving direct deposition of these metals.[36]

An important aspect of hepatic function is its role as the major site of Cu excretion (due to bile formation) in vertebrates.[45] In Cu-loaded animals and among those with naturally impaired Cu metabolism,[45,49,50] the overburdening of bilary excretion pathways is typified by the subcellular localization of Cu in "particulate" or noncytosolic fractions (especially lysosomes and tertiary lysosomes). Therefore, the subcellular distribution of metals can be used to infer metal-induced stress.

In the 1986 sucker populations, the proportion of hepatic Cu in the cytosol increased with distance from the smelter (Table 6), and in fish from all lakes, Cd was associated with cellular particulate fractions. The low proportion of cytosolic Cu in fish from Hook and White Lakes, therefore, could be indicative of active excretory processes occurring in the livers of fish from these lakes. Although the subcellular distributions of metals were only circumstantial (metal concentrations in bile were not analyzed), when considered in conjunction with the elevated levels of Cu in sediment still present in 1986 and the reduction in hepatic and renal Cu between 1981 and 1986, metal excretion must have been active in suckers collected in 1986.

Metallothionein (MTN) has been used as an indicator of heavy metal exposure in fish.[16,17] Recently, reductions in hepatic MTN have been used to demonstrate recovery of metal-stressed fish after metal loadings were reduced in Buttle Lake, British Columbia.[51] White suckers from Hamell Lake in 1981 had higher concentrations of hepatic and renal MTN than those observed in suckers from Thompson Lake, and Hamell Lake suckers were more resistant to Cd toxicity. This resistance and elevated MTN concentrations did not appear to be capable of sustaining this population, because suckers could not be captured from Hamell Lake in 1986. Previous laboratory studies[52] demonstrated that acclimation to Cd toxicity does occur in white suckers after exposure to sublethal concentrations of Cd, Hg, and Zn. For example, tolerance to Cd-induced lethality increased 2.3- and 3.6-fold in suckers exposed to Cd and Hg, respectively. Further studies on the potencies and organ specificities of these metals in producing increased MTN concentrations demonstrated correlations between MTN induction and acclimation to Cd toxicity.[23] Correlations of elevated MTN concentrations produced in fish by metals in groups IB and IIB of the periodic table to acclimation to toxicity produced by these metals have frequently been observed.[16,17]

Under conditions of acute metal exposure, the induction of MTN synthesis may produce metal-binding capacity in excess of that required.[53] This phenomenon of "overproduction" has been observed by others[54,55] and may be responsible for transient metal tolerance of exposed fish.[55] This apparent excess of metal-binding capacity has been correlated with the general induction of sulfhydryl-containing compounds[56] and is the major evidence supporting general indices of sulfhydryl status in fish as a measurement of metal-induced stress. In fathead minnows,[57] hepatic MTN contained less than 75% of complete metal saturation after a 2-week exposure to sublethal concentrations of Zn. In approaches similar to the one used to determine saturation indices in this study, Cousins[44] describes as "appreciable" the amounts of metal-free, thiol-containing polypeptides during both steady- and induced-states of synthesis. Artificial increases in percentages of free- thiols (i.e., saturation indices) due to proteolytic digestion were prevented in the 1986 samples by using liquid nitrogen to freeze fresh tissues immediately after their dissection and by including the protease inhibitor, PMSF, in the gel filtration analyses.

Thomas et al.[58,59] observed a dose- and time-dependent elevation of liver and posterior kidney AST in the striped mullet *(Mugil cephalus)* in response to a 6-week Cd exposure. This elevation was found to be due mainly to increased MTN within the first 4 weeks of exposure, followed by elevations in glutathione (GSH) after 4 weeks.[56] In mammals, Cd is bound to liver and kidney MTN[60] and some is excreted in the bile bound to GSH.[61,62] It appears that similar mechanisms were utilized by Nesootao Lake suckers. These fish had increased concentrations of MTN in liver and posterior kidney and increased AST in posterior kidney, which were coincident with the highest concentrations of Cd in liver and kidney.

Another interesting adaptation by white suckers to metal pollution in proximity to the smelter is the increasing level of posterior kidney AsA in relation to decreasing distance from the smelter. This response appears to be related more to the degree of metal exposure, as indicated by metal concentrations in the sediments, than to the degree of tissue metal loading in these fish. A similar graded response to metal accumulation was observed for kidney AST. Both of these biochemicals (AsA and AST) are antioxidants and are involved in detoxifying processes[58,59,63] by terminating free radicals and protecting cellular and subcellular membranes.[64] Cd produces adverse effects on membranes in fish;[65-69] and the antioxidants, AsA and AST, respond to low concentrations of Cd,[64] presumably to protect the integrity of renal tubules.[70] Andersson et al.[71] found a dose-dependent elevation in liver AsA in perch exposed to bleached Kraft mill effluents. They attributed this response to increased synthesis of AsA reflecting an enhanced demand for AsA in processes such as detoxification and wound healing. This is especially likely in cyprinids, such as white suckers, which have the ability to synthesize AsA.[72,73] Increased levels of AsA in the kidney may also serve to maintain ion regulation by preserving Ca homeostasis in metal-exposed suckers. Both individual and mixed metal exposures are known to alter Ca metabolism in freshwater fish.[37,74,75] Thomas et al.[58] point out that Ca uptake from the gut is dependent on AsA. The stable plasma Ca levels observed in these white suckers may be due to adequate tissue stores of AsA, although bone Ca levels were not measured in this study. Furthermore, the lack of a hyperglycemic response in white suckers exposed to environmental metal pollution may be due to the increased utilization of glucose for the synthesis of AsA.

Increased antioxidant levels in the kidney may play a role in maintaining ionoregulation in fish exposed to metal pollution. The modifying effect of major water ions on ionoregulation, however, is also a factor in this study. Higher levels of ambient Na, Cl, and Ca could serve to partially ameliorate the effects of metal exposure by helping to maintain physiological conditions (i.e., iono-regulation) in these fish by decreasing the permeability of surface membranes to metals. Hook Lake (0.89 nmol/L Ca; 0.092 nmol/L Na; 0.073 nmol/L Cl) located within 10 km of the smelter had approximately 13 times the sediment load of total metals when compared to the reference lake (Twin Lake — 0.29 nmol/L Ca; 0.053 nmol/L Na; 0.014 nmol/L Cl), located in 31 km from the smelter. There were no significant differences in kidney or liver total metal burden in suckers from the six lakes sampled. The highest kidney and liver Cd concentrations, however, were found in Nesootao Lake white suckers, and the highest liver Cu, and second highest kidney Cd concentrations were found in Neso Lake suckers. These two lakes had among the lowest ambient Ca concentrations, of the six lakes studied (0.22 and 0.21 nmol/L, respectively), and resident suckers showed significant, but not pronounced, ionic changes. The loss of plasma ions (Na, Cl, K) in Neso Lake suckers may indicate hemodilution due to decreased gill permeability to water or possible kidney damage resulting in decreased ion retention. This type of response is symptomatic of Cu toxicity in the striped bass[76] and bluegill.[77]

Nesootao Lake suckers showed a decrease in plasma Na and K, but no hemodilution or altered glucose metabolism was apparent. Larsson et al.[78] found slight decreases in plasma K and no effect on glucose or Ca metabolism in perch from three Cd-contaminated stations in the Eman River system. These authors suggest that the perch may have acquired an increased tolerance to Cd.

It is apparent that the biochemical parameters measured in white suckers are affected by sediment and tissue burdens of metals. The ability of these clinical measurements to integrate environmental variables (i.e., ambient ions and metal deposition) is also evident, since the greatest ionic disturbances were found in Nesootao and Neso Lake white suckers. Suckers from these two lakes had the highest and second highest kidney Cd concentrations, and the lakes had approximately one-fourth the ambient Ca concentrations found in Hook Lake (9.8 km from the smelter). Three of the biochemical variables (AsA, AST, and MTN) measured in sucker kidneys demonstrated a graded, dose-dependent response, indicative of fish populations adapting to a pollution gradient. These findings are in agreement with effects previously observed in perch in the vicinity of a smelter[79,80] and a pulp mill[71] and corroborate the observations of McFarlane and Franzin[7] with respect to the effects of metal deposition on the accumulation of metals in fish in the Flin Flon area.

In conclusion, this study provides biological and sediment data which indicate that electrostatic precipitators have had little or no effect 4 years after their installation. Four years, however, may not be sufficient time for reducing the metal burdens in these heavily contaminated sediments. Diagenetic remobilization processes, for example, may cause metals to move out of deeper sediments into pore water and migrate up to surficial sediment segments.[81,82] In another study,[51] however, MTN was a sensitive indicator for evaluating the effectiveness of controlling metal mining discharges into freshwater lakes. Two years after installation of controls, which reduced metal discharges by 80%, hepatic MTN concentrations in rainbow trout were about 24% of concentrations reported before controls were used. The reduced MTN concentrations in the livers of those trout were not different from those observed in trout from reference lakes. In this Flin Flon study, similar reductions in MTN of white suckers were not observed between 1981 and 1986, although renal and hepatic MTN concentrations were generally lower in lakes receiving less metals. In this study, rigorous comparisons between 1981 and 1986 cannot be made because different lakes were sampled and additional biochemical indicators of stress were measured in 1986. Despite this fact, it is apparent that white sucker populations near Flin Flon continue to be under metal-induced stress. The biochemical data set from 1986 is the first such set of "sensitive" stress indicators that has been collected from aquatic systems near Flin Flon in the 60 years the smelter has been in operation. These biochemical parameters, including MTN, and metal concentrations in surficial sediments should be measured in the near future in samples obtained from lakes studied in both years to confirm the lack of effectiveness of metal pollution controls.

ACKNOWLEDGMENTS

We gratefully acknowledge the excellent assistance of Carol Catt in typing the manuscript. We are also grateful to an anonymous referee for providing constructive comments and literature references on diagenic migration of metals in sediments. These studies were supported by the Canadian Department of Fisheries and Oceans and by a Wildlife Toxicology grant from the World Wildlife Fund.

REFERENCES

1. Franzin, W. G. "Aquatic Contamination in the Vicinity of the Base Metal Smelter of Flin Flon, Manitoba, Canada - A Case History," in *Environmental Impacts of Smelters,* J. O. Nriagu, Ed. (New York: John Wiley & Sons, 1984), pp. 523-550.

2. Heywood, W. W., "Ledge Lake Area, Manitoba and Saskatchewan," *Geol. Surv. Can. Memoir No.* 337 (1966).

3. Strachan, L. "Environmental Report. Hudson Bay Mining and Smelting Company Limited, Flin Flon, Manitoba," Dept. Mines, Nat. Resour. Environ., Environ. Manage. Div., Environ. Control Branch Environ. Rep., 12 pp. attachments. Presentation to the Manitoba Clean Environment Commission (1979).

4. Frazer, W. W. "Presentation with Respect to Discharges to the Atmosphere from Flin Flon Metallurgical Operations of Hudson Bay Mining and Smelting Co. Ltd." Presentation to Manitoba Clean Environment Commission, December 1983, p. 38.

5. Phillips, S. F., D. L. Wotton, and D. B. McEachern. "Snow Chemistry in the Flin Flon Area of Manitoba," *Water Air Soil Pollut.* 30:253-261 (1986).

6. Van Loon, J. C. and R. J. Beamish. "Heavy Metal Contamination by Atmospheric Fallout of Several Flin Flon Area Lakes and the Relation to Fish Populations," *Can. J. Fish. Aquat. Sci.* 34:899-906 (1977).

7. McFarlane, G. A. and W. G. Franzin. "An Examination of Cd, Cu, and Hg Concentrations in Livers of Northern Pike, *Esox lucius,* and White Sucker *Catostomus commersoni,* from Five Lakes Near a Base Metal Smelter at Flin Flon, Manitoba," *Can. J. Fish. Aquat. Sci.* 37:1573-1578 (1980).

8. Franzin, W. G., G. A. McFarlane, and A. Lutz. "Atmospheric Fallout in the Vicinity of a Base Metal Smelter at Flin Flon, Manitoba," *Can. J. Fish. Aquat. Sci.* 37:1573-1578 (1979).

9. Zoltai, S. C. "Distribution of Base Metals in Peat near a Smelter at Flin Flon, Manitoba," *Water Air Soil Pollut.* 37:217-228 (1988).

10. Harrison, S. E. and J. F. Klaverkamp. "Metal Contamination in Liver and Muscle of Northern Pike *(Esox lucius)* and White Sucker *(Catostomus commersoni)* and in Sediments from Lakes Near the Smelter at Flin Flon, Manitoba," *Environ. Toxicol. Chem.* 9:941-956 (1990).

11. Dixon, D. G., P. V. Hodson, J. F. Klaverkamp, K. M. Lloyd, and J. R. Roberts. "The Role of Biochemical Indicators in the Assessment of Aquatic Ecosystem Health—Their Development and Assessment," *Nat. Res. Counc.,* Ottawa, 119 pp. (1985).

12. Neff, J. M. "Use of Biochemical Measurements to Detect Pollutant-Mediated Damage to Fish," in *Aquatic Toxicology and Hazard Assessment: 7th Symp.* R. D. Cardwell, R. Purdy, and R. C. Bahnes, Eds. (Philadelphia, PA: American Society Testing Materials, 1985), pp. 155-183.

13. Adams, S. M., K. L. Shepard, M. S. Greeley, Jr., B. D. Jimenez, M. G. Ryon, L. R. Shugart, J. F. McCarthy, and D. E. Hinton. "The Use of Bioindicators for Assessing the Effects of Pollutant Stress on Fish," *Mar. Environ. Res.* 28:459-464 (1989).

14. Brown, S. B., R. E. Evans, H. S. Majewski, G. B. Sangalang, and J. F. Klaverkamp. "Responses of Plasma Electrolytes, Thyroid Hormones and Gill Histology in Atlantic Salmon *(Salmo salar)* to Acid and Limed River Waters," *Can. J. Fish. Aquat. Sci.* 47:2431-2440 (1990).

15. Majewski, H. S., S. B. Brown, R. E. Evans, H. C. Freeman, and J. F. Klaverkamp. "Responses of Kidney, Liver, Muscle and Bone in Atlantic Salmon *(Salmo salar)* to Diet and Liming in Acidic Nova Scotia Rivers," *Can. J. Fish. Aquat. Sci.* 47:2441-2450 (1990).

16. Klaverkamp, J. F., W. A. MacDonald, D. A. Duncan, and R. Wagemann, in *Contaminant Effects on Fisheries,* V. W. Cairns, P. V. Hodson, and J. O. Nriagu, Eds. (New York: John Wiley & Sons, 1984), pp. 99-113.

17. Hamilton, S. J. and P. M. Mehrle. "Metallothionein in Fish: Review of Its Importance in Assessing Stress from Metal Contaminants," *Trans. Am. Fish. Soc.* 115:596-609 (1986).

18. Haux, C. and L. Forlin. "Biochemical Methods for Detecting Effects of Contaminants on Fish," *Ambio* 17:376-380 (1988).

19. Hamer, D. H. "Metallothionein," *Annu. Rev. Biochem.* 55:913-951 (1986).

20. Kagi, J. H. R. and Y. Kojima. "Chemistry and Biochemistry of Metallothionein," *Experimentia Suppl.* 52:25-61 (1987).

21. Palmiter, R. D. "Molecular Biology of Metallothionein Gene Expression," *Experimentia Suppl.* 52:63-80 (1987).

22. Fowler, B. A. "Intracellular Compartmentation of Metals in Aquatic Organisms: Roles in Mechanisms of Cell Injury," *Environ. Health Perspect.* 71:121 (1987).

23. Klaverkamp, J. F. and D. A. Duncan. "Acclimation to Cadmium Toxicity by White Suckers: Cadmium Binding Capacity and Metal Distribution in Gill and Liver Cytosol," *Environ. Toxicol. Chem.* 6:275-289 (1987).

24. Webb, M. "Toxicological Significance of Metallothionein," *Experimentia Suppl.* 52:109-134 (1987).

25. Kagi, J. H. R. and A. Schaffer. "Biochemistry of Metallothionein," *Biochemistry* 27:8509-8515 (1988).

26. Stainton, M. P., M. J. Capel, and F. A. J. Armstrong. "The Chemical Analysis of Freshwater," 2nd ed., *Can. Fish. Mar. Serv. Misc. Publ.* 25:1-180 (1977).

27. Schindler, D. W., R. Wagemann, R. B. Cook, T. Ruszczynski, and J. Prokopowich. "Experimental Acidification of Lake 223 Experimental Lakes Area: Background Data and the First Three Years of Acidification," *Can. J. Fish. Aquat. Sci.* 37:342-354 (1980).

28. Sturgeon, R. E., J. A. H. Desaulniers, S. S. Berman, and D. S. Russel. "Determination of Trace Metals in Estuarine Sediments by Graphite-Furnace Atomic Absorption Spectrometry," *Anal. Chim. Acta* 134:283-291 (1982).

29. Harrison, S. E., M. D. Dutton, R. V. Hunt, J. F. Klaverkamp, A. Lutz, W. A., MacDonald, H. S. Majewski, and L. J. Wesson. "Metal Concentrations in Fish and Sediment From Lakes near Flin Flon, Manitoba," *Can. Data Rep. Fish. Aquat. Sci.* 747:iv + 74 (1989).

30. Dutton, M. D., M. Stephenson, and J. F. Klaverkamp. "A Modified Mercury Displacement Assay for Measuring Metallothionein in Fish," *Environ. Toxicol. Chem.,* submitted for publication.

31. Moron, M. S., J. W. Depierre and B. Mannervik. "Levels of Glutathione, Glutathione Reductase and Glutathione S-transferase Activities in Rat Lung and Liver," *Biochim. Biophys. Acta* 582:67-78 (1979).

32. Jagota, S. K. and H. M. Dani. "A New Colorimetric Technique for The Estimation of Vitamin C Using Folin Phenol Reagent," *Anal. Biochem.* 127:178-182 (1982).

33. Peterson, G. L. "A Simplification of the Protein Assay of Lowry et al., Which is More Generally Applicable," *Anal. Biochem.* 83:346-356 (1977).

34. SAS Institute, Inc. *SAS User's Guide: Basics* (Cary, NC: SAS Institute, Inc., 1982), 921 pp.

35. Dutton, M. D. and J. F. Klaverkamp. "Metal-binding Proteins in Livers and Kidneys of Northern Pike *(Esox lucius)* and White Suckers *(Catostomus commersoni)* from Lakes Receiving Atmospheric Deposition of Cu, Zn and Cd," *Environ. Toxicol. Chem.,* submitted for publication.

36. Johnson, M. G., L. R. Culp, and S. E. George. "Temporal and Spatial Trends in Metal Loadings to Sediments of the Turkey Lakes, Ontario," *Can. J. Fish. Aquat. Sci.* 43:754-762 (1986).

37. Bendell-Young, L. and H. H. Harvey. "Concentrations and Distribution of Fe, Zn, and Cu in Tissues of the White Sucker *(Catostomus commersoni)* in Relation to Elevated Levels of Metals and Low pH," *Hydrobiologia* 176/177:349-354 (1989).

38. Johnson, M. G. "Trace Element Loadings to Sediments of Fourteen Ontario Lakes and Correlations with Concentrations in Fish," *Can. J. Fish. Aquat. Sci.* 44:3-13 (1987).

39. Chereminisoff, P. M. and R. A. Young. *Pollution Engineering Practice Handbook* (Ann Arbor, MI: Ann Arbor Sci. Publ., 1975).

40. Yan, N. D. and G. E. Miller. "Effects of Deposition of Acids and Metals on Chemistry and Biology of Lakes near Sudbury, Ontario," in *Environmental Impacts of Smelters,* J. O. Nriagu, Ed. (New York: John Wiley & Sons, 1984), pp. 234-282.

41. Gale, G. H., R. C. Somerville, J. Chornoby, B. Haystead, N. Provins, D. Braun, D. Munday, and A. Walker. "Geological Setting of Mineral Deposits at Ruttan, Thompson, Snow Lake, and Flin Flon, Manitoba." Geol. Assoc. Can., Winnipeg Section, p. 59 (1982).

42. Syme, E. C., A. H. Bailes, D. P. Price, and D. V. Ziehlke. "Flin Flon Volcanic Belt: Geology and Ore Deposits at Flin Flon and Snow Lake, Manitoba," Geol. Assoc. Can., Winnipeg Section. 79 pp. (1982).

43. Jeffries, D. S. "Atmospheric Deposition of Pollutants in The Sudbury Area," in *Environmental Impacts of Smelters,"* J. O. Nriagu, Ed. (New York: John Wiley & Sons, 1984), pp. 117-154.

44. Cousins, R. J. "Metallothionein—Aspects Related to Copper and Zinc Metabolism," *J. Inher. Metab. Dis.* 6 (Suppl. 1):15-21 (1983).

45. Sternlieb, I. 1988. "Copper and Zinc," in *The Liver: Biology and Pathobiology,* 2nd ed., I. M. Arias, W. B. Jakoby, H. Popper, D. Schachter, and D. A. Shafritz, Eds. (New York: Raven, 1985), pp. 525-533.

46. Cousins, R. J. and M. L. Failla. "Cellular and Molecular Aspects of Mammalian Zn Metabolism and Homeostasis," in *Zinc in the Environment. Part 2: Health Effects,* J. O. Nriagu, Ed. (New York: Interscience, 1980), pp. 121-136.

47. Bradley, R. W. and J. R. Morris. "Heavy Metals in Fish From a Series of Metal Contaminated Lakes near Sudbury, Ontario," *Water Air Soil Pollut.* 27:341-354 (1986).

48. Bendell-Young, L. I. and H. H. Harvey. "Metal Concentration and Calcification of Bone of White Sucker *(Catostomus commersoni)* in Relation to Lake pH," *Water Air Soil Pollut.* 30:657-664 (1986).

49. Bunton, T. E., S. M. Baksi, S. G. George, and J. M. Frazier. "Abnormal Hepatic Copper Storage in a Teleost Fish *(Morone americana),*" *Vet. Pathol.* 24:515-524 (1987).

50. Gross, J. B., Jr., B. M. Myers, L. J. Kost, S. M. Kuntz, and N. F. LaRusso. "Biliary Copper Excretion by Hepatic Lysosomes in the Rat. Major Excretory Pathway in Experimental Copper Overload," *J. Clin. Invest.* 83:30-39 (1988).

51. Deniseger, J., L. J. Erickson, A. Austin, M. Roch, and M. J. R. Clark. "The Effects of Decreasing Heavy Metal Concentrations on the Biota of Buttle Lake, Vancouver Island, British Columbia," *Water Res.* 24:403-416 (1990).

52. Duncan, D. A. and J. F. Klaverkamp. "Tolerance and Resistance to Cadmium in White Suckers *(Catostomus commersoni)* Previously Exposed to Cadmium, Mercury, Zinc or Selenium," *Can. J. Fish. Aquat. Sci.* 40:128-138 (1983).

53. Cousins, R. J. "Absorption, Transport, and Hepatic Metabolism of Copper and Zinc: Special Reference to Metallothionein and Ceruloplasmin," *Physiol. Rev.* 65:238-309 (1985).

54. Petering, D. H., S. Krezoski, J. Villalobos, C. F. Shaw, III, and J. D. Otvos. "Cd-Zn Interactions in the Ehrlich Cell: Metallothionein and Other Sites," *Experimentia Suppl.* 52:573-580 (1987).

55. McCarter, J. A. and M. Roch. "Hepatic Metallothionein and Resistance to Copper in Juvenile Coho Salmon," *Comp. Biochem. Physiol.* 74C:133-137 (1983).

56. Wofford, H. W. and P. Thomas. "Interactions of Cadmium with Sulfhydryl-Containing Compounds in Striped Mullet *(Mugil cephalus L.),*" *Mar. Environ. Res.* 14:119-137 (1984).

57. Hobson, J. F. and W. J. Birge. "Acclimation-Induced Changes in Toxicity and Induction of Metallothionein-Like Proteins in the Fathead Minnow Following Sublethal Exposure to Zinc," *Environ. Toxicol. Chem.,* 8:157-169 (1989).

58. Thomas, P., M. Bally, and J. M. Neff. "Ascorbic Acid Status of Mullet, *Mugil cephalus* Linn., Exposed to Cadmium," *J. Fish. Biol.* 20:183-196 (1982a).

59. Thomas, P., H. W. Wofford, and J. M. Neff. "Effects of Cadmium on Glutathione Content of Mullet *(Mugil cephalus)* Tissues," in *Physiological Mechanisms of Marine Pollutant Toxicity,* W. B. Vernberg, A. Calabrese, F. P. Thurberg, and F. J. Vernberg, Eds. (San Diego: Academic Press, 1982), p. 109-125.

60. Noël-Lambot, F., C. Gerday, and A. Disteche. "Distribution of Cd, Zn and Cu in Liver and Gills of the Eel *Anguilla anguilla* with Special Reference to Metallothioneins," *Comp. Biochem. Physiol.* 61C:177-187 (1978).

61. Cherian, M. G. and J. J. Vostal. "Bilary Excretion of Cadmium in Rat. I. Dose-Dependent Biliary Excretion and the Form of Cadmium in the Bile," *J. Toxicol. Environ. Health* 2:945-954 (1977).

62. Elinder, C. G. and M. Pannone. "Biliary Excretion of Cadmium," *Environ. Health Perspect.* 28:123-126 (1979).

63. Zannoni, V. G. and M. M. Lynch. "The Role of Ascorbic Acid in Drug Metabolism," *Drug Metab. Rev.* 2:57-69 (1973).

64. Palace, V. P., H. S. Majewski, and J. F. Klaverkamp. "Interactions Between Cellular Antioxidant Defenses and Cadmium Accumulation and Toxicity in the Rainbow Trout," *Environ. Toxicol. Chem.*, submitted for publication.
65. Slack, A. V. and G. A. Holliden. *Sulfur Dioxide Removal From Waste Gases*, 2nd ed. (Park Ridge, NJ: Noyes Data Corp., 1975).
66. Wani, G. P. and A. N. Latey. "The Effect of Cadmium on the Gills and Kidney of a Cyprinid Fish *Garra mullya* (Sykes)," *Biovigyanam* 10:35-40 (1984).
67. Stromberg, P. C., J. G. Ferrante, and S. Carter. "Pathology of Lethal and Sublethal Exposure of Fathead Minnows, *Pimephales promelas*," to Cadmium: A Model for Aquatic Toxicity Assessment," *J. Toxicol. Environ. Health* 11:247-259 (1983).
68. Saxema, O. P. "Cadmium-Induced Neoplasia in *Channa punctatus* (Bloch)," *Curr. Sci.* 50:735-736 (1981).
69. Koyama, J. and Y. Itazawa. "Effects of Oral Administration of Cadmium on Fish. II. Results of Morphological Examination," *Bull. Jpn. Soc. Sci. Fish.* 43:527-533 (1977).
70. Hawkins, W. E., L. G. Tate, and T. G. Sarphie. "Acute Effects of Cadmium on the Spot *Leiostomus xanthurus* (Teleostei): Tissue Distribution and Renal Ultrastructure," *J. Toxicol. Environ. Health* 6:283-295 (1980).
71. Andersson, T., L. Forlin, J. Hardig, and A. Larrson. "Physiological Disturbances in Fish Living in Coastal Water Polluted with Bleached Kraft Pulp Mill Effluents," *Can. J. Fish Aquat. Sci.* 45:1525-1536 (1988).
72. Yamamoto, Y., M. Sato, and S. Ikeda. "Existence of L-Gulonolactone Oxidase in Some Teleosts," *Bull. Jpn. Soc. Sci. Fish.* 44:775-779 (1978).
73. Sato, M., R. Yoshinaka, Y. Yamamoto, and S. Ikeda. "Non-Essentiality of Ascorbic Acid in the Diet of Carp," *Bull. Jpn. Soc. Sci. Fish.* 44(10):1151-1156 (1978).
74. Dwyer, F. J., C. J. Schmitt, S. E. Finger, and P. M. Mehrle. "Biochemical Changes in Long Ear Sunfish, *Lepomis megalotis*, Associated with Lead, Cadmium and Zinc from Mine Tailings," *J. Fish. Biol.* 33:307-317 (1988).
75. Roch, M. and E. J. Maly. "Relationship of Cadmium-Induced Hypocalcemia with Mortality in Rainbow Trout *(Salmo gairdneri)* and the Influence of Temperature on Toxicity," *J. Fish. Res. Bd. Can.* 36:1297-1301 (1979).
76. Courtois, L. A. and R. D. Meyerhoff. "Effects of Copper Exposure on Water Balance," *Bull. Environ. Contam. Toxicol.* 14:221 (1975).
77. Heath, A. G. "Changes in Tissue Adenylates and Water Content of Bluegill, *Lepomis macrochirus*, Exposed to Copper," *J. Fish. Biol.* 24:299 (1984).
78. Larsson, A., C. Haux, and M.-L. Sjobeck. "Fish Physiology and Metal Pollution: Results and Experiences from Laboratory and Field Studies," *Ecotoxicol. Environ. Safety* 9:250-281 (1985).
79. Larsson, A., C. Haux, M.-L. Sjobeck, and G. Lithner. "Physiological Effects of an Additional Stressor on Fish Exposed to a Simulated Heavy Metal Containing Effluent from a Sulfide Ore Smeltery," *Ecotoxicol. Environ. Safety* 8:118-128 (1984).
80. Sjobeck, M.-L., C. Haux, A. Larsson, M. Edgren, and G. Lithner. "Testing of Physiological Methods in a Field Study on Perch, *Perca fluviatus*, from a Heavy Metal Contaminated Area of the Gulf of Bothnia," Rep. 147, Swedish Environment Protection Board, Brackish Water Toxicology Laboratory, 1984.

81. Ridgway, I. M. and N. B. Price. "Geochemical Associations and Post-Depositional Mobility of Heavy Metals in Coastal Sediments: Loch Etive, Scotland," *Mar. Chem.* 21:229-248 (1987).

82. McKee, J. D., T. P. Wilson, D. T. Long, and R. M. Owen. "Geochemical Partitioning of Pb, Zn, Cu, Fe and Mn across the Sediment-Water Interface in Large Lakes," *J. Great Lakes Res.* 15:46-58 (1989).

3

Toxic Metal Uptake and Essential Metal Regulation in Terrestrial Invertebrates: A Review

Alan Beeby

Department of Biotechnology, South Bank Polytechnic,
London, United Kingdom

OVERVIEW

A small number of metals can properly be described as "essential," for which organisms have mechanisms of uptake and regulation. Terrestrial animals rely on their digestive tracts for essential metal assimilation and have several mechanisms for storage and excretion. Some of these may have evolved as a detoxification system to remove excess metal from sensitive sites. There is evidence that toxic metals may be lost by the same route, possibly due to similarities in their chemical properties. Such pathways may have to cope with large metal burdens if toxic effects are to be minimized in highly polluted habitats. This chapter examines the evidence that such regulatory processes have become adapted for toxic metal excretion and the forms this adaptation has taken in the various groups of terrestrial invertebrates.

INTRODUCTION

For plants and animals to survive in polluted habitats, they have to distinguish between the metals they need and those likely to poison them. The mechanisms by which essential metals are assimilated, however, are not always sufficiently selective. Chemical similarities between metals may also allow potentially toxic elements to follow these routes of uptake. Even novel elements like Pu trace pathways of nutrient ion absorption in higher plants.[1] In vertebrates, Pb and Ca are known to interact at the biochemical, cellular, and system levels,[2] and there are complex interactions by which Cd and Pb interfere with Cu and Zn metabolism.[3]

The most important route for metal assimilation in terrestrial invertebrates is usually assumed to be the digestive tract. Despite the aerial nature of much metal pollution, respiratory uptake remains largely unquantified, and cutaneous uptake has only rarely been evaluated.[4] The assimilation of toxic metals varies not only with diet,[5] but also with an organism's essential metal demand. Williamson and Evans[6] record the highest lead concentrations in those roadside invertebrates with a calcified exoskeleton. Similarly, in a soil community, Carter[7] notes elevated Cu and Zn concentrations in millipedes "with their highly calcareous exoskeletons." Near metal smelters, millipedes, slugs, and worms have more Pb than insects,[8] and Cu and Cd are higher in isopods, oligochaetes, and collembolans[9] than in other litter dwellers. Thus, those invertebrates with a high Ca requirement are commonly shown to have higher toxic metal concentrations.

Toxic metals interact and substitute with both macronutrient and trace metals. These interactions are inevitably a consequence of similarities in the coordination chemistry of the two elements, reflecting their positions in the periodic table. Venugopal and Luckey[10] note that most metals in the fourth period are essential, while the common toxic metals are in periods 5 (soluble) and 6 (less soluble, but potentially more toxic). This has been extended by Nieboer and Richardson:[11] metals confined to the left hand side of the periodic table comprise macronutrients which tend to bind to oxygen sites in biological ligands (their "class A"). "Class B" metals form strong bonds with N or S and exert their toxicity by affecting protein structure and enzyme function (Table 1). Within a period, this toxicity rises with increasing electropositivity.[10] Trace elements and some toxic metals with catholic binding affinities are deemed borderline between the two classes. These may displace class A or other borderline metals. For example, Ni displaces Zn in carbonic anhydrase causing a loss of activity.[11]

Substitution is more frequently reported for the essential metals of group IIA which play a structural role (Ca and Mg) and less so for the more mobile metals of group IA (Na and K). Cells maintain K and Mg inside the cell, and Na and Ca outside, using metabolic energy. Binding sites for specific metals are selective on the basis of cation size, down to 0.2 Å,[12] but isomorphous replacement of metal ions depends on charge, size, and geometric demand.[12] Simkiss and Taylor[13] suggest that the principal interferences will occur between class A metals (such

Table 1
A Biological Classification of Metals[a]

Period	Class"A"		"Borderline"						Class"B"		
2	Li	Be									
3	Na	Mg									
4	K	Ca	Cr	Mn	Fe Co Ni	Cu	Zn		Cu		
5	Rb	Sr					Cd	Sn	Ag		
6	Cs							Pb	Au	Hg	Pb

[a] Derived from Nieboer and Richardson[11] and Venugopal and Luckey.[10] This is a corruption of part of the periodic table, based upon the biological ligands favored by these metals. The macronutrient metals are found within "class A", the trace and some toxic metals within the "borderline" class. Note that Cu and Pb, which share characteristics of both "class B" and the "borderline" class, appear twice, as do the groups to which the "class B" metals belong (IB, IIB, IV).

as Sr displacing Ca in bone) or between class B metals (Cd and Hg interfering with Cu metabolism). However, Cd will substitute for Ca due to its similar ionic radius in the protein concanavalin A,[11] and may inhibit phosphoglucose isomerase by displacing Mg in *Monodonta*.[14]

Marine mollusks and arthropods show protein polymorphisms which allow for enzyme viability under different metal insults.[15] However, there is little equivalent work on terrestrial invertebrates. A variety of adaptive responses have been proposed, and many implicate mechanisms of essential metal regulation, but the relative roles of acclimation, adaptation, or their combination are largely unknown. Few studies have demonstrated inherited tolerance in a generation removed from the pollution source, thereby distinguishing between phenotypic or genotypic responses in a terrestrial invertebrate.

This chapter aims to demonstrate the significant role that essential metal regulation may play in toxic metal assimilation by terrestrial invertebrates, and how, in turn, this may determine their adaptation to life in a metal-contaminated environment. It begins by considering toxic metal interactions with Ca, and to a lesser extent, Mg. Ca is a class A metal required in varying amounts by different invertebrate phyla. Yet, despite its role in some structural tissues, many groups isolate excess intracellular Ca in concretions. Metal interactions in these intracellular granules and also in structural tissues are considered here. By way of contrast, the effect of toxic metals on Cu and Zn, two borderline trace metals which themselves become toxic in excess, are also reviewed. In each case, the evidence for an adaptive response to the toxic metal insult by terrestrial invertebrates is considered.

METAL INTERACTIONS IN INTRACELLULAR GRANULES

Ca may be in excess in many biological systems, and mechanisms to reduce its intracellular concentration are relatively well understood. Precipitation as intracellular granules is likely to be the least demanding energetically,[16] and a variety of granule types have been described from all major invertebrate groups.[17] Two varieties of Ca-rich granule are found: extracellular calcium carbonate granules acting as a Ca reserve,[18,19] and insoluble calcium phosphate granules used for metal retention[18,20] or a phosphate reserve.[21] The carbonate granules are believed to buffer gut pH in earthworms.[22] Although mollusks have both types of granule, tissue buffering is also aided by ammonia generation.[23,24]

Phosphate granule production may increase in response to some metals — possibly with Pb in earthworms[25] and with Zn, but not Cu, in mollusks.[26,27] Total granule volume increases with dietary Cu and Fe in *Musca*.[28] Toxic metals are carried as a protein-bound fraction in the blood of the snail, *Helix aspersa*,[29] and are eventually added to these granules, although their organic content is relatively low.[30] The high Ca concentration may inhibit the enzyme pyrophosphatase, inducing phosphate precipitation of the metals on the granule surface.[30] Class A metals are most likely to precipitate, while class B metals (Cd, Cu, and Hg) will tend to remain protein-bound.[13] Ca can be released from these granules. Simkiss[31] shows how Mn entering the cell can lead to corrosion of the granule surface and the liberation of Ca in *Helix aspersa*. This excess Ca induces toxic effects, including necrosis of the hepatopancreas.

Arthropods with a digestive gland produce calcium phosphate granules: Pb-rich granules are known in spiders,[32] and *Dysdera* has separate granules with large amounts of Cu and Zn,[33] although Hopkin and Martin[34] detect no net uptake of Zn, Cd, or Pb by this spider on a contaminated diet. Some insects produce granules in the mid-gut[35] or in Malpighian tubules, both of which are sites of toxic metal accumulation.[36] These granules may not be lost and thus serve as a permanent site for storage in *Musca*.[28]

Earthworms differ from arthropods by producing calcium phosphate granules with a higher organic content.[37] These "chloragosomes" are confined to the chloragogenous tissue surrounding the posterior alimentary canal. Binding of Pb to chloragosomes probably incurs an energetic cost[38,39] since Pb will displace Ca and Zn.[40,22] Although Pb is thought to displace Ca in binding to the phospholipid of lichens,[11] the capacity of chloragosomes for the metal is undiminished by removal of their phospholipid.[40] In addition, the negative correlation between Pb and Ca, Zn and Pb, and Pb and P in the chloragosomes of *Dendrobaena rubida*[22] suggests antagonistic binding between these elements. The chloragosomes may also be the principal reserve for physiological Zn.[41]

Alternative granule types are produced from different routes of accumulation. In mollusks, Cd is initially bound by specific binding proteins resembling mammalian metallothionein (MTN),[42,43] and is later incorporated into a further granule type.[17] Cd-rich granules ("cadmosomes") have been described for earthworms.[22] No calcium phosphate granules have been identified in the hepatopancreas of woodlice,[33] where Pb is mostly confined to other, membrane-bound granules.

These "cuprosomes"[44] are rich in S and are primarily used for Cu storage; Cu is largely protein-bound, but Pb is precipitated as phosphate.[45] These granules are probably end products of the lysosomal system.[46] While the cuprosomes are a principal storage site for Cu in the isopod hepatopancreas, an additional granule type, found in different cells, stores Fe and may also be used for toxic metal accumulation under high insults.[33] The lack of cuprosomes in gastropods[47] may account for their greater susceptibility to Cu poisoning.[48] A more comprehensive granule classification is given by Hopkin.[33]

Granules do not account for the whole metal burden. A significant proportion of Pb, Cd, and Cu is found beyond membrane-bound vesicles in the woodlouse, *Porcellio scaber*,[45] and in lead-polluted worms.[39] One third of the Zn and most of the Cd in the hepatopancreas of the snail, *Helix pomatia,* is associated with a soluble protein.[49] Only one half of the total Pb burden is held in the hepato-pancreatic granules of the slug, *Arion,* on a dosed diet, and Ireland[50] suggested that granule binding sites for the metal had become saturated.

METAL INTERACTIONS WITHIN THE SOFT TISSUES

Concentrations of Ca are elevated in the soft tissues of many terrestrial animals exposed to toxic metal pollution, including annelids and mollusks (Figure 1). Compared to an uncontaminated population, slugs with elevated Zn in their calcium phosphate granules show high Ca levels in the hepatopancreas and the rest of the tissues.[51] In addition, high Ca concentrations in the soft tissues of juvenile snails *(Helix aspersa),* probably required for shell building, are matched by the highest lead levels of any age group.[52] The interaction of Pb and Ca is not simply a product of diet selection, at least in mollusks.[53] The correlation of

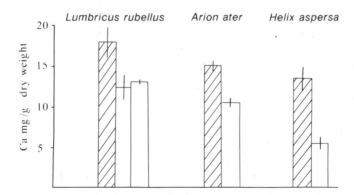

FIGURE 1. The increase of Ca concentrations in the soft tissues of earthworms and gastropod mollusks from habitats polluted with lead (redrawn from Ireland[51,97] and Beeby and Richmond[63]). In each case, the population from the contaminated site (hatched boxes) is shown to have higher mean Ca levels. The bars show the standard error term associated with each mean.

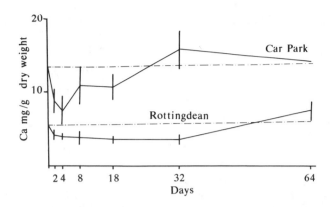

FIGURE 2.   The loss and restoration of Ca concentrations in the soft tissues of two populations of *Helix aspersa* with time on a high lead diet. The Rottingdean site has no significant metal pollution, but the car snails come from a central London site that is grossly polluted. This latter population has much higher Ca levels in its soft tissues, and will restore to these field levels (denoted by the broken lines) more rapidly after initial exposure to Pb. The standard error of each mean is shown by the bars (where *n* >3).

elevated Pb and Sr observed in *Arion*[54] may be due to Sr, which is a chemical analog of Ca.

High toxic metal levels in the tissues appear to increase essential metal turnover. In slugs there is a periodic loss of Cu from the hepatopancreas apparently following a cycle of Ca loss.[27] The mantle and hepatopancreas of *Helix pomatia* may show the same pattern.[55] The rate of cell turnover is increased in the hepatopancreatic epithelium of *Arion* at high Cu doses,[56] and the same authors suggest that Ca excretion is accelerated to speed the removal of Cu. However, Cu is not concentrated in the hepatopancreas, and more than one half of the total copper in the snail, *Arianta,* is held in the foot/mantle tissue.[57]

A competitive interaction between Zn and Ca is thought to explain an equimolar correspondence of the two metals in the oyster, *Ostrea.*[58] Zn is lost with Ca granules and also with the excretion of lipofuscin from the hepatopancreas of *Arion.*[59,60] A loss and restoration of Ca and Zn follow addition of Ag to the diet of the snail, *Achatina,*[61] but there is no equivalent response in the slug, *Arion.* Dietary Cd causes Zn and Cd loss from the intestinal epithelium in *Arion,*[2] and a high Pb diet will induce Ca loss from the hepatopancreas and intestine.[62] Pb has the same effect on whole soft tissue concentrations of Ca in *Helix aspersa.*[63] However, snails from a polluted car park fed a high Pb diet will restore their Ca levels and maintain a higher concentration of Ca compared to a population from an unpolluted site (Figure 2). The car park snails also retain less Pb in the soft tissues.[64] This level of Ca may be an adaptive response to high ambient Pb, perhaps representing an increased concentration of calcium phosphate granules to provide additional binding sites and a more rapid excretion of Pb. Within the mollusks, this increase in the turnover of Ca appears to be a response to both borderline and class B metals, even in animals without previous exposure to toxic metals. Perhaps this represents an increase in the period of a Ca cycle in the soft tissues.

High dietary Mn may facilitate Mg assimilation by the woodlouse, *Porcellio spinicornis,* possibly due to its role in activating enzymes responsible for Ca and Mg assimilation,[65] although why this effect appears to exclude the hepatopancreas is not explained. Prosi and Back[66] suggest that the majority of the Pb added to the hepatopancreas may arrive via the hemolymph in *Porcellio scaber,* and this is also the most significant route for Ca additions to the hepatopancreas of mollusks.[67] *Gammarus,* a freshwater amphipod, shows a strong inverse relationship between hemolymph Cd and total body Ca, such that Cd uptake is more a function of the crustacean's Ca status than the external concentration of Cd.[68] Zn and Cd may inhibit Ca uptake, and Wright[68] concludes there is "accidental active uptake" of Cd.

Different species of woodlice show different rates of metal accumulation. Cd and Pb concentrations are higher in the hepatopancreas of *Oniscus* than in *Porcellio* from the same site;[69] possibly a higher proportion of these metals are held in the more heavily calcified exoskeleton of *Porcellio.* Differences in gut pH[69] or gut flora between these species[70,71] might also be responsible. Bengtsson et al.[72] note the possible significance of such microbial interactions in the metal assimilation of litter feeders and Simkiss[73] shows enhanced Cu uptake by *Helix* with sulfate-reducing bacteria in their crop. As Simkiss notes, such enhancement will be greater in animals lacking an acidic gut.

Pb uptake by woodlice is related to their Ca assimilation.[74] While Zn is the primary cause of mortality in *Porcellio* near a smelter in Pennsylvania, this is also partially determined by the presence of other metals (Cu, Cd, Pb, Ca), other elements (S, P), and the pH of the litter.[75] The lead burden of the mother also reduces the amount of Ca she provides to the developing young in her brood pouch,[76] although a depression of fertility is achieved only at high Pb (12,800 ppm dry wt) and Zn (1600 ppm dry wt) doses.[77] Joosse et al.[78] also recorded reductions in progeny numbers, growth, and food consumption in Zn-rich diets. Cd and Zn are shown to have a synergistic effect, reducing brood size and extending gestation time in *Porcellio.*[48] However, a reduction in respiration rate attributable to Zn in *Porcellio* from an uncontaminated site is absent in a population from the vicinity of a Zn smelter, and this may indicate tolerance.[78]

Centipedes from a metal-contaminated site survive longer than those from an uncontaminated site when fed a high metal diet. Hopkin and Martin[79] suggest that this population may be resistant to Cd, possibly through some form of adaptation of their midgut. The oribatid mite, *Platynothrus peltifer,* will avoid

Pleurococcus spiked with Cd above certain levels,[80] just as some woodlice may reject diets with high Zn or Cu.[78,81] The collembolan *Orchesella cincta* is claimed to show tolerance by rejecting Pb-rich foods,[82] although this is not demonstrated on individuals reared in the absence of Pb. Populations of the same species from polluted and unpolluted sites will both reject algae rich in Fe.[83]

Fecal losses account for most of the Pb consumed by the collembolan *Orchesella cincta*,[84] but the majority of that assimilated is lost with exfoliation of the gut lining. This insect may show tolerance to Pb and Cd; a population from a polluted habitat had a small, but significantly increased, excretion efficiency for Pb and Cd.[85] This represents an increase in the metal content of gut linings lost at molt, those metals being bound to granules within the intestinal epithelium. Interestingly, van Straalen et al.[85] speculate that the adaptation may not be complete due to a lack of sufficient binding capacity in the granules.

Earthworms need Ca solely for metabolic processes, such as extracellular buffering and muscular contraction. Ca-rich tissues were thought to play a part in Pb excretion in earthworms nearly 100 years ago,[86] although the interaction between the two metals was first established by Ireland, who describes an elevation of tissue Ca in *Dendrobaena rubida* from soils polluted with Pb.[87,88]

Morris and Morgan[89] suggest how the correlation between Ca and Pb might operate in earthworms: an increase in available Pb saturates the granule-binding capacity (as Ireland[50] suggested for *Arion*), and subsequent leakage of Pb causes damage to cell membranes allowing an accumulation of intracellular Ca.[90] Why this should induce a correspondence between whole body Ca and Pb or enhance Ca uptake is unclear. Nevertheless, the absence of high Ca levels in *Lumbricus rubellus* from a contaminated site may indicate adaptation; without any cytotoxic effect of Pb, no overall increase of Ca in the worms occurs.[91]

Pb may exert other selective pressures on worms. It impairs Ca regulation by the calciferous glands.[91] Ca is the main agent of muscular depolarization in mollusks and worms, rather than Na,[12] and Weilgus-Serafinska[92] suggests that the presence of Pb in the muscles impairs earthworm locomotory function by interference with their calcium ion transport.

Again, differences between species occur within a site.[93] *Lumbricus* are shown to accumulate more Pb than *Allolobophora*,[94,95] although the contrary was found by Ash and Lee.[96] *Lumbricus rubellus* has more Ca and Zn than *Dendrobaena rubida* but only one third the Pb and one half the Cd.[22] Some earthworms may regulate their gut pH using calcium carbonate secretions from the calciferous glands in the anterior gut, conferring a tolerance to acidic conditions. High Ca in the gut of these species may produce a competitive interaction between Pb and Ca uptake,[22] and such worms are known to have chloragosomes with a higher Ca content.[41] Earthworms with inactive calciferous glands show higher levels of Pb in the whole tissues compared to acid-tolerant species.[89] Even so, some acid-tolerant worms have a greater resistance to toxic metals than others: *Eisenia* is more able to survive Pb additions to the soil than either *Lumbricus* or *Den-*

drobaena,[97] and *Eisenoides* also survives with very high levels of Pb in its tissues.[75]

Raising the gut pH will effectively reduce the availability of dietary Pb. In this case, the role of Ca is largely incidental. For the same reason, Ca-rich soils have lower levels of available Pb. Ca demand will also be reduced in such soils and Ma[98] proposes that both Ca and Pb uptake would therefore be reduced. There is consequently a range of possible interactions between Pb and Ca which determine tissue Pb concentrations; these are summarized in Table 2.

Zn and Ca are correlated in the whole tissues of *Dendrobaena* in the absence of significant soil pollution,[38] but a diet dosed with Cd had no effect on the Ca status of two species of earthworm.[99] The major route for Zn and Pb loss in *Dendrobaena* is via the feces,[100] but Andersen and Laursen[95] believe Pb is also lost through the calciferous glands. They suggest that Pb levels are higher in the coelomic waste nodules of species lacking this route because their calciferous glands are inactive. However, Morgan and Morris[22] found no toxic metals in the calcium carbonate granules of *Dendrobaena rubida* or *L. rubellus* and conclude that this is not a significant route for toxic metal excretion. Instead, the Pb:Ca interaction is confined to the posterior alimentary canal,[25] the site of the chloragogenous tissue. Because of its role in glycogen storage, this tissue has been likened to the liver of vertebrates[39] and, indeed, like the digestive gland of mollusks, it is the primary site for calcium phosphate granules or chloragosomes. Earthworm mucus and urine have also been implicated in the excretion of Zn and Pb.[40]

A summary of the principal interactions between Ca and Mg and the more common toxic metals in terrestrial invertebrates is given in Table 3. The primary response by either essential metal is shown for each metal deemed to be in excess for each group. An increase in essential metal turnover appears to be a common feature in most of these responses.

Table 2
The Major Environmental Determinants of Pb Uptake by Earthworms[a]

Soil	Primary Effect	Secondary Effect	Pb Status
High soil acidity	1. High Pb availability	—	+
	2. High Ca demand	High Pb uptake	+
	3. Acid-tolerant species add Ca to gut	Ca competes with Pb for binding sites	−
	4. Acid-intolerant species unable to regulate gut pH	High Pb availability	+
High soil Pb	1. High Pb in chloragosomes	High Ca in anterior alimentary canal	—
	2. Saturation of chloragosomes	Ca levels elevated due to cytoxicity	+
High CEC[b]	1. Low Pb availability	—	

[a] A particular soil parameter will have a primary effect on metal uptake or some other physiological response, which, in turn, may induce the secondary effects shown.
[b] Cation-exchange capacity.

Table 3
The Principal Interactions of Ca and Mg with Various Toxic Metals in the Soft Tissues of Different Terrestrial Invertebrates[a]

Principal Reponse	Toxic Metal	Group	Reference
Increased Ca level	Ag	Mollusca	87
	Mg		61
	Pb		63
	Zn		51
	Pb	Annelida	97
			89
	Zn		88
	Pb	Isopoda	74
Increased Mg level	Ag	Mollusca	61
	Pb		63
	Mn	Isopoda	65
Increased granule production	Cu	Mollusca	27
	Zn		26
	Pb	Annelida	22
	Cu and Fe	Insecta	28
Increased Ca loss from the soft tissues	Ag	Mollusca	61
	Pb		4
			63
	Cd		4
	Cu		27
	Zn		59,60
Ca displacement from granules	Pb	Mollusca	22
			40
	Zn		51
Mg displacement	Be	Mollusca	118
	Pb		114

[a] The response associated with a range of metals is shown, including trace and essential metals in excess.

METAL INTERACTIONS IN THE STRUCTURAL TISSUES

The lower rates of metal assimilation by insects are probably a function of low demand, due in part to a lack of respiratory pigments.[33] In addition, the nature of the exoskeleton means less Ca is needed for structural tissues in most insects[101] and Zn may be added to some parts of the insect cuticle to harden it.[102] While Mg, K, and Zn are required in a basal diet for proper growth in *Tribolium*, no major Ca need is detected.[103] Equally, demand for Mg may decline in the presence of other minerals, especially Na.[103,104] The toxicity of Cd, Mn, and Sr are increased in the absence of Ca in *Blatella*, although this is not true of Pb and Zn.[105] Gordon suggests that Sr and Mn may interfere with Ca metabolism in this insect. *Formica* larvae retain toxic metals in tissues rich in Ca, and there may be no distinction between Ca and toxic metal uptake from the insect lumen.[36]

Large losses of metals occur with the meconium left after pupation.[106,107] Cd and Zn in the exuviae of developing chironomids reflect the concentration of the solution in which they have been cultured,[108] although surface adsorption onto the castes is not quantified in this study. Most Zn assimilated by the centipede *Lithobius* is retained in the exoskeleton and subcuticular tissues, with Cu largely held in the midgut,[79] although Hopkin[33] refers to there being no detectable Zn in their exuviae. Up to 80% of the Pb burden is held in the exoskeleton of surface-washed carabid beetles.[109] Molting may provide one route of detoxification for arthropods shedding their gut lining.[110] A higher molting rate may be an adaptation to elevated Mn in the collembolan *Orchesella cincta.*[83]

Woodlice use their exoskeleton as a storage-excretion device,[101] with Ca being moved out of the carapace at each molt for retention and reutilization. Besides Ca, the crustacean exoskeleton contains magnesium carbonates and phosphates,[65,101] and Pb is deposited there in freshwater isopods.[111] Alikhan[112] suggests that Mg and Mg levels in the exuviae reflect dietary concentrations. In this case, the molted exoskeleton could represent a detoxification route. However, the exuviae are usually consumed by the owner, and further experiments are needed to show that toxic metals may indeed be lost by this route.

A resistant population of the freshwater isopod *Asellus meridianus* assimilates dietary Pb and Cu more readily than a nontolerant population, with Pb added to the exoskeleton and associated tissues.[111] In *Gammarus* the exoskeleton is a major binding site for Cd, and the hepatopancreas only becomes important when the exoskeleton is "saturated."[68] By coating some specimens with grease, this study attempts to distinguish adsorption onto the carapace from the external media from dietary intake and suggests that surface adsorption is unimportant in this amphipod.

The molluskan shell also serves as a depository of toxic metals.[62,113,114] The potential for Ca being replaced in the shell may depend on the ionic radius of the substituting metal.[115] Sr and Ba will displace Ca[116,117] and, in doing so, may determine the shell's crystalline structure.[115] High dietary Ca inhibits Be retention by the shell and the hepatopancreas of *Achatina* and also causes loss of Zn and Mg from both sites.[118] In contrast, Ag was not deposited in the shell of this species fed a dosed diet.[61] Pb displaces Ca more readily than Cd in *Mytilus,* although Pb is largely held in the calcium carbonate fraction and Cd in the organic matrix of the shell.[119,120] Mg is lost from the shells of adult *Helix* accumulating Pb,[114] and a similar loss of Mg and Ca occurs in the shells of juveniles (A. Beeby, unpublished data, 1989). In addition, snails from a polluted habitat show an increased capacity to add Pb to the shell, and again, juveniles from this site show the same characteristic as adult snails. This may indicate an adaptation by this population.[114]

The role of Ca and Mg in structural tissues probably accounts for the deposition of the toxic metals at these sites. Indeed, the retention time of a toxic metal in any tissue may reflect the rate of turnover of Ca or Mg. By virtue of this, these structural tissues are relatively inert sites which serve as long-term repositories of toxic metals. Calcium phosphate granules represent sites for short-term accumulation. Perhaps the chloragosomes and waste nodules fulfill the equivalent roles in earthworms. In those animals able to shed their cuticle, toxic metal retention in the exuviae may be an important route for loss if they are not subsequently consumed. The extent to which woodlice, in particular, can recover Ca from their exoskeleton while losing other metals with the exuviae has yet to be quantified.

INTERACTIONS IN TRACE METAL ASSIMILATION

The low concentration of trace elements in the natural environment requires effective uptake and retention mechanisms in the biota. Isopods need Cu for hemocyanin production but will store far more in the hepatopancreas than their respiratory needs;[121] this is perhaps an adaption to high demand and low dietary levels in nonpolluted habitats.[33]

The interrelations between trace and toxic metals can be very complicated: Pb and Zn interfere with hemoglobin synthesis in mammals, and excess Cu may compete with intestinal binding sites for Zn.[3] The antagonism between Cd and Zn may alter Zn and Cu relations which, in turn, disrupts Cu and Fe interactions.[3] Cd has a major effect on egg production by *Platynothrus peltifer,* and van Straalen et al.[80] suggest that this is due to disruption of the role of Zn in the mite's reproductive metabolism. In *Helix* high levels of Cd causes tissue damage and reduced shell growth and reproductive activity;[122] Cd is one of several metals able to block the Ca current of *Helix* neurons.[123]

Like insects, woodlice regulate their Zn assimilation,[78] although concentrations differ between species - *Porcellio* has four times as much Zn in its hepatopancreas as *Oniscus* from the same site.[69] This may be due to its role in the enzyme carbonic anhydrase and the greater precipitation of Ca in the exoskeleton of *Porcellio*.[69] Since Cd is often presumed to follow uptake routes of Cu and Zn, Cd:Zn and Cd:Cu ratios ought to be similar between species. Hopkin et al.[69] suggest that, because they differ, *Ligia oceanica* and *Porcellio scaber* may have become adapted to reduce their Cd uptake, retaining less than *Oniscus asellus*. The same authors suggest that the large size of the hepatopancreas of *Oniscus asellus* may represent an adaptation to high Zn.[124]

In a population of the freshwater isopod *Asellus meridianus* tolerant to Cu, Pb displaces Cu from the cuprosomes of the hepatopancreas, while a Pb-tolerant population is able to restrict the uptake of both Pb and Cu.[125] However, in a closely related species, *A. aquaticus,* acclimation is shown to be a component of Pb resistance.[126] *Porcellio* collected from near a Zn smelter accumulate more Zn than a control population, and this appeared to involve an energetic cost, at least in the males,[127] although again, acclimation may well play some role in this response. Joosse and Verhoef[48] report an increased capacity to retain Cd in *Porcellio* deemed to be tolerant. Most significantly, this has been shown for a laboratory-reared generation, presumably without exposure to the metal.

Zn or Se in the soil depress earthworm Cd uptake. The most significant determinant of their Cd assimilation appears to be the Zn:Cd ratio in the soil.[128] Cd may have a low relative toxicity to worms because it is held by specific binding proteins.[98,129] Possibly the range of Cd concentrations normally encountered in most habitats are effectively bound by such proteins in both annelids and mollusks, with tissue damage only occurring at very high exposures. Cu has a high toxicity to earthworms,[98] but, at low levels, Cu may enhance growth in adult worms and may be necessary for successful cocoon production.[130] Failure to meet this Cu demand may be compensated for by Pb,[130] and notably high levels of Cu appear to suppress Pb uptake in this group.[97] Both metals may serve to counterbalance low pH effects, reduce cocoon parasites or, in some way, induce "high-quality" cocoons.[130] The same may be true of the collembolan *Onychiurus* with either metal ridding it of a gut parasite.[131]

Pb and Cd may be assimilated by the same physiological pathway in the collembolan *Orchesella*.[132] Hg is bound by the same protein which binds Zn and Cu in *Blatella* and will displace Zn from the lysosomes, the remainder of which become granules.[133] Christie et al.[134] note that resistance to Hg in some *Drosophila* strains is associated with an increased sensitivity to Cd but offer no reason why the two should be linked. The same authors note that the gene coding for Cd resistance is located on the X chromosome, but its effects are augmented by genes located on a second chromosome.

Cu levels fall within a narrow range in most insects (9 to 20 μg/g dry wt) indicating regulation.[104] Timmermans and Walker[108] suggest there is an efficient excretory mechanism for Cu between the larval and pupal stage of *Chironomus riparius* to maintain a low concentration in imagines. The trace elements Cu and Zn show more evidence of regulation than Cd in the grasshopper, *Chorthippus*.[135] Cd assimilation depresses Cu uptake in males and Zn uptake in both sexes of *Chorthippus*[135] (Figure 3) and in *Chironomus* larvae from water spiked with high levels of Cd.[136] In this latter case, removing Cd allowed Zn but not Cu levels to be restored. A protein binding Cd in *Sarcophaga* larvae may also be responsible for Cu and Zn regulation; Cd competes with Cu for binding sites, but Zn is largely unaffected by Cd administration.[107]

METAL INTERACTIONS IN INDUCIBLE PROTEINS

Both Cu and Cd bind to particular cells in the midgut region of *Drosophila*, but, while a significant proportion of the Cu is in a particulate fraction, nearly all Cd remains in solution.[137] Cd induces a specific binding protein with several characteristics of mammalian MTN, and which will bind Cu but not Zn.[137] Cd can be readily displaced from a low-molecular weight protein induced in *Chironomus yoshimatsui* by Cu.[136] Of the two varieties of this protein described for *Drosophila* by Mokdad et al.,[138] only one is also induced by Cu. These authors could detect no inducible ligand for Zn and, for this reason, suggest that this protein ordinarily serves to regulate Zn assimilation.

That Cd retention varies with age and developmental stage in *Drosophila* may imply some phenotypic component to the response, possibly acclimation. Even so, the greater Cd resistance in some strains of *Drosophila* results from an increased capacity to produce a specific binding protein,[139] with more of the toxic metal being retained. Suzuki et al.[140] also show that tolerance to Cd in the worm *Eisenia foetida* is due to the induction of a specific binding protein. In this case, three components are induced, the most important of which has a very high molecular weight. This component also binds most Cd in *Dendrodrilus rubidus* and *L. rubellus,* although, while the latter has two types, *Dendrodrilus* has a solitary binding protein.[129]

Cd binding proteins appear to have a long evolutionary history, although the inducible components probably arose separately in the insects and crustaceans.[138] An increase in the accumulation of Cd and Zn by *Porcellio* from a polluted site is attributed to an increased production of specific binding protein, possibly indicating tolerance in this population.[141] Joosse and Verhoef[48] report a Cd binding protein which appears to be inducible by both Cu and Cd from tolerant and nontolerant *Porcellio*. However, no such protein could be detected in the hepatopancreas of *Porcellio* examined by Dallinger and Prosi.[46]

Cooke et al.[142] suggest that in snails Cd competition for protein ligands in the hepatopancreas accounts for the similarities in Cd and Zn assimilation. In addition, the displacement of Zn by Cd from the ligand is thought to account for its appearance in granules. Cd is associated with a protein in *Arion* which, like *Drosophila,* has an inducible low-molecular weight component which seems to have a specific role in detoxification.[42] The greater accumulation of Zn and Cd by the molluskan kidney[143] may be due to a recycling of the metal-binding protein there.[144,145]

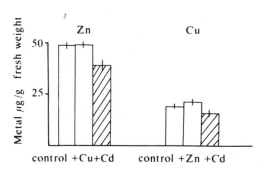

FIGURE 3.   The effect of dietary Ca on Zn and Cu concentrations in *Chorthippus brunneus* (redrawn from Johnson[135]). Neither Cu or Zn appear to affect soft tissue concentrations of each other in this grasshopper, but cadmium significantly depresses Zn but not Cu. Mean concentrations are shown with standard error bars.

Table 4
**The Range of Adaptations Suggested for Invertebrates Resistant to the
Effects of Various Toxic Metals and for Which There May Be a Genetic
Component[a]**

Adaptation	Metal	Species	Reference
Ca regulation	Pb	*Eisenia*	97
	Pb	*Lumbricus rubellus*	25
	Pb	*Helix aspersa*	63
Increased excretion	Mn and Fe	*Orchesella cincta*	82
	Pb and Cd	*Orchesella cincta*	83
	Pb	*Helix aspersa*	85
Increased storage	Pb	*Helix aspersa* (shell)	114
	Cd	*Drosophila*	138
Food avoidance	Zn	*Porcellio scaber*	78
	Pb	*Orchesella cincta*	82
Increased ligand production	Cd	*Drosophila*	137
	Cd, Zn	*Porcellio scaber*	140
	Cd	*Eisenia foetida*	139
	Cd	*Arion lusitanicus*	43
	Cu, Cd	*Drosophila*	147

[a] With the exception of food avoidance and specific protein production, these may
simply represent an acceleration of existing processes.

CONCLUSION

An understanding of the relationships between essential and toxic metal uptake
helps to explain why toxic metals are assimilated at all. While most attention
here has been focused on the common metallic pollutants and the essential metals
of group IIA, the potential range of interactions is large. These are a product
not only of similarities in the chemistry of the metals but also result from a lack
of specificity in biological ligands. It may be that in populations without a past
exposure to toxic metals there has been no selective pressure for ligands to be
more specific.

The recurrence of metallothionein-like proteins in the major phyla suggests
that these have a long evolutionary history and are possibly derived from a Zn-
binding protein used to regulate tissue levels of this trace metal. Equally, the
prevalence of an inducible Cd-binding protein suggests that many groups have
found it prudent to protect themselves against accidental Cd uptake by producing
a more competitive ligand for the metal. The relative uniformity of the protein
across these phyla may indicate that only a small genetic change has been required
for each adaptation. Tolerance to other metals may also involve only a small
number of genetic changes. Indeed, the various responses which have been
proposed largely appear to be an acceleration of existing processes, such as
excretion, Ca and Mg turnover, molt frequency, and granule production.

The association of a number of toxic metals with Ca results from both its
structural role and from its precipitation inside and outside the cell. The conditions
under which Ca is deposited appear to favor the precipitation of other metals,
be they class A, B, or borderline (Table 4). Tolerance may exploit this association

by accelerating aspects of the organism's Ca metabolism; an increase in Ca retention or loss under toxic metal insults have both been shown for a number of groups. Within the soft tissues, there appear to be three possible strategies: an increase in the shedding of granules, the provision of extra binding sites (by producing more granules), or competitive inhibition by Ca for binding sites. However, such responses have to be distinguished from cytotoxic effects of the pollutant[90] or the liberation of Ca from the granules.[31]

Not all options are open to all groups; woodlice may protect themselves from excess Cu by storing it in cuprosomes, a strategy not available to gastropods that lack these granules and to whom the metal is more toxic.[47,48] Avoidance by diet selection may also be a form of tolerance, although the mechanism by which the metal content of the food is detected has yet to be elucidated. In many cases, contamination will consist of surface particulates which are rarely simulated in laboratory diets. Collembolans and woodlice appear to be able to distinguish contaminated food,[82] while snails do not avoid a high Pb diet.[53] Woodlice will ordinarily consume their exuviae after molting, and it would be interesting to see if populations using them as a storage-excretion device consume contaminated exuviae as readily as uncontaminated ones.

Serving as a large-capacity sink, the exoskeleton and the shell provide sites where toxic metals can be isolated from more sensitive tissues. Borderline and class B metals are found here, along with class A metals — associations which invertebrates share with the skeleton of vertebrates.[146] In the absence of these sites, alternative long-term sinks are possible: earthworms may use the Ca-rich waste nodules in the posterior segments in this way,[95] or if the animal has a shorter life-span, granules may be retained for the duration.[28] The capacity of these sites for storage of toxic metals may depend upon their scope for essential metal storage and their rate of essential metal turnover. It would be interesting to compare the biological half-life of Ca and Mg with those of toxic metals resident in the same tissues.

Cd uptake follows Zn and Cu assimilation, probably through its capacity to bind to a protein originally associated with Zn regulation. The assimilation of Cu and Zn may be limited by provision of a fixed amount of carrier protein within the organism. In uncontaminated environments, ligand availability would regulate the uptake of these trace metals and might thus explain the low inducibility of the proteins by Cu and Zn in most groups. The capacity to produce a low-molecular weight variant of this protein while on a contaminated diet confers a resistance to Cd in several groups (although in earthworms it is a high-molecular weight component that is inducible). The extent to which tolerance has produced a novel protein in each group will provide valuable insights into their origins.

Certainly, the best evidence to date of a genotypic response by terrestrial invertebrates to toxic metals are these inducible proteins, which appear to be relatively common among the phyla. Further work on other forms of suspected tolerance needs to focus on generations removed from any contamination to establish whether genotypic change has occurred in a population with a history of exposure to metal pollution. Manipulating their essential metal levels will yield clues to the nature of any such adaptation.

ACKNOWLEDGMENTS

My thanks to Sharon Holmes, Andy Badenoch and Jackie Beeby for typing the manuscript and several anonymous reviewers for their invaluable comments.

REFERENCES

1. Cataldo, D. A., R.E. Wildung, and T. R. Garland. "Speciation of Trace Inorganic Contaminants in Plants and Bioavailability to Animals: An Overview," *J. Environ. Qual.* 16:289-295 (1987).
2. Pounds, J. G. "Effect of Lead Intoxication on Calcium Homeostasis and Calcium-Mediated Cell Function: A Review," *Neurotoxicology* 5:295-332 (1984).
3. Petering, H. G. "The Effect of Cadmium and Lead on Copper and Zinc Metabolism," in *Trace Element Metabolism in Animals, Vol. II,* W. G. Hoekstra, et al., Eds. (London: University Park Press, 1974), pp. 311-325.
4. Ireland, M. P. "Sites of Water, Zinc and Calcium Uptake and Distribution of These Metals after Cadmium Administration in *Arion ater* (Gastropoda: Pulmonata)," *Comp. Biochem. Physiol.* 73A:217-221 (1982).
5. Kratz, W., H. Grutte, and G. Weigmann. "Cadmium Accumulation of Soil Fauna after Artificial Application of Cadmium Nitrate in a Ruderal Ecosystem," in *Heavy Metals in the Environment, Vol. 1,* T. D. Lekkas, Ed. (Edinburgh: CEP, 1985), pp. 667-669.
6. Williamson, P. and P. R. Evans. "Lead Levels in Roadside Invertebrates and Small Mammals," *Bull. Environ. Contam. Toxicol.* 8:280-288 (1972).
7. Carter, A. "Cadmium, Copper and Zinc in Soil Animals and Their Food in a Red Clover System," *Can. J. Zool.* 61:2751-2757 (1983).
8. Beyer, W. N., O. H. Pattee, L. Sileo, D. J. Hoffman, and B. M. Mulhearn. "Metal Contamination in Wildlife Living near Two Zinc Smelters," *Environ. Pollut.* 38A:63-86 (1985).
9. Hunter, B. A., M. S. Johnson, and D. J. Thompson. "Ecotoxicology of Copper and Cadmium in a Contaminated Grassland Ecosystem. II. Invertebrates," *J. Appl. Ecol.* 24:587-599 (1987).
10. Venugopal, B. and T. D. Luckey. *Environmental Quality and Safety Supplement, Vol. 1,* F. Coulston and F. Korte, Eds. (San Diego: Academic Press/G. Thieme, 1975), pp. 4-73.
11. Nieboer, E. and D. H. S. Richardson. "The Replacement of The Nondescript Term "Heavy Metals" by a Biologically and Chemically Significant Classification of Metal Ions," *Environ. Pollut.* 1B:3-26 (1980).
12. Williams, R. J. P. "The Biochemistry of Sodium, Potassium, Magnesium and Calcium," *Quart. Rev. Chem. Soc.* 24:331-365 (1970).
13. Simkiss, K. and M. Taylor. "Cellular Mechanisms of Metal-Ion Detoxification and Some New Indices of Pollution," *Aquat. Toxicol.* 1:279-290 (1981).

14. Lavie, B. and E. Nevo. "Genetic Selection of Homozygote Allozyme Genotypes in Marine Gastropods Exposed to Cadmium Pollution," *Sci. Total Environ.* 57:91-98 (1986).

15. Nevo, E., R. Noy, B. Lavie, A. Beiles, and S. Muchtar. "Genetic Diversity and Resistance to Marine Pollution," *Biol. J. Linn. Soc.* 29:139-144 (1986).

16. Simkiss, K. "Biomineralization and Detoxification," *Calcif. Tiss. Res.* 24:199-200 (1977).

17. Brown, B. E. "The Form and Function of Metal-Containing "Granules" in Invertebrate Tissues," *Biol. Rev.* 57:621-667 (1982).

18. Simkiss, K. "Intracellular and Extracellular Routes in Biomineralization," *Symp. Soc. Exp. Biol.* 30:423-444 (1976).

19. Mason, A. Z. and J. A. Nott. "The Role of Intracellular Biomineralized Granules in the Regulation and Detoxification of Metals in Gastropods with Special Reference to the Marine Prosobranch *Littorina littorea,*" *Aquat. Toxicol.* 1:239-256 (1981).

20. Burton, R. F. "The Storage of Calcium and Magnesium Phosphates and of Calcite in the Digestive Glands of the Pulmonata (Gastropoda)," *Comp. Biochem. Physiol.* 43A:655-663 (1972).

21. Campbell, J. W. and B. D. Boyan. "On the Acid-Base Balance of Gastropod Molluscs," in *Mechanisms of Mineralization in the Invertebrates and Plants.* N. Watabe and K. Wilbur, Eds. (Columbia, S.C.: University of South Carolina Press, 1976), pp. 109-133.

22. Morgan, A. J. and B. Morris. "The Accumulation and Intracellular Compartmentation of Cadmium, Lead, Zinc, and Calcium in two Earthworm Species *(Dendrobaena rubida* and *Lumbricus rubellus)* Living in a Highly Contaminated Soil," *Histochemie* 25:269-285 (1982).

23. Loest, R. "Ammonia Volatilization and Absorption by Terrestrial Gastropods: A Comparison Between Shelled and Shell-less Species," *Physiol. Zool.* 52:461-469 (1979).

24. Loest, R. "Ammonia-Forming Enzymes and Calcium-Carbonate Deposition in Terrestrial Pulmonates," *Physiol. Zool.* 52:470-483 (1979).

25. Morgan, J. E. and A. J. Morgan. "Calcium-Lead Interactions Involving Earthworms. Part 2: The Effect of Accumulated Lead on Endogenous Calcium in *Lumbricus rubellus,*" *Environ. Pollut.* 55:41-54 (1988).

26. Schoettli, G. and H. G. Seiler. "Uptake and Localization of Radioactive Zinc in the Visceral Complex of the Land Pulmonate *Arion rufus,*" *Experientia* 26:1212-1213 (1970).

27. Marigomez, J. A., E. Angulo, and J. Moya. "Copper Treatment of the Digestive Gland of the Slug *(Arion ater* L. 1. Bioassay Conduction and Histochemical Analysis," *Bull. Environ. Contam. Toxicol.* 36:600-607 (1986).

28. Sohal, R. S. and R. E. Lamb. "Intracellular Deposition of Metals in the Midgut of the Adult Housefly *Musca domestica,*" *J. Insect. Physiol.* 23:1349-1354 (1977).

29. Howard, B. and K. Simkiss. "Metal Binding by *Helix aspersa* Blood," *Comp. Biochem. Physiol.* 70A:559-561 (1981).

30. Howard, B., P. C. H. Mitchell, A. Ritchie, K. Simkiss, and M. Taylor. "The Composition of Invertebrate Granules from the Metal-Accumulating Cells of the Common Garden Snail *(Helix aspersa),*" *Biochem. J.* 194:507-511 (1981).

31. Simkiss, K. "Surface Effects in Ecotoxicology," *Functional Ecol.* 4:303-308 (1990).

32. Ludwig, M. and G. Alberti. "Mineral Congregations 'Spherites' in The Midgut Gland of *Coelotes terrestris* (Araneae): Structure, Composition and Function," *Protoplasma* 143:43-50 (1981).

33. Hopkin, S. P. *Ecophysiology of Metals in Terrestrial Invertebrates* (London: Elsevier Applied Science, 1989), p. 366.

34. Hopkin, S. P. and M. H. Martin. "Assimilation of Zinc, Cadmium, Lead, Copper and Iron by the Spider *Dysdera crocata,* A Predator of Woodlice," *Bull. Environ. Contam. Toxicol.* 34:183-187 (1985).

35. Humbert, W. "The Mineral Concentrations in the Midgut of *Tomocerus minor* (Collembola): Microprobe Analysis and Physio-ecological Significance," *Rev. Ecol. Biol. Soc.* 14:71-80 (1977).

36. Jeantet, A. Y., C. Ballan-Dufrancais, and R. Martoja. "Insects Resistance to Mineral Pollution. Importance of Spherocrystal in Ionic Regulation," *Rev. Ecol. Biol. Soc.* 14:563-582 (1977).

37. Prento, P. "Metals and Phosphate in the Chloragosomes of *Lumbricus terrestris* and Their Possible Physiological Significance," *Cell Tissue Res.* 196:123-134 (1979).

38. Ireland, M. P. and K. S. Richards. "The Occurrence and Localisation of Heavy Metals and Glycogen in the Earthworm *Lumbricus rubellus* and *Dendrobaena rubida* from a Heavy Metal Site," *Histochemie* 51:153-166 (1978).

39. Richards, K. S. and M. P. Ireland. "Glycogen-Lead Relationship in the Earthworm *Dendrobaena rubida* from a Heavy Metal Site," *Histochemie* 56:55-64 (1978).

40. Ireland, M. P. "Heavy Metal Binding Properties of Earthworm Chloragosomes," *Acta Biol. Acad. Sci. Hung.* 29:388-394 (1978).

41. Morgan, A. J. "The Elemental Composition of the Chloragosomes of Nine Species of British Earthworms in Relation to Calciferous Gland Activity," *Comp. Biochem. Physiol.* 73A:207-216 (1982).

42. Dallinger, R., B. Berger, and A. Bauer-Hilty. "Purification of Cadmium-Binding Proteins from Related Species of Terrestrial Helicidae (Gastropoda, Mollusca): A Comparative Study," *Mol. Cell Biochem.* 85:135-145 (1989).

43. Dallinger, R., H. H. Janssen, A. Bauer-Hilty, and B. Berger. "Characterization of an Inducible Cadmium-Binding Protein from Hepatopancreas of Metal-Exposed Slugs (Arionidae, Mollusca)," *Comp. Biochem. Physiol.* 92C:355-360 (1989).

44. Wieser, W. and J. Klima. "Compartmentalization of Copper in the Hepatopancreas of Isopods," *Mikroskopie* 22:1-9 (1969).

45. Prosi, F. and R. Dallinger. "Heavy Metals in the Terrestrial Isopod *Porcellio scaber* Latreille. 1. Histochemical and Ultrastructural Characterization of Metal-Containing Lysosomes," *Cell Biol. Toxicol.* 4:81-96 (1988).

46. Dallinger, R. and F. Prosi. "Heavy Metals in The Terrestrial Isopod *Porcellio scaber* Latreille. II. Subcellular Fractionation of Metal Accumulating Lysosomes from the Hepatopancreas," *Cell Biol. Toxicol.* 4:97-109 (1988).

47. Moser, H. and W. Wieser. "Copper and Nutrition in *Helix pomatia* (L.)," *Oecologia (Berl.)* 42:241-251 (1979).

48. Joosse, E. N. G. and S. C. Verhoef. "Developments in Ecophysiological Research on Soil Invertebrates," *Adv. Ecol. Res.* 16:175-248 (1987).

49. Dallinger, R. and W. Wieser. "Molecular Fractionation of Zn, Cu, Cd and Pb in the Midgut Gland of *Helix pomatia, L.*," *Comp. Biochem. Physiol.* 79C:125-129 (1984).

50. Ireland, M. P. "Effect of Chronic and Acute Lead Treatment in the Slug *Arion ater* on Calcium and Delta-Aminolaevulinic Acid Dehydratase Activity," *Comp. Biochem. Physiol.* 79C:287-290 (1984).

51. Ireland, M. P. "Distribution of Essential and Toxic Metals in the Terrestrial Gastropod, *Arion ater*," *Environ. Pollut.* 20:271-278 (1979).

52. Beeby, A. N. and S. L. Eaves. "Short-Term Changes in Ca, Pb, Zn and Cd Concentrations of the Garden Snail, *Helix aspersa* Muller from a Central London Car Park," *Environ. Pollut.* 30A:233-244 (1983).

53. Beeby, A. N. "The Role of *Helix aspersa* as a Major Herbivore in the Transfer of Lead Through a Polluted Ecosystem," *J. Appl. Ecol.* 22:267-275 (1985).

54. Popham, J. D. and J. M. D'Auria." *Arion ater* (Mollusca: Pulmonata) as an Indicator of Terrestrial Environmental Pollution," *Water Air Soil Pollut.* 14:115-124 (1980).

55. Dallinger, R. and W. Wieser. "Patterns of Accumulation, Distribution and Liberation of Zn, Cu, Cd and Pb in Different Organs of the Land Snail *Helix pomatia, L.*," *Comp. Biochem. Physiol.* 79C:117-124 (1984).

56. Marigomez, J. A., E. Angulo, and J. Moya. "Copper Treatment of the Digestive Gland of the Slug *Arion ater* L. 2. Morphometrics and Histopathology," *Bull. Environ. Contam. Toxicol.* 36:608-615 (1986).

57. Berger, B. and R. Dallinger. "Accumulation of Cadmium and Copper by the Terrestrial Snail *Arianta arbustorum* L: Kinetics and Budgets," *Oecologia (Berl.)* 79:60-65 (1989).

58. Coombs, T. L. "The Distribution of Zinc in the Oyster *Ostrea edulis* and Its Relation to Enzyme Activity and Other Metals," *Mar. Biol.* 12:170-178 (1972).

59. Recio, A., J. A. Marigomez, E. Angulo, and J. Moya. "Zinc Treatment of the Digestive Gland of the Slug *Arion ater* L. 1. Cellular Distribution of Zinc and Calcium," *Bull. Environ. Contam. Toxicol.* 41:858-864 (1988).

60. Recio, A., J. A. Marigomez, E. Angulo, and J. Moya. "Zinc Treatment of the Digestive Gland of the Slug *Arion ater* L. 2. Sublethal Effects at the Histological Level," *Bull. Environ. Contam. Toxicol.* 41:865-871 (1988).

61. Ireland, M. P. "A Comparative Study of the Uptake and Distribution of Silver in a Slug *Arion ater* and a snail *Achatina fulica*," *Comp. Biochem. Physiol.* 90C:189-194 (1988).

62. Ireland, M. P. "Effects of Wound Healing on Zinc Distribution and Alkaline Phosphatase Activity of *Helix aspersa* (Gastropoda: Pulmonata)," *J. Molluscan Stud.* 52:169-173 (1986).

63. Beeby, A. N. and L. Richmond. "Calcium Metabolism in Two Populations of the Snail *Helix aspersa* on a High Lead Diet," *Arch. Environ. Contam. Toxicol.* 17:507-511 (1988).

64. Beeby, A. N. and L. Richmond. "Adaptation by an Urban Population of the Snail *Helix aspersa* to a Diet Contaminated with Lead," *Environ. Pollut.* 46:73-82 (1987).

65. Bercovitz, K. and M. A. Alikhan. "Effect of Forced Fasting on Magnesium and Manganese Regulation in a Terrestrial Isopod, *Porcellio spinicornis* Say (Porcellionidae, Isopoda, Crustacea)," *Bull. Environ. Contam. Toxicol.* 43:151-158 (1989).

66. Prosi, F. and H. Back. "Indicator Cells for Heavy Metal Uptake and Distribution in Organs from Selected Invertebrate Animals," in *Heavy Metals in the Environment, Vol 2*, T. D. Lekkas, (Edinburgh: CEP, 1985), pp. 242-245.

67. Fournie, J. and M. Chetail. "Calcium Dynamics in Land Gastropoda," *Am. Zool.* 24:857-870 (1984).

68. Wright, D. A. "Cadmium and Calcium Interactions in the Freshwater Amphipod *Gammarus pulex*," *Freshwater Biol.* 10:123-133 (1980).

69. Hopkin, S. P., M. H. Martin, and S. J. Moss. "Heavy Metals in Isopods from the Supra-Littoral Zone on the Southern Shore of the Severn Estuary. U.K.," *Environ. Pollut.* 9B:239-254 (1985).

70. Coughtrey, P. J., M. H. Martin, J. Chard, and S. N. Shales. "Microorganisms and Metal Retention in the Woodlouse *Oniscus asellus*," *Soil Biol. Biochem.* 112:23-27 (1980).

71. Wood, S. and B. S. Griffiths. "Bacteria Associated with the Hepatopancreas of the Woodlice *Oniscus asellus* and *Porcellio scaber* (Crustacea, Isopoda)," *Pedobiologia* 31:89-94 (1988).

72. Bengtsson, G., S. Nordstrom, and S. Rundgren. "Population Density and Tissue Metal Concentration of Lumbricids in Forest Soils Near a Brass Mill," *Environ. Pollut.* 30A:87-108 (1983).

73. Simkiss, K. "Prokaryote-Eukaryote Interactions in Trace Element Metabolism," *Experientia* 41:1195-1197 (1985).

74. Beeby, A. N. "Interaction of Lead and Calcium Uptake by The Woodlouse, *Porcellio scaber* (Isopoda, Porcellionidae), *Oecologia* (Berl.) 32:255-262 (1978).

75. Beyer, W. N., G. W. Miller, and E. J. Cromartie. "Contamination of the O2 Soil Horizon by Zinc Smelting and Its Effects on Woodlouse Survival," *J. Environ. Qual.* 13:247-251 (1984).

76. Beeby, A. N. "Lead Assimilation and Brood-Size in The Woodlouse *Porcellio scaber* (Crustacea, Isopoda) Following Oviposition," *Pedobiologia* 20:360-365 (1980).

77. Beyer, W. N. and A. Anderson. "Toxicity to Woodlice of Zinc and Lead Oxides Added to Soil Litter," *Ambio* 14:173-174 (1985).

78. Joosse, E. N. G., K. J. Wulffraat, and H. P. Glas. "Tolerance and Acclimation to Zinc of the Isopod *Porcellio scaber* Latr.," in *Heavy Metals in the Environment*, T. D. Lekkas, Ed. (Edinburgh: CEP, 1981), pp. 425-428.

79. Hopkin, S. P. and M. H. Martin. "Assimilation of Zinc, Cadmium, Lead and Copper by The Centipede, *Lithobius variegatus* (Chilopoda)," *J. Appl. Ecol.* 21:535-546 (1984).

80. van Straalen, N. M., J. H. M. Schobben, and R. G. M. De Goede. "Population Consequences of Cadmium Toxicity in Soil Microarthropods," *Ecotoxicol. Environ. Safety* 17:190-204 (1989).

81. Dallinger, R. "The Flow of Copper Through a Terrestrial Food Chain III. Selection of an Optimum Copper Diet by Isopods," *Oecologia (Berl.)* 30:273-276 (1977).

82. Joosse, E. N. G. and S. C. Verhoef. "Lead Tolerance in Collembola," *Pedobiologia* 25:11-18 (1983).

83. Nottrot, F., E. N. G. Joosse, and N. M. van Straalen. "Sublethal Effects of Iron and Manganese Soil Pollution on *Orchesella cincta* (Collembola)," *Pedobiologia* 30:45-53 (1987).

84. Joosse, E. N. G. and J. B. Buker. "Uptake and Excretion of Lead by Litter-Dwelling Collembola," *Environ. Pollut.* 18:235-240 (1979).

85. van Straalen, N. M., G. M. Groot, and H. R. Zoomer. "Adaptation of Collembola to Heavy Metal Soil Contamination," in *Heavy Metals in the Environment, Vol. 1,* T. D. Lekkas, Ed. (Edinburgh: CEP, 1986), pp. 16-20.

86. Hogg, T. W. "Immunity of Some Low Forms of Life from Lead Poisoning," *Chem. News* 71(1850):223-224 (1895).

87. Ireland, M. P. "Distribution of Pb, Zn, and Ca in *Dendrobaena rubida* (Oligochaeta) Living in Soil Contaminated by Base Metal Mining in Wales," *Comp. Biochem. Physiol.* 52B:551-555 (1975).

88. Ireland, M. P. and R. J. Wooton. "Variations in the Lead, Zinc and Calcium Content of *Dendrobaena rubida* (Oligochaeta) in a Base Metal Mining Area," *Environ. Pollut.* 10:201-208 (1975).

89. Morris, B. and A. J. Morgan. "Calcium-Lead Interactions in Earthworms: Observations on *Lumbricus terrestris* L. Sampled from a Calcareous Abandoned Leadmine Site," *Bull. Environ. Contam. Toxicol.* 33:226-233 (1986).

90. Morgan, A. J. "Calcium-Lead Interactions Involving Earthworms: An Hypothesis," *Chem. Ecol.* 2:251-261 (1986).

91. Morgan, J. E. and A. J. Morgan. "Calcium-Lead Interactions Involving Earthworms. Part 1: The Effect of Exogenous Calcium on Lead Accumulation by Earthworms Under Field and Laboratory Conditions," *Environ. Pollut.* 54:41-53 (1988).

92. Wielgus-Serafinska, E. "Influence of Lead Poisoning and Ultrastructural Changes in the Body Wall of *Eisenia foetida* (Savigny), Oligochaeta. 1. Short Action of Different Concentrations of Lead and Ultrastructural Changes in the Cells of the Body Wall," *Folia Histochem. Cytochem.* 17:181-188 (1979).

93. Beyer, W. N. and E. J. Cromartie. "Survey of Pb Cu Zn Cd Cr As and Se in Earthworms and Soil from Diverse Sites," *Environ. Monitor Assessm.* 8:27-36 (1987).

94. Andersen, C. "Cadmium, Lead and Calcium Content, Number and Biomass in Earthworms (Lumbricidae) from Sewage Sludge Treated Soil," *Pedobiologia* 19:309-319 (1979).

95. Andersen, C. and J. Laursen. "Distribution of Heavy Metals in *Lumbricus terrestris, Aporrectodea longa* and *A. Rosea* Measured by Atomic Absorption and X-ray Fluorescence Spectrometry," *Pedobiologia* 24:347-356 (1982).

96. Ash, C. P. J. and D. L. Lee. "Lead, Cadmium, Copper and Iron in Earthworms from Roadside Sites, *Environ. Pollut.* 22a:59-67 (1980).

97. Ireland, M. P. "Metal Accumulation by the Earthworms *Lumbricus rubellus, Dendrobaena veneta* and *Eiseniella tetraedra* Living in Heavy Metal Polluted Sites," *Environ. Pollut.* 19:201-206 (1979).

98. Ma, W. C. "Biomonitoring of Soil Pollution: Ecotoxicology Studies of the Effect of Soil-Borne Heavy Metals on Lumbricid Earthworms," *Annu. Rep. Res. Inst. Nature Manage.,* (Arnham Netherlands, 1983), pp. 83-87.

99. Ireland, M. P. and K. S. Richards. "Metal Content, after Exposure to Cadmium, of Two Species of Earthworms of Known Differing Calcium Metabolic Activity," *Environ. Pollut.* 26A:69-78 (1981).

100. Ireland, M. P. "Excretion of Lead, Zinc and Calcium by the Earthworm *Dendrobaena rubida* Living in Soil Contaminated with Zinc and Lead," *Soil Biol. Biochem.* 8:347-350 (1976).

101. Hackman, R. H. "Cuticle: Biochemistry," in *Biology of the Integument,* J. Bereiter-Hahn et al., Eds. (Berlin: Springer-Verlag, 1984).

102. Hillerton, J. E. and J. F. V. Vincent. "The Specific Location of Zinc in Insect Mandibles," *J. Exp. Biol.* 101:333-336 (1982).

103. Medici, J. C. and M. W. Taylor. "Mineral Requirements of the Confused Flour Beetle *Tribolium confusum* (Duval)," *J. Nutr.* 88:181-186 (1966).

104. Levy, R. and H. L. Cromroy. "Concentration of Some Major and Trace Elements in Forty-one Species of Adult and Immature Insects Determined by Atomic Absorption Spectroscopy," *Ann. Entomol. Soc. Am.* 66:523-526 (1973).

105. Gordon, H. T. "Minimal Requirements of the German Roach *Blatella germanica* L.," *Ann. N.Y. Acad Sci.* 77:290-351 (1959).

106. Nourteva, P. and S. L. Nourteva. "The Fate of Mercury in Sarcosaprophagous Flies and in Insects Eating Them," *Ambio* 11:34-37 (1982).

107. Aoki, Y. and K. T. Suzuki. "Excretion of Cadmium and Change in the Relative Ratio of Iso-Cadmium-Binding Proteins During Metamorphosis of Fleshfly *(Sarcophaga peregrina),"* *Comp. Biochem. Physiol.* 78C:315-317 (1984).

108. Timmermans, K. R. and P. A. Walker. "The Fate of Trace Metals During the Metamorphosis of Chironomids (Diptera, Chironomidae)," *Environ. Pollut.* 62:73-85 (1989).

109. Roberts, R. D. and M. S. Johnson. "Dispersal of Heavy Metals from Abandoned Mine Workings and Their Transference through Terrestrial Food Chains," *Environ. Pollut.* 16:293-310 (1978).

110. van Straalen, N. M. and J. H. van Meerendonk. "Biological Half-Lives of Lead in *Orchesella cinta* (L.) (Collembola)," *Bull. Environ. Contam. Toxicol.* 38:213-219 (1987).

111. Brown, B. E. "Uptake of Copper and Lead by A Metal Tolerant Isopod *Asellus meridianus* Rac.," *Freshwater Biol.* 7:235-244 (1977).

112. Alikhan, M. A. "Magnesium and Manganese Regulation During Moult-Cycle in *Porcellio spinicornis.* Say. (Procellionidae Isopoda)," *Bull. Environ. Contam. Toxicol.* 42:699-706 (1989).

113. Everard, M. and D. Denny. "The Transfer of Lead by Freshwater Snails in Ullswater, Cumbria," *Environ. Pollut.* 35A:299-314 (1984).

114. Beeby, A. N. and L. Richmond. "The Shell as a Site of Lead Deposition in the Snail *Helix aspersa,"* *Arch. Environ. Contam. Toxicol.* 18:623-628 (1989).

115. Watabe, N. "Shell," in *Biology of the Integument.* J. Bereiter-Hahn et al., Eds. (Berlin: Springer-Verlag, 1984), pp. 448-485.

116. Sandi, E. "Radioactivity in Snail Shell Due to Fallout," *Nature* 193:290 (1962).

117. Gardenfors, U., T. Westermark, U. Emanuelsson, H. Mutuei, and H. Walden. "Use of Land-Snail Shells as Environmental Archives," *Ambio* 17:347-349 (1988).

118. Ireland, M. P. "Studies on The Effects of Dietary Beryllium at Two Different Calcium Concentrations in *Achatina fulica* (Pulmonata)," *Comp. Biochem. Physiol.* 83C:435-438 (1986).

119. Sturesson, U. "Lead Enrichment in Shells of *Mytilus edulis,"* *Ambio* 5:253-256 (1976).

120. Sturesson, U. "Cadmium Enrichment in Shells of *Mytilus edulis*," *Ambio* 7:122-125 (1978).

121. Wieser, W. "Ecophysiological Adaptations of Terrestrial Isopods: A Brief Review," *Symp. Zool. Soc. London* 53:247-265 (1984).

122. Russell, L. K., J. I. Denhaven, and R. P. Botts. "Toxic Effects of Cadmium on the Garden Snail *(Helix aspersa),*" *Bull. Environ. Contam. Toxicol.* 26:634-640 (1981).

123. Akaike, N., K. S. Lee, and A. M. Brown. "The Calcium Current of *Helix* neuron," *J. Gen. Physiol.* 71:509-531 (1978).

124. Hopkin, S. P. and M. H. Martin. "The Distribution of Zinc, Cadmium, Lead and Copper within the Woodlouse *Oniscus asellus* (Crustacea, Isopoda)," *Oecologia (Berl.)* 54:227-232 (1982).

125. Brown, B. E. "Lead Detoxification by A Copper Tolerant Isopod," *Nature* 276:388-390 (1978).

126. Fraser, J. "Acclimation to Lead in the Freshwater Isopod *Asellus aquaticus,*" *Oecologia (Berl.)* 45:419-420 (1980).

127. Joosse, E. N. G., H. E. van Capelleveen, L. H. van Dalen, and J. van Diggelen. "Effects of Zinc, Iron and Manganese on Soil Arthropods Associated with Decomposition Processes," in *Heavy Metals in the Environment, Vol. 1,* T. D. Lekkas, Ed. (Edinburgh: CEP, 1983), pp. 467-470.

128. Beyer, W. N., R. L. Chaney, and B. M. Mulhern. "Heavy Metal Concentrations in Earthworms from Soil Amended with Sewage Sludge," *J. Environ. Qual.* 11:381-385 (1982).

129. Morgan, J. E., C. G. Norey, A. J. Morgan, and J. Kay. "A Comparison of the Cadmium-Binding Proteins Isolated from the Posterior Alimentary Canal of the Earthworms *Dendrodrilus rubidus* and *Lumbricus rubellus,*" *Comp. Biochem. Physiol.* 92C:15-21 (1989).

130. Bengtsson, G., T. Gunnarsson, and S. Rundgren. "Effects of Metal Pollution on the Earthworm *Dendrobaena rubida* (Say.) in Acidified Soils," *Water Air Soil Pollut.* 28:361-383 (1986).

131. Bengtsson, G., T. Gunnarsson, and S. Rundgren. "Influence of Metals on Reproduction Mortality and Population Growth in *Onychiurus armatus* (Collembola)," *J. Appl. Ecol.* 22:967-978 (1985).

132. van Straalen, N. M., T. B. A. Burghoots, M. J. Doornhoff, G. M. Groot, M. P. M. Janssen, E. N. G. Joosse, J. H. van Meerendonk, J. P. J. J. Theeuwen, H. A. Verhoef, and H. R. Zoomer. "Efficiency of Lead and Cadmium Excretion in Populations of *Orchesella cincta* (Collembola) from Various Contaminated Soils," *J. Appl. Ecol.* 24:953-968 (1987).

133. Jeantet, A. Y., C. Ballan-Dufrancais, and J. Ruste. "Quantitative Electron-Probe Microanalysis on Insects Exposed to Mercury. II. Involvement of the Lysosomal System in Detoxification Processes," *Biol. Cell.* 39:325-334 (1980).

134. Christie, N. T., M. W. Williams, and K. B. Jacobson. "Genetic and Physiological Parameters Associated with Cadmium Toxicity in *Drosophila melanogaster,*" *Biochem. Genet.* 23:571-583 (1985).

135. Johnson, M. S. "Consumer-Producer Relationships for Trace Metals in *Chorthippus brunneus* (Thunberg)," *Bull. Environ. Contam. Toxicol.* 37:234-238 (1980).

136. Yamamura, M., K. T. Suzuki, S. Hatakeyama, and K. Kubota. "Tolerance to Cadmium and Cadmium-Binding Proteins Induced in the Midge Larva *Chironomus yoshimatsui* (Diptera, Chioronomidae)," *Comp. Biochem. Physiol.* 75C:21-24 (1983).

137. Maroni, G. and D. Watson. "Uptake and Binding of Cadmium, Copper and Zinc by *Drosophila melanogaster* Larvae," *Insect Biochem.* 15:55-63 (1985).

138. Mokdad, R., A. Debec, and M. Wegnez. "Metallothionein Genes in *Drosophila melanogaster* Constitute A Dual System," *Proc. Natl. Acad. Sci. U.S.* 84:2658-2662 (1987).

139. Jacobson, K. B., M. W. Williams, J. E. Turner, and N. T. Christie. "Cadmium Toxicity in *Drosophila:* Genetic and Physiological Parameters," in *Heavy Metals in the Environment, Vol. 2,* T. D. Lekkas, Ed. (Edinburgh: CEP, 1985), pp. 239-241.

140. Suzuki, K. T., M. Yamamura, and T. Mori. "Cadmium-Binding Proteins Induced in The Earthworm," *Arch. Environ. Contam. Toxicol.* 9:415-424 (1980).

141. van Capelleveen, E. H. E. "The Ecotoxicity of Zinc and Cadmium for Terrestrial Isopods" in *Heavy Metals in The Environment, Vol. 2,* T. D. Lekkas, Ed. (Edinburgh: CEP, 1985), pp. 245-247.

142. Cooke, M., A. Jackson, G. Nickless, and D. J. Roberts. "Distribution and Speciation of Cadmium in the Terrestrial Snail *Helix aspersa*," *Bull. Environ. Contam. Toxicol.* 23.445-451 (1979).

143. Coughtrey, P. J. and M. H. Martin. "The Distribution of Pb, Cd, Zn and Cu within the Pulmonate Mollusc *Helix aspersa* Muller," Oecologia (Berl.) 23:315-322 (1976).

144. Robinson, W. E. and D. K. Ryan. "Metal Interactions within the Kidney, Gill, and Digestive Gland of the Hard Clam, *Mercenaria mercenaria* Following Laboratory Exposure to Cd," *Arch. Environ. Contam. Toxicol.* 15:23-30 (1986).

145. Simkiss, K., M. Taylor, and A. Z. Mason. "Metal Detoxification and Bioaccumulation," *Mar. Biol. Lett.* 3:87-201 (1982).

146. van Barneveld, A. and C. J. A. van den Hamer. "Influence of Ca and Mg on the Uptake and Deposition of Pb and Cd in Mice," *Toxicol. Appl. Pharmacol.* 79:1-10 (1985).

147. Maroni, G., J. Wise, J. E. Young, and E. Otto. "Metallothionein Gene Duplications and Metal Tolerance in Natural Populations of *Drosophila melanogaster*," *Genetics* 117:739-744 (1987).

4

Allometry of Metal Bioaccumulation and Toxicity

Michael C. Newman[1] and Mary Gay Heagler[2]

[1]University of Georgia, Savannah River Ecology Laboratory, P.O. Drawer E, Aiken, South Carolina 29802
and
[2]Rutgers University, Department of Environmental Sciences, New Brunswick, New Jersey 08903

OVERVIEW

Many factors influence bioaccumulation and toxicity of inorganic contaminants. One such factor which varies within and between populations and species is animal size. Despite the prevalence and magnitude of scaling effects on accumulation and toxicity, relatively few studies have adequately quantified these effects or clearly elucidated underlying mechanisms. The purpose of this chapter is to synthesize the present literature on ecotoxicological allometry, to define potential techniques, to identify errors and their remedies, and to suggest future avenues for research. It is not the purpose of this chapter to provide an extensive review of such studies.

INTRODUCTION

Advances in this area have been made in a variety of disciplines. Considerations of size in toxicokinetics stem from early research on effective dosages of drugs and poisons.[1,2] Allometric considerations for bioaccumulation have a more complex origin than those for toxicological allometry. Allometry studies became common in radiation epidemiology[3,4] and radioecology[5-7] after nuclear weapons appeared at the end of World War II. General results from such studies prompted assessment of radiotracers to measure metabolic processes in the field including the confounding effects of animal size.[8-10] The growing need for heavy metal biomonitoring and relative ease of heavy metal quantification (atomic absorption spectrophotometry) by the mid-1960s fostered the emergence of another large body of literature involving allometry and bioaccumulation.

By the 1970s, the emerging field of heavy metal ecotoxicology was generally characterized by an overabundance of data and a paucity of paradigms. Boyden[11,12] was one of the first individuals in this field to attempt to define a general quantitative model of size effects on bioaccumulation and to formulate falsifiable hypotheses of underlying mechanisms. Boyden's power model was used primarily to redescribe[13*] or "fit" a data set to a simplifying model. This allowed a certain degree of description and limited extrapolation. Since the 1970s, many workers have attempted to move the power equation from the "redescription" to "generative representation" status (e.g., a model that describes the data and explains the phenomenon). For example, Boyden's suggestion that linkage to metabolic rate dictates the behavior of one class of power relations has been examined in attempts to increase the confidence in and generality of power models.[14-17] Such desirable model qualities are becoming increasingly important as the field becomes more intently focused on cause/effects models, especially bioenergetically based models.[18-22]

A similar process has been taking place relative to allometry and metal toxicity. In 1975, Anderson and Weber[23] formulated a relationship for the effects of body size on toxicity of poisons by modifying established size-effective dosage models.[1,2] The associated models remain in a state of transition between redescription and generative representation.[24-26]

Concepts and techniques applied in ecotoxicological allometry were borrowed exclusively from those of physiological and morphological allometry. This linkage to classic allometry provided a rapid infusion of techniques and concepts, and continuity among disciplines. Full advantage of this continuity was taken in the last decade with the emergence of bioenergetic models of metal accumulation and toxicity. However, several conceptual and procedural errors from

---

* Taylor's classification of models as redescriptions or generative representations is used throughout this chapter. A redescriptive model summarizes observations and permits prediction "on the basis that past patterns might continue." Generative representations contain sufficient detail "to explain the phenomena observed [such that] we can make confident predictions for situations not yet observed."

physiological and morphological allometry were directly transplanted to ecotoxicology. Further, many concepts were overextended in attempts to link physiological or morphological relationships with ecotoxicological ones.

Despite the importance of scaling effects, a general review of ecotoxicological allometry has not been developed to date. The purpose of this chapter is to synthesize the present literature on ecotoxicological allometry, to define potential techniques, to identify significant errors and their remedies, and to suggest future avenues for research. It is not the purpose of this chapter to provide an extensive review of such studies.

ALLOMETRY

"Although the curve fits well, it does not follow that the formula from which it is derived is the unique description of the relationships . . . and the theory from which it is based consequently 'true'."[27]

Overview

Allometry is the study of size and its consequences.[28] Characters examined are most often morphological,[28-32] physiological,[28,33,34] or biochemical.[35-38] Huxley[39] is largely credited for firmly establishing the use of power equations to describe these relationships. Indeed, Huxley's Law of Simple Allometry is a central paradigm of allometry. By 1987, more than 750 published allometric power relationships had been described.[40] Despite their clearly empirical nature, an enormous literature has been generated in an attempt to identify the "basic factor"[29] underlying allometric relationships. Numerous hypotheses now exist, but the underpinnings for this law remain ambiguous. For example, Rubner's Law (metabolic rate is linked to size-dependent change in surface:volume ratio through its influence on heat loss in warm-blooded animals) failed to explain scaling of metabolism because protozoans and cold-blooded metazoans also conform to this relationship. A more recent example involves explanations derived from dimensional analysis[41] which are actively being debated at this time.[42]

Quantification

The general power function can be written as follows:

$$Y = aX^b \qquad (1)*$$

where a and b = constants
Y = size of body part
X = some standard such as body weight

In Huxley's work,[39] the constant b was the ratio of growth of Y to growth of

---

* Throughout this chapter, the original constant and variable designations have been changed from those given in the source publications for the sake of uniformity.

X. It has been referred to as the constant specific growth ratio, coefficient of relative growth, or growth-partition coefficient. When physiological variables are scaled to weight using this relationship, the exponent is referred to as the mass exponent. When $b = 1$, there is a simple proportionality between Y and X (body size). When $b < 1$, Y increases more slowly than X; when $b > 1$, Y increases more rapidly than X. The constant a was attributed no biological or general significance by Huxley. It is the value of Y when $X = 1$. Consequently, it is often given such interpretations as "the rate of oxygen uptake for unit weight" or "mass independent metabolism" in studies of metabolic rate.[28,40] Schmidt-Nielsen[43] refers to it as the proportionality coefficient. In dimensional analysis of allometric relationships, this mass coefficient is also compared to Meeh's constants for solids expressed as mass and is interpreted in terms of geometric similitude.[40] Gould[28] suggests that vague assignment of biological meaning to this constant should be avoided as, depending on the units employed, it can lead to absurd units, e.g., brain weight of a 1-mm tall human. Biological interpretation of a is also complicated by its partial dependence on the value of b.[28] Gould[28] recommends that, until techniques are formulated that remove the influence of b on a, no biological meaning be attributed to this term.

Techniques for describing the allometric relationships between more than two variables have also been outlined. Adolph[44] suggested that several heterogonic equations (equations expressing extent of disproportionality) can be multipled provided each contained a common variable. He outlined the following relationships. Given the two allometric relationships,

$$Y = a_1 X^{b1} \tag{2}$$

$$Z = a_2 X^{b2} \tag{3}$$

where X = body size
Y and Z = biological variables

Then

$$Y = a_1 (Z/a_2)^{(b1/b2)} \tag{4}$$

$$Z = a_2 (Y/a_1)^{(b2/b1)} \tag{5}$$

$$X = (Y/a_1)^{(1/b1)} \tag{6}$$

He further explained that allometric equations [Equation (1)] can be regarded as the algebraic relationship between two separable, simultaneous exponential relationships:

$$Y = ce^{-\beta t} \tag{7}$$

$$X = de^{-\alpha t} \tag{8}$$

With

$$b = \beta/\alpha$$

$$a = c/(d^{\beta/\alpha})$$

Equation (2) can be rewritten in terms of these exponential equations [Equations (7) and (8)] as follows:

$$Y = [c/(d^{\beta/\alpha})]X^{\beta/\alpha} \tag{9}$$

This reexpression of the power relationship could be useful in metal ecotoxicology, a discipline which draws heavily upon exponential relationships from pharmacokinetics and toxicokinetics.

Frequently, biological rates are expressed on a per unit mass basis in allometric studies. The allometric relationship [Equation (1)] for such mass specific rates (e.g., specific metabolic rate) can be expressed with the equation,

$$Y = aX^{(b-1)} \tag{10}$$

where Y = independent variable expressed on a per unit mass basis
X, a, and b = same as defined for Equation (1)

Although the power equation [Equation (1)] dominates allometry, it is an empirical relationship without a clearly defined, underlying mechanism.[28] It is often used when simpler models would fit the data more accurately.[27,28,45] For example, Smith[45] reanalyzed 60 sets of allometric data by fitting them to both a simple linear relationship and a power relationship. Only in 12 cases did the power model provide a better fit (higher correlation coefficient) than the linear model. All 12 exceptions involved interspecific comparisons where the range of sizes was large (broad allometry[45]). He concluded that there was little statistical justification for the general application of a power relationship to allometric data. Heusner[40] suggested that the use of the power equation for intraspecies comparisons with narrow size ranges (narrow allometry[45]) may be particularly unwarranted. Uncritical use of this empirical relationship can compromise data description and subsequent interpretation.[28,44-46]

Techniques
The same approach has been used to analyze data in an overwhelming majority

of allometric analyses. The power relationship is converted to a linear one by taking the logarithms of X and Y. The relationship [Equation (1)] is reexpressed as the following:

$$\text{Log } Y = b\text{Log } X + \text{Log } a \tag{11}$$

Least-squares regression techniques are performed on the Log X and Log Y variables. The intercept (Log a) and slope (b) are used to estimate the mass coefficient and mass exponent, respectively. The correlation coefficient (r) and a bilogarithmic plot are used to support the goodness-of-fit for the resulting model.

There are several advantages and disadvantages of this approach. The transformation often enhances the statistical qualities of the model by improving the normality of the data, and it reduces the influence of outliers on the results. However, sole reliance on the correlation coefficient and bilogarithmic plot for assessing fit to the model can be unjustified.[45] The correlation coefficient is influenced by the ranges of X and Y as well as the fit of the data to the model.[46] Also, the visual appearance of the bilogarithmic plot is highly distorted.[46]

Several errors are frequently made during these statistical manipulations. The regression model associated with the transformed data includes an error term missing from Equation (11).

$$\text{Log } Y = b\text{Log } X + \text{Log } a + \epsilon \tag{12}$$

where ϵ = random error of the model

After regression of the log-transformed variables, the results are usually back-transformed to Equation (1) without consideration of the error term. The model may then be used to make predictions. This can lead to a bias in predicted Y values (original units) as Equation (12) becomes

$$Y_p = aX^b 10^\epsilon \tag{13}$$

where Y_p = the predicted value of Y

Unless the data fit the model perfectly ($\epsilon = 0$), predicted values will be biased by [Equation (14)][47]

$$10^{\sigma_e^2/2} \tag{14}$$

where σ_e^2 = error variance of the Log-Log regression model

An estimate of

$$\sigma_e^2 = \sum_{i=1}^{N} r_i^2/(N - 2) \tag{15}$$

where r_i = the ith regression residual
N = the number of data pairs

The reader should refer to Beauchamp and Olson[48] for a more complete discussion of bias estimation.

The median response, not the mean response, is predicted from the model if this bias is ignored.[47] Assuming a normal distribution of residuals from the Log-Log regression, the unbiased, predicted Y can be estimated to be

$$Y_{unbiased} = \hat{Y}_p(10^{\sigma_e^2/2}) \tag{16}$$

where $\hat{Y}$ = biased prediction

If the assumption of a normal distribution of residuals is rejected using statistics such as the Kolmogorov D Statistic ($N > 50$) or Shapiro and Wilk W Statistic ($N < 50$) the above bias correction is inappropriate. Koch and Smillie[47] suggest a "smearing estimate of bias" in such cases.

$$Y_{unbiased} = \hat{Y}_p(1/N \sum_{i=1}^{N} 10^{r_i}) \tag{17}$$

Bias correction procedures similar to Equation (16) have been outlined in other publications,[48] including those involving allometry.[49] Unfortunately, there has been a general failure to implement these corrections when necessary. Further, the residual distribution is not examined in most instances. Failure to present the complete regression model, including the model error [Equation (12)], in descriptive studies can compromise later use in predictive modeling. These shortcomings in allometric data analysis have been transfered to studies of scaling in ecotoxicology.

The regression technique described above assumes no measurement error or inherent variation in the independent variable, X. The technique minimizes the sum of the squares of deviations on the Y-axis only. In many allometric studies, this is an assumption of convenience, not a conclusion reached after careful examination of the data set.[50] When this assumption is not valid, functional regression techniques are more appropriate than the predictive regression techniques described above. In such techniques, the deviations in the Y and X directions are considered by minimizing "the sum of the products of the vertical and horizontal distance of each point from the line."[50] Ricker[50] outlines functional regression techniques relative to allometry with the following set of equations:

$$Y = u + vX \tag{18}$$

where u = the intercept of the model
v = geometric mean estimate of the model

In terms of the predictive regression variables, the estimate of v is

$$v = \pm b/r \tag{19}$$

where r = correlation coefficient of the predictive regression [Equation (11)]
b = slope from the predictive regression

The sign of v is the same as those of b and r.
The 95% confidence limit for v was described by Ricker[50] as the following:

$$\text{C.I.} = t\sqrt{[v^2(1 - r^2)/(N - 2)} \tag{20}$$

where t = Student t statistic for 0.95 and $N-2$ d.f.

The intercept of the functional regression is estimated as follows:

$$u = \text{Log}Y_m - b\text{Log}X_m \tag{21}$$

where $\text{Log}Y_m$ and $\text{Log}X_m$ are means of Log Y and Log X values

The functional regression is more appropriate when X is measured with error and/or X is subject to inherent variation. However, the results cannot be used to test for significant association between the two variables. The reader is directed to Ricker[50] for a more thorough discussion of this point. The major significance attributed to the allometric exponent (b) in allometry suggests that more critical evaluation of predictive regression assumptions regarding X is warranted. Such inattention to assumptions regarding X has also been characteristic of ecotoxicological allometry research.

ALLOMETRY AND BIOACCUMULATION
" . . . there is no clear understanding of the mechanism relating body size to metal concentrations, and the current confusion in the literature still confounds attempts to rationalize such data."[51]

General Factors Influencing Elemental Allometry
A variety of factors influence the relationship between body burden and size. Morgan[5] suggests that, considering the complex effects of metabolic rate, respiratory rate, pulse rate, pulse volume, and relative surface on body burdens,

exponents for allometric relations should not be considered "anything more than multiple process statistics."

Temporal factors such as age, growth, and duration of exposure clearly influence body burden allometry. Often the relationship between size and age is so complex or difficult to quantify that it is left undefined. However, Williamson[52] was able to use multivariate techniques to identify the opposite effects of age and size on Cd concentrations in the snail, *Cepaea hortensis*. The effective retention of Cd in this snail resulted in a gradual accumulation of Cd with age. Such accumulation of Cd could be facilitated by incorporation into intracellular granules as described by Simkiss[53] (see also, Chapter 3). The biological half-life for this metal was long relative to the life-span of the snail; therefore, the body burden was not in equilibrium with environmental concentrations. The size of individuals had the opposite effect on Cd concentration. This effect was related by Williamson[52] to the higher specific ingestion rate for smaller snails relative to large snails. Hg, another group IIB metal with the potential for high lipid solubility, has shown a similar trend with fish age.[54] The influence of age was assessed indirectly for the last group IIB metal, Zn. Using shell dimension changes with age as a covariate with size, Lobel and Wright[51] demonstrated the importance of considering mussel age and size as covariates in biomonitoring efforts. Assuming different exposure duration for a range of mosquitofish sizes and size-dependent accumulation kinetics, b values were shown to increase slowly from 0.32 (10 days of exposure) to 0.53 (350 days of exposure).[55]

Jeffree's[56] examination of alkaline earth metals in freshwater mussels provides a clear example of one mechanism contributing to age effects on body burden. He examined body burdens of Ca, Ba, Mg, and ^{226}Ra. Ca and its two analogs (^{226}Ra and Ba) increased with age and shell size. These strong relationships suggested that the concentrations in the animals were not in equilibrium with environmental concentrations. The relationship between Mg and age was not as clear as those for the other three elements. He gave the following explanation.

These metals were found associated with intracellular, calcium-magnesium pyrophosphate granules. The stability constants for the respective hydrogen phosphates of these metals were used to indicate their relative insolubilities when in association with granules (Figure 1) and suggest one potential mechanism for the age-dependent behavior of these metals. Mg, the most soluble of the four metals examined, had the poorest correlation with age. The least soluble metals had clearer trends with age, as they were less prone to dissolution from the granules and consequent clearance. As a result, the age-dependencies of Ca, Ba and ^{226}Ra concentrations were clearer than that of Mg.

Age of an individual is more than an indication of exposure duration. Physiological, cytological, and biochemical changes associated with reproduction or early development can have significant effects on size-dependent body burdens. Changes associated with sexual maturation can produce abrupt slope changes in the bilogarithmic plot of body burden versus size.[57] Seasonal changes associated

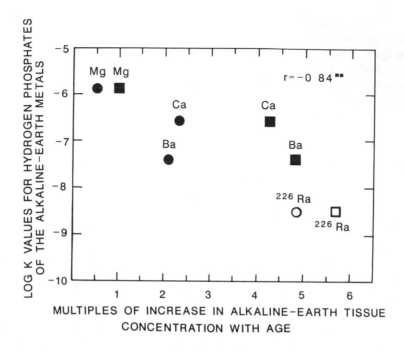

FIGURE 1.   Multiples of increase in tissue concentration with mussel age versus log of stability constants for the hydrogen phosphates of Mg, Ca, Ba, and ^{226}Ra. The symbols (○ and □) refer to results from different locations. (Modified from Jeffree 1988.[56])

with gametogenesis can also influence size-body burden relationships.[58-60] More subtle effects associated with reproduction are also likely. For example, Cs excretion by women is modified significantly during pregnancy by a shift in aldosterone, a hormone involved in K regulation.[61]

Biochemical changes associated with ontogeny can also influence body burdens. For example, dusky shiners (Figure 2) displayed a decrease in Hg concentration with increasing size. This is contrary to many studies that have demonstrated an increase in Hg concentration with fish size.[18,19,54,62-65] Under the assumption that most of the accumulated Hg was present as methylmercury, this difference can be attributed to biochemical shifts associated to life history phenomena. The lipid content of this species during sampling decreased with fish size as described by a power function with a b value of 0.62 ± 0.04. The increase in Hg concentration with decreasing fish size was likely linked to the lipid content in fish cohorts as they developed and entered reproductive status. This apparently anomalous relationship for size-dependent Hg concentration could be attributed to lipid dynamics such as those described by Roberts et al.[66] for methylmercury and chlordane bioaccumulation.

Growth dynamics can also be important in the relationship between body burden and size. An apparent dilution will occur if growth is significant within the range of animal sizes examined.[18,60,67] Interactions between toxicant body burden and growth (inhibition[26] or hormesis[68]) have not been examined in studies of body burden allometry in contaminated environments. Feeding efficiency and ingestion rate also vary with size and could have a significant influence on these relationships.[34,69] Such effects of feeding can be confounded by many factors such as food quality.[10,34]

Surveys and Redescription Models

Table 1 is a summary of regression models for elemental body burdens versus animal size. It is a retabulation of Boyden's[11,12] data (131 regressions from 13 marine molluskan species) which has been supplemented with 15 additional regressions from his papers[11,12] and 69 entries from other studies. As a result, the data remain biased toward marine mollusks. Boyden used b values from 131 of the regression models contained in Table 1 to identify different types of body burden relationships. When b values for suspect regression models* were excluded from his consideration, three types of relationships were identified (Figure 13 of Boyden[12]). The two groupings containing most of the relationships had b values in the regions of 0.77 (generally between 0.70 and 0.90) and 1.03 (typically between 0.90 and 1.10). A third type contained a small number of the

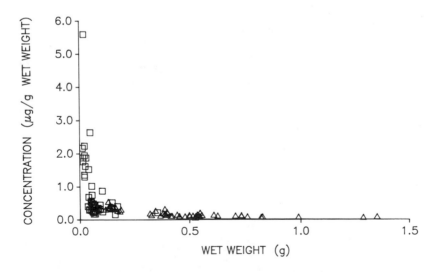

FIGURE 2. Mercury concentration in various-sized dusky shiners *(Notropis cummingae)* from a southeastern U.S. stream.

* Regression models generated from data sets with an inadequate range in animal size and those with associated significance of greater than $P = 0.001$ were not included.

Table 1
Summary of Allometric Body Burden Relationships

Element	Species	Type[a]	N	Size Range (g)	Conc. Range (μg/g)	Regression a(se)	b(se)	r	v	Reference
As	Helisoma trivolvis	WB-D	50	0.0009-0.0413	48.0-138.0	27.2(0.1)	0.75(0.05)	0.91	0.83	70
	H. trivolvis	WB-D	27	0.0030-0.0199	3.0-6.1	7.2(0.1)	1.11(0.07)	0.96	1.16	70
Cd	Buccinum undatum	WB-D	20	0.06-21.4	8.3-12.7	6.5	1.18(0.14)	0.98	1.21	12
	Cerastoderma edule	WB-D					0.77(0.04)			11
	Chlamys opercularis	WB-D	20	0.2-3.7		7.3	0.96(0.11)	0.97	0.98	12
	Cottus gobio	WB-D	36			0.018	1.40(0.12)			71
	C. gobio	WB-D	20			0.030	1.38(0.08)			71
	Crassostrea gigas	WB-D	39	0.01-4.3		6.3	0.85(0.04)	0.99	0.86	12
	C. gigas	WB-D	22	0.02-4.0		25.9	0.86(0.09)	0.90	0.90	12
	C. gigas	WB-D	30	0.07-11.6		11.8	0.85(0.13)	0.92	0.92	12
	Crepidula fornicata	WB-D	21	0.05-0.78		11.5	1.12(0.17)	0.95	1.18	12
	Donax trunculus	WB-D	29		0.6(0.1)[b]	0.8	1.29(0.14)	0.88	1.47	72
	Gambusia holbrooki	WB-D	94	0.0039-0.298	0.24-1.34	0.52(0.05)	1.01(0.03)	0.96	1.05	76
	H. trivolvis	WB-D	50	0.0009-0.041	21.5-51.4	30.3(0.07)	0.99(0.04)	0.96	1.03	70
	H. trivolvis	WB-D	27	0.0030-0.0199	0.43-0.72	0.78(0.12)	1.09(0.06)	0.96	1.13	70
	Littorina littorea	WB-D	37	0.01-0.7		2.6	0.97(0.11)	0.96	1.00	12
	Mercenaria mercenaria	WB-D					0.77(0.04)			11
	M. mercenaria	WB-D	35	0.02-302		1.4	0.81(0.09)	0.97	0.83	12
	M. edulis	WB-D					1.03(0.02)			11
	M. edulis	WB-D	21	0.26-1.94		3.7	0.97(0.12)	0.97	1.00	12
	M. edulis	WB-D	17	0.18-1.3		5.8	1.05(0.28)	0.93	1.13	12
	M. edulis	WB-D	22	0.08-1.09		64.7	1.02(0.17)	0.94	1.09	12
	M. edulis	WB-D	20	0.17-1.03		94.2	1.08(0.30)	0.87	1.24	12
	M. edulis	WB-D	40	0.04-3.5		1.6	0.95(0.06)	0.95	0.99	12
	M. edulis	WB-D	53			1.86	0.9	0.51	1.76	74
	M. edulis	WB-D	126	0.004-0.92		2.49(0.08)	0.65(0.06)	0.98	0.96	59
	Ostrea edulis	WB-D	38	0.02-2.5		5.2	0.94(0.07)	0.98	0.96	12
	O. edulis	WB-D	24	0.34-6.63		6.4	0.96(0.15)	0.94	1.02	12
	Patella intermedia	WB-D	14	0.021-1.71		4.7	1.35(0.38)	0.92	1.47	12
	P. intermedia	WB-D	16	0.021-1.71		6.3	1.49(0.21)	0.97	1.53	12
	Patella vulgata	WB-D	35	0.02-0.80		401.0	2.05(0.25)	0.96	2.13	12

Element	Species	Tissue	n	Range 1	Range 2	a	b	r	ratio	N
	P. vulgata	WB-D	29	0.01–0.80		686.4	1.98(0.28)	0.95	2.08	12
	P. vulgata	WB-D	30	0.01–0.85		716.9	1.96(0.11)	0.92	2.12	12
	P. vulgata	WB-D	34	0.02–1.10		704.2	1.70(0.11)	0.98	1.73	12
	P. vulgata	WB-D	32	0.01–1.10		190.7	1.37(0.09)	0.99	1.39	12
	P. vulgata	WB-D					2.00			11
	Scaphander lignarius	WB-D	20	0.09–3.8		6.8	0.94(0.04)	0.996	0.94	12
	Venerupis decussata	WB-D	30	0.17–2.77		8.9	0.77(0.22)	0.83	0.93	12
	V. decussata	WB-D					0.77(0.04)			11
Co	*B. undatum*	WB-D	20	0.06–21.4		7.3	0.67(0.05)	0.99	0.68	12
	S. lignarius	WB-D	20	0.09–3.8		26.4	1.07(0.06)	0.99	1.08	12
Cr	*H. trivolvis*	WB-D	50	0.009–0.041	1.26–7.15	0.80(0.13)	0.638(0.08)	0.77	0.83	70
	H. trivolvis	WB-D	27	0.003–0.0199	0.42–1.11	0.41(0.21)	0.888(0.10)	0.86	1.03	70
¹³⁷Cs	*G. holbrooki*	WB-W	28	0.12–1.20	(0.80–3.37)[c]	0.39(0.15)	0.96(0.12)	0.85	1.13	67
	G. holbrooki	WB-W	55	0.15–0.96	(0.90–2.98)[c]	0.35(0.09)	0.88(0.08)	0.85	1.04	67
	Acanthopagrus butcheri	WM-W	88	4.3–799.0		1.750	0.92			75
	Aldrichetta forsteri	WM-W	79	7.8–385.0		2.377	0.88			75
	Amniataba caudavittatus	WM-W	96	4.0–319.9		2.506	0.89			75
Cu	*B. undatum*	WB-D	20	0.06–21.4	118–187	122.6	1.10(0.22) / 0.77(0.04)	0.94	1.17	11
	C. edule	WB-D	20	0.02–3.7		35.0	0.96(0.10)	0.99	0.97	12
	C. opercularis	WB-D	39	0.01–4.3	447.4–809.5	446.4	0.82(0.04)	0.99	0.83	12
	C. gigas	WB-D	22	0.02–4.0		260.6	0.80(0.13)	0.94	0.85	12
	C. gigas	WB-D	21	0.05–0.73		198.5	1.13(0.23)	0.93	1.22	12
	C. fornicata	WB-D	33		11.8(1.5)[b]	8.8	0.83(0.09)	0.86	0.95	72
	D. trunculus	WB-D	94	0.0039–0.298	4.3–21.6	11.26(0.04)	1.04(0.03)	0.97	1.07	76
	G. holbrooki	WB-D	50	0.0009–0.041	45–98	55.11(0.07)	0.95(0.04)	0.93	1.02	70
	H. trivolvis	WB-D	27	0.003–0.0199	7–16	7.02(0.18)	0.92(0.09)	0.90	1.02	70
	H. trivolvis	WB-D	37	0.01–0.7		642.2	1.07(0.20)	0.97	1.10	12
	L. littorea	WB-D					0.77(0.04)			11
	M. mercenaria	WB-D	35	0.02–3.02		23.4	0.79(0.14)	0.94	0.89	12
	M. mercenaria	WM-W	163	2.4–832.0		10.26	0.571			75
	Mugil cephalus	WB-W	50		0.96–1.45		0.80			73
	M. edulis	WB-D	21	0.26–1.94		7.3	0.80(0.21)	0.90	0.89	12
	M. edulis	WB-D	17	0.18–1.3		6.9	0.77(0.19)	0.93	0.82	12
	M. edulis	WB-D	22	0.08–1.09		11.2	0.81(0.05)	0.99	0.82	12
	M. edulis	WB-D	20	0.17–1.03		12.7	0.83(0.22)	0.88	0.94	12

Table 1 (continued)
Summary of Allometric Body Burden Relationships

Element	Species	Type[a]	N	Size Range (g)	Conc. Range (µg/g)	a(se)	b(se)	r	v	Reference
	M. edulis	WB-D	96				0.77(0.04)			11
	M. edulis	WB-W	119			14.95	0.89	0.65	1.37	74
	M. edulis	WM-W	160	0.004-0.92		3.69(0.04)	0.86(0.03)			59
	Nematalosa vlaminghi	WB-D	38	4.3-449.1		3.66	0.812			75
	O. edulis	WB-D		0.02-2.5	405.5-423.0	391.9	1.05(0.09)	0.97	1.08	12
	P. intermedia	WB-D					0.77(0.04)			11
	P. intermedia	WB-D	14	0.02-1.71		19.2	0.81(0.11)	0.99	0.82	12
	P. intermedia	WB-D	16	0.02-1.71		19.9	0.73(0.10)	0.97	0.75	12
	P. vulgata	WB-D	35	0.02-0.80		26.6	0.81(0.10)	0.94	0.96	12
	P. vulgata	WB-D	29	0.01-0.80		31.2	0.79(0.06)	0.98	0.81	12
	P. vulgata	WB-D	30	0.01-0.85		36.9	0.80(0.12)	0.99	0.81	12
	P. vulgata	WB-D	34	0.02-1.10		24.0	0.70(0.12)	0.89	0.78	12
	P. vulgata	WB-D	32	0.01-1.10		25.4	0.76(0.07)	0.97	0.79	12
	P. vulgata	WB-D					0.77(0.04)			11
	Pecten maximus	WB-D	37	0.16-8.30		22.3	0.65(0.10)	0.99	0.66	12
	S. lignarius	WB-D	20	0.09-3.80		44.5	0.98(0.15)	0.99	0.99	12
	V. decussata	WB-D	30	0.17-2.77		11.7	0.76(0.13)	0.89	0.85	12
Fe	A. butcheri	WM-W	88	4.30-799.0		23.3	0.82			75
	A. forsteri	WM-W	79	7.80-385.0		29.1	0.84			75
	A. caudavittatus	WM-W	96	4.00-319.9		20.4	0.99			75
	B. undatum	WB-D	20	0.06-21.4		65.2	1.09(0.11)	0.98	1.00	12
	C. edule	WB-D					1.03(0.02)			11
	C. gigas	WB-D	39	0.01-4.3		236.5	0.70(0.12)	0.91	0.77	12
	C. gigas	WB-D	22	0.02-4.0		313.0	0.80(0.04)	0.99	0.81	12
	C. gigas	WB-D	30	0.70-11.6		365.3	0.80(0.05)	0.98	0.82	12
	D. trunculus	WB-D	33		663(146)[6]	377.0	0.64(0.14)	0.63	1.02	72
	G. holbrooki	WB-D	94	0.0039-0.298	10.8-155.6	40.2(0.06)	0.90(0.04)	0.93	0.97	76
	H. trivolvis	WB-D	50	0.0009-0.041	1956-3875	2035.2(0.06)	0.92(0.03)	0.97	0.95	70
	H. trivolvis	WB-D	27	0.0030-0.019	756-2103	3031.4(0.17)	1.22(0.09)	0.94	1.29	70
	L. littorea	WB-D	37	0.01-0.7		364.3	0.74(0.20)	0.85	0.87	12
	M. mercenaria	WB-D	35	0.02-3.02		108.7	0.77(0.18)	0.95	0.81	12

	Species	Tissue	n	Range	Range (2)	Mean	Mean (SD)	r		n
Mn	M. mercenaria	WB-D	163	2.4-832.0		83.4	0.77(0.04)	0.96	0.72	11
	M. cephalus	WM-W	122	0.004-0.92		5.57(0.07)	0.68			75
	M. edulis	WB-D	21	0.26-1.94		91.1	0.65(0.06)	0.90	0.81	59
	M. edulis	WB-D	22	0.08-1.09		152.0	0.77(0.04)	0.80	1.00	11
	M. edulis	WB-D	20	0.17-1.03		228.1	0.69(0.09)	0.96	0.80	12
	M. edulis	WB-D	40	0.04-3.5		201.7	0.73(0.16)	0.97	0.87	12
	M. edulis	WB-D	160	4.3-449.1		67.5	0.80(0.35)	0.95	0.85	12
	M. edulis	WM-W		0.02-2.5			0.77(0.06)	0.99	0.76	75
	N. vlaminghi	WB-D		0.34-6.63			0.75			12
	O. edulis	WB-D		0.02-1.71			0.85(0.07)	0.97	0.72	12
	O. edulis	WB-D		0.02-1.71			0.81(0.12)	0.93	0.73	12
	P. intermedia	WB-D	38	0.01-0.85		223.3	0.76(0.04)	0.90	0.92	12
	P. intermedia	WB-D	24	0.02-1.10		216.0	0.70(0.11)	0.97	0.76	12
	P. vulgata	WB-D	14	0.01-1.10		1195.9	0.68(0.19)	0.85	1.21	12
	P. vulgata	WB-D	16	0.17-2.77		1504.2	0.83(0.17)	0.93	0.84	11
	P. vulgata	WB-D	30	0.06-21.4		2594.2	0.74(0.07)	0.95	1.05	12
	V. decussata	WB-D	34	0.01-4.3		1259.5	1.03(0.02)	0.97	1.10	75
	V. decussata	WB-D	32	0.02-4.0		1368.0	1.03(0.31)	0.91	1.15	75
	A. butcheri	WM-W	30		4.3-799	367.9	0.67			
	A. forsteri	WM-W	88		7.8-385	6.84	0.71			
	A. caudavittatus	WM-W	79		4.0-319.9	2.60	0.58			
	Arenicola marina	WB-D	96		2.5-10	8.70	0.61			
	B. undatum	WB-D	20	0.07-11.6		6.0	0.78(0.12)	0.79	1.00	77
	C. gigas	WB-D	39	0.05-0.78		33.0	1.00(0.11)	0.70	0.80	12
	C. gigas	WB-D	22	0.0039-0.2980		22.2	1.06(0.13)	0.99	1.13	12
	C. gigas	WB-D	30	0.0009-0.0413		16.9	1.05(0.20)	0.86	0.87	12
	C. fornicata	WB-D	21	0.0030-0.0199		22.6	0.79(0.29)	0.77	1.21	12
	D. trunculus	WB-D	18		12.4(2.0)[b]	6.2	0.56(0.14)			72
	G. holbrooki	WB-D	94		8.8-43.9	34.31	1.12(0.02)			76
	H. trivolvis	WB-D	50		65-276	60.85(0.11)	0.75(0.06)			70
	H. trivolvis	WB-D	27		131-501	202.89(0.31)	0.94(0.15)			70
	L. littorea	WB-D	37	0.01-0.7		34.1	0.61(0.18)	0.83	0.73	12
	M. cephalus	WM-W	163	2.4-832.0		17.42	0.45			75
	M. edulis	WB-D	119	0.004-0.92		3.77(0.06)	0.70(0.05)	0.82	0.82	59
	M. edulis	WB-D	22	0.08-1.09		4.5	0.80(0.26)	0.81	0.95	12
	M. edulis	WB-D	17	0.18-1.3		6.9	0.77(0.35)			12

Table 1 (continued)
Summary of Allometric Body Burden Relationships

Element	Species	Type[a]	N	Size Range (g)	Conc. Range (µg/g)	Regression a(se)	b(se)	r	v	Reference
	M. edulis	WB-D	40	0.04-3.5		5.4	0.73(0.17)	0.91	0.80	12
	N. vlaminghi	WM-W	160	4.3-449.1		11.54	0.74			75
	O. edulis	WB-D	24	0.34-6.63		10.5	0.95(0.13)	0.95	1.00	12
	O. edulis	WB-D	38	0.02-2.5		17.9	1.04(0.14)	0.93	1.12	12
	P. maximus	WB-D	37	0.16-8.3		24.1	0.87(0.11)	0.93	0.93	12
	S. lignarius	WB-D	20	0.09-3.8		7.9	1.19(0.11)	0.98	1.21	12
Ni	B. undatum	WB-D	20	0.06-21.4		5.6	0.68(0.07)	0.98	0.69	12
	C. edule	WB-D					1.03(0.02)			11
	C. opercularis	WB-D	20	0.2-3.7		5.5	0.78(0.11)	0.97	0.80	12
	C. gigas	WB-D	30	0.07-11.6		5.4	0.72(0.07)	0.97	0.75	12
	C. fornicata	WB-D	21	0.05-0.73		11.5	1.04(0.22)	0.92	1.13	12
	M. mercenaria	WB-D	35	0.02-3.02		9.4	1.05(0.20)	0.93	1.13	12
	M. mercenaria	WB-D					1.03(0.02)			11
	M. edulis	WB-D					1.03(0.02)			11
	M. edulis	WB-D	21	0.26-1.94		5.3	0.67(0.12)	0.83	0.81	12
	M. edulis	WB-D	17	0.18-1.3		7.6	0.75(0.26)	0.88	0.85	12
	M. edulis	WB-D	22	0.08-1.09		11.3	0.76(0.15)	0.92	0.82	12
	M. edulis	WB-D	40	0.04-3.5		3.5	0.75(0.25)	0.73	1.02	12
	M. edulis	WB-D	118	0.04-0.92		3.12(0.05)	0.73(0.04)			59
	O. edulis	WB-D	38	0.02-2.5		6.0	0.63(0.10)	0.90	0.70	12
	O. edulis	WB-D	24	0.34-6.63		6.1	0.67(0.11)	0.94	0.71	12
	S. lignarius	WB-D	20	0.09-3.8		54.0	1.00(0.11)	0.99	1.00	12
	V. decussata	WB-D	30	0.02-3.02		22.5	1.05(0.20)	0.94	1.12	12
	V. decussata	WB-D						1.03(0.02)			11
Pb	A. marina	WB-D			10-40		0.74			77
	B. undatum	WB-D	20	0.06-21.4		9.1	0.71(0.09)	0.93	0.76	12
	Campeloma decisum	WB-D	57	0.0046-0.278		2.21	0.75	0.90	0.83	78
	C. edule	WB-D						0.77(0.04)			11
	C. opercularis	WB-D	20	0.2-3.7		32.1	0.98(0.10)	0.99	0.99	12
	C. gobio		36			0.00008	1.43(0.13)			71
	C. gigas	WB-D	22	0.02-4.0		20.7	0.75(0.10)	0.93	0.80	12

Element	Species	Method	Range	n	Range					n
Zn	C. gigas	WB-D	0.07-11.6	30		9.1	0.78(0.08)	0.97	0.81	12
	C. fornicata	WB-D	0.05-0.73	21		11.4	1.00(0.17)	0.94	1.06	12
	L. littorea	WB-D	0.01-0.79	37		6.5	0.73(0.18)	0.84	0.87	12
	M. mercenaria	WB-D					1.03(0.02)			11
	M. mercenaria	WB-D	0.02-3.02	35		7.8	1.01(0.18)	0.93	1.09	12
	M. edulis	WB-D					0.77(0.04)			11
	M. edulis	WB-D	0.08-1.09	22		19.0	0.83(0.15)	0.93	0.89	12
	M. edulis	WB-D	0.04-3.5	40		14.4	0.72(0.13)	0.88	0.81	12
	M. edulis	WB-D		88		25.14	0.73	0.57	1.28	74
	P. intermedia	WB-D	0.02-1.71	16		9.6	0.73(0.21)	0.93	0.80	12
	P. intermedia	WB-D					0.77(0.04)			11
	P. vulgata	WB-D					0.77(0.04)			11
	P. vulgata	WB-D	0.02-0.80	35		20.4	0.79(0.15)	0.90	0.88	12
	P. vulgata	WB-D	0.01-0.80	29		15.9	0.74(0.17)	0.83	0.89	12
	P. vulgata	WB-D	0.01-0.85	30		14.4	0.67(0.16)	0.87	0.77	12
	P. maximus	WB-D	0.15-8.3	37		13.2	0.68(0.06)	0.99	0.69	12
	Physa integra	WB-D	0.0007-0.0070	16		46.7	0.97	0.92	1.05	78
	O. edulis	WB-D	0.02-2.5	38		6.2	0.65(0.11)	0.90	0.72	12
	O. edulis	WB-D	0.34-6.63	24		7.8	0.76(0.12)	0.94	0.81	12
	S. lignarius	WB-D	0.09-3.8	20		44.5	0.97(0.03)	0.99	0.97	12
	V. decussata	WB-D	0.17-2.77	30		7.8	0.81(0.14)	0.93	0.87	12
	V. decussata	WB-D					0.77(0.04)			11
	A. butcheri	WM-W	4.3-799.05	88		46.45	0.86			75
	A. forsteri	WM-W	7.8-385.0	79		27.93	0.89			75
	A. caudavittatus	WM-W	4.0-319.9	96		61.80	0.79			75
	A. marina	WB-D			40-320		0.72			77
	B. undatum	WB-D	0.66-21.4	20	450-1040	508.4	1.20(0.10)	0.97	1.23	12
	C. edule	WB-D					0.77(0.04)			11
	C. opercularis	WB-D	0.2-3.7	20		980.4	0.97(0.17)	0.96	1.01	12
	C. gigas	WB-D	0.01-4.3	39		4293.4	0.95(0.03)	0.99	0.95	12
	C. gigas	WB-D	0.02-4.0	22		2570.4	1.0(0.10)	0.98	1.03	12
	C. fornicata	WB-D	0.05-0.78	21		81.9	0.94(0.18)	0.94	1.00	12
	D. trunculus	WB-D		33		68.0	0.74(0.09)	0.82	0.90	72
	G. holbrooki	WB-D	0.0039-0.2980	94	165-594	203.9(0.03)	0.899(0.02)	0.97	0.93	76
	H. trivolvis	WB-D	0.0009-0.0413	50	137-223	154.1(0.04)	0.97(0.02)	0.98	0.98	70
	H. trivolvis	WB-D	0.0030-0.0199	27	57-105	77.7(0.14)	1.008(0.07)	0.95	1.06	70

Table 1 (continued)
Summary of Allometric Body Burden Relationships

Element	Species	Type[a]	N	Size Range (g)	Conc. Range (µg/g)	Regression a(se)	b(se)	r	v	Reference
	L. littorea	WB-D	37	0.01-0.7		185.0	0.75(0.08)	0.97	0.77	12
	M. mercenaria	WB-D	35	0.02-3.02		177.3	1.04(0.16)	0.99	1.05	12
	M. mercenaria	WB-D					1.03(0.02)			11
	M. cephalus	WM-W	163	2.4-832.0		53.95	0.749			75
	M. edulis	WB-D	50	0.1-10	21.0-25.9		0.85			73
	M. edulis						0.77(0.04)			11
	M. edulis	WB-D	21	0.26-1.94		96.8	0.86(0.10)	0.97	0.87	12
	M. edulis	WB-D	17	0.18-1.3		147.0	0.84(0.31)	0.87	0.97	12
	M. edulis	WB-D	22	0.08-1.09		148.5	0.86(0.20)	0.89	0.97	12
	M. edulis	WB-D	20	0.17-1.03		227.3	0.86(0.25)	0.86	1.00	12
	M. edulis	WB-D	40	0.04-3.5		264.5	0.81(0.06)	0.97	0.83	12
	M. edulis	WB-D	48			189.67	0.41(0.28)	0.35	1.17	51
	M. edulis	WB-D	43			20.99	0.67(0.12)	0.41	1.62	51
	M. edulis	WB-D	20			706.32	0.13(0.26)	0.62	0.20	51
	M. edulis	WB-D	98	36.8-348.1	249-393	1.85	1.19(0.04)	0.43		51
	M. edulis	WB-D	98	238.4-4660		1.55	1.15(0.03)	0.46	2.50	51
	M. edulis	WB-D	96			218.0	0.187	0.69	1.26	74
	M. edulis	WB-D	119	0.004-0.92		4.95(0.04)	0.86(0.04)			59
	N. vlaminghi	WM-W	160	4.3-449.1		48.87	0.81			75
	O. edulis	WB-D	38	0.02-2.5	1846.5-5823.6	3437.2	1.03(0.07)	0.98	1.05	12
	O. edulis	WB-D	24	0.34-6.63	1816.0-11185.5	4358.1	0.98(0.16)	0.93	1.05	12
	P. intermedia	WB-D					0.77(0.04)			11
	P. intermedia	WB-D	14	0.02-1.71		237.8	0.71(0.08)	0.99	0.72	12
	P. intermedia	WB-D	16	0.02-1.71		193.9	0.76(0.07)	0.99	0.77	12
	P. vulgata	WB-D	35	0.02-0.80		358.3	0.84(0.06)	0.98	0.86	12
	P. vulgata	WB-D	29	0.01-0.80		354.9	0.91(0.05)	0.99	0.92	12
	P. vulgata	WB-D	30	0.01-0.85		433.6	0.93(0.06)	0.98	0.95	12
	P. vulgata	WB-D	34	0.02-1.10	52	388.2	0.84(0.07)	0.97	0.86	12
	P. vulgata	WB-D	32	0.02-1.10	27	302.5	0.89(0.05)	0.99	0.90	12
	P. vulgata	WB-D					0.77(0.04)			11
	P. maximus	WB-D	37	0.15-8.3		672.5	0.72(0.10)	0.96	0.75	12

Species								
S. lignarius	WB-D	20	0.09-3.8	984.0	1.06(0.11)	0.98	1.00	12
V. decussata	WB-D	30	0.17-2.77	92.2	0.97(0.14)	0.96	1.01	12
V. decussata	WB-D				1.03(0.02)			11

Note: Boyden (1974) reports the same b value for several species for the metals Cd, Cu, Fe, Pb, Ni, and Zn. The b values are reported for the individual species in Table 1 but were used once in Figures 3 and 4.

[a] WB, whole body; WM, white muscle tissue; D, dry weight; W, wet wt. WB in mollusks does not include shell.

[b] Range indicated by standard deviation.

[c] Bq/g.

relationships and was characterized by b values of greater than 1.30 [in the range of 2 (Reference 11)]. Although he was later found to be incorrect,[60] Boyden hypothesized that b values were generally constant for some species-metal combinations. This simplifying assumption is often made in bioaccumulation models (see Bergner[79]).

Boyden hypothesized that some process linked to metabolic rate fostered a relationship with a b value in the range of 0.77. This suggestion was based on the similarity to b values for size-dependent metabolic rate. Although this may seem a reasonable suggestion, it is inappropriate to assume a common mechanism based on similarity between b values alone.[40] Boyden himself suggests that alternative allometric relationships could produce similar b values. For example, the amount of gill or general body surface for influx per unit mass of tissue to bind the element decreases with increasing size[30-32] and the clearance rate could decrease with increasing size.[7,8,55] The combined effects of these two relationships could generate a power relationship with a b value less than unity. However, Boyden rejected the surface-to-volume ratio hypothesis based on the false assumption of an isometric relationship (b = 0.67) between surface area and mass for most organisms.[12] He suggested that more research was needed to effectively assess these potential explanations. Boyden's hypothesized linkage to metabolism has received considerable criticism but, to date, it has not been rigorously tested.

The second set of relationships was characterized by b values in the range of unity. Boyden suggested that these relationships were determined by the number of tissue sites available to bind the element. This also seems to be a reasonable hypothesis. However, Fagerström[15] developed the following counterargument that a b value of unity would imply linkage to metabolism. If steady state is assumed for a biologically indeterminant element,* the rate of its turnover will be directly proportional to the animal's energy metabolism. Under such conditions, the following relationships can be defined:

$$T_{1/2} \alpha X^{1-b} \tag{22}$$

$$C \, \alpha X^0 \tag{23}$$

$$\Theta \, \alpha X^{1-b} \tag{24}$$

$$Y \, \alpha X^1 \tag{25}$$

where $T_{1/2}$ = biological half-life

* An element is indeterminant if, at steady state, its concentration in the organism is directly proportional to the concentration in the environment.[6] Uptake and elimination of an indeterminant element are dominated by simple mass equilibria relations. An element is defined as biologically determinant if its concentration in the organism is relatively constant over a wide range of environmental concentrations.[6] Many essential elements or their analogs seem to be determinant.

C = whole body concentration
Θ = metabolic turnover rate
Y = whole body burden
X and b as defined previously

Equation (25) suggests that linkage to metabolic processes will produce a b value of approximately unity for a biologically indeterminant element. It should be kept in mind that the conclusions of this argument may not be applicable for systems under nonequilibrium conditions or for biologically determinant elements. Indeed, studies of elimination kinetics for essential elements (Zn and K)[4,8-10] or their analogs (^{134}Cs and ^{137}Cs)[3-5,7,61] suggest that this [Equation (22)] may not be accurate for biologically determinant elements. It [Equation (23)] is certainly not true for concentrations of major elements such as Ca, P, N, and S which vary with size-dependent, relative proportions of bone to protein in fish.[80]

Finally, a small number of regressions were characterized by b values greater than 1 and approximating 2. Such relationships were thought to be due to a high affinity of the metal to some binding component and to consequent, rapid removal from circulation. Cd accumulation in several species had such high b values.

Although many of Boyden's suggestions have been found to be false or remain unsubstantiated, his work is cited and used to interpret most subsequent studies of metal body burdens and size. Further, his hypothesis are clearly stated and are amenable to the process of falsification. The central importance of this work warrants reexamination of associated data and conclusions.

Boyden[12] estimated slopes for both predictive (b) and functional regression (v) models. He used a histogram of b values to define the three different relationships which he clearly stated were somewhat arbitrary. When all of Boyden's data and the supplemental data were reexamined (Figure 3), the present authors found no discrete distributions of b values. Rather, a skewed distribution with a median of 0.83 was suggested. When, as suggested by Boyden, only samples with sufficiently broad size ranges* were used (Figure 4), the same skewed distribution was noted. There is only a slight indication of bimodality in these data. The median b value from this distribution was 0.80. One must conclude that clear evidence for three types of discrete relationships is still lacking more than a decade after Boyden's preliminary attempts to clarify the allometry of metal accumulation. An additional aspect of the regression results confounds interpretation of the b values. As discussed previously, there is a covariance between the r (correlation coefficient) and b such that, as the r becomes smaller, the b value will tend to be biased increasingly downward. When the b values are plotted against r (Figure 5), this bias is clear even for the data with a wide size range (Figure 5 inset).

Further, the b value will be strongly influenced by the range in concentrations.

---

* In this chapter, this requirement was formalized to a size range of no less than 50-fold.

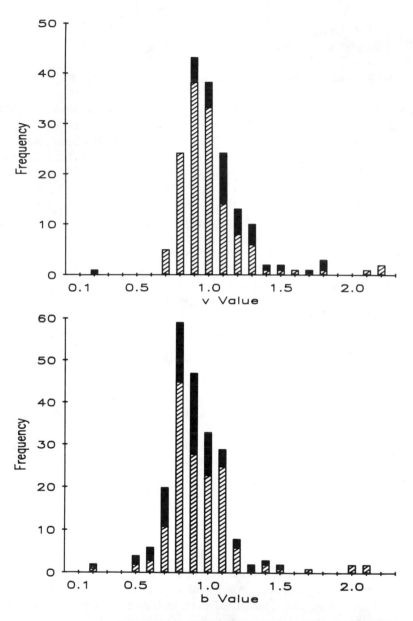

FIGURE 3.   The frequency of b and v values from Boyden[11,12] (cross-hatching) and other studies (solid) regardless of weight range. B values: N = 214; median = 0.83; range = 0.13 to 2.05. V values: N = 170, median = 0.95, range = 0.20 to 2.13

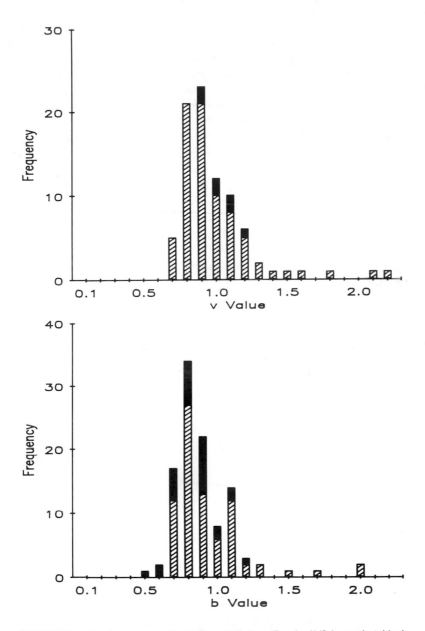

FIGURE 4. The frequency of b and v values from Boyden[11,12] (cross-hatching) and other studies (solid) when the weight range was 50-fold or more. B values: N = 107; median = 0.80; range = 0.45 to 1.98. V values: N = 85; median = 0.85; range = 0.66 to 2.12.

Figure 2 shows Hg concentrations for dusky shiners *(Notropis cummingsae)* sampled from a southeastern U.S. stream.[81] Table 2 is a tabulation of a, b, and v estimates when the concentration range is progressively truncated. As the individuals with the highest concentrations of Hg (smaller fish) are discarded, the estimates of b increase. If a power model is to be used, it is essential to clearly define the ranges of body burdens or concentrations as well as the animal size range. The widest possible ranges for X and Y should be obtained.

Regardless, it was questionable if b values were the most appropriate statistics to tabulate for such discussion. The assumptions of no measurement error or no inherent variability in the surrogate measure of "size" (wet or dry wt) are not warranted in many studies. For example, Boyden and others, including the senior author, have estimated weights for smaller individuals by pooling similarly sized animals and taking an average weight. There is an inherent error associated with this technique. Even in the absence of such procedures, there can be significant error associated with estimating wet or dry wt. The confounding effects of age, growth, reproduction, mass of food in the gut, and a variety of factors make the measurment of "size" susceptible to variation. Consequently, the slope of the functional regression (v) would seem more appropriate than b as the basis for such discussion. Subsequent comparison to the allometric literature becomes confused as this body of information also contains many estimates of b but few estimates of v. Fortunately, Equation (19) can be used to estimate v when r is given.

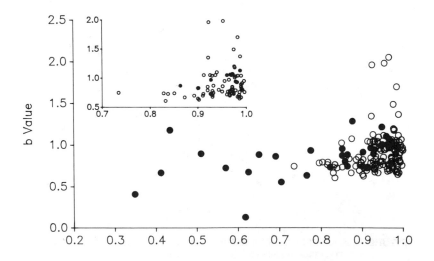

FIGURE 5.   The covariance of b value and correlation coefficient (r) for all re-lationships, e.g., Figure 3, or those from relationships with 50-fold or wider range in size, e.g., Figure 4 (inset). Open circles are those used in Boyden's original Figure 13.

Table 2
Concentration Range Effects on Regression Results

Concentration (µg/g)	N	b̂	â	r	v
≦5.56	96	0.321	0.082	0.86	0.37
<5.00	95	0.347	0.085	0.86	0.40
<2.00	92	0.381	0.092	0.86	0.44
<1.00	83	0.505	0.101	0.81	0.62
<0.80	82	0.512	0.101	0.82	0.62
<0.70	81	0.521	0.102	0.82	0.64
<0.60	80	0.528	0.103	0.81	0.65
<0.50	75	0.550	0.103	0.81	0.68
<0.40	64	0.601	0.105	0.77	0.78

When v values for body burden power equations are used (Figures 3 and 4), the suspected bimodality becomes even less apparent (Figure 4). Further, the medians of the associated slopes are closer to 0.90 than 0.77.

Although there was wide variation in values for each element and there was no clear bimodality (Figures 3 and 4), b or v values for some elements (As, Cd, and ^{137}Cs) did tend toward 1 and those for other elements (Co, Cu, Fe, Ni, Pb, and Zn) tended to be slightly less than 1 (Table 3). Median slopes for the essential elements (Co, Cr, Cu, Fe, Mn, Ni, and Zn) were only slightly lower than those for nonessential elements (As, Cd, and Pb). There are no further discernible patterns in median values for the tabulated elements. There is an obvious need for carefully designed experiments to clarify the ambiguity arising from interpretation of redescription models of field data. Further, when conclusions are drawn from this survey, the bias toward mollusks should be kept in mind. These invertebrates have mechanisms for uptake, sequestration, detoxification, and elimination that are common to all animals; however, the major role of such mechanisms as metal incorporation into intracellular granules[53] may not be applicable to certain phyla. Such characteristics which strongly influence body burden can restrict the generalization of findings from one species to another.

Despite attempts to define the mechanism(s) underlying physiological or morphological allometric equations, they remain empirical relationships. Certainly, the same may be said for body burden allometry. Consequently, there is no reason why alternate models should not be explored in redescription of body burden data. Such models would be most useful if they were amenable to interpretation using allometric or pharmacokinetic theory.

If warranted, a simple linear model may be as good or better than a power model for normalization of biomonitoring data. The argument associated with this statement is identical to that given above during discussion of narrow allometry. This approach was taken by Ashraf and Jaffar[82] relative to As concentrations in various sized tuna. Strong and Luoma[60] also found no advantage to transforming the body concentration and size data in their examination of body size effects on metal concentrations in a marine bivalve. Kumagai and Saeki[83] used an exponential model to relate Cu concentration in a marine mollusk to shell height. Williamson[52] used a multivariate approach which incorporated estimated age into the power model. Although an overextension of the model, Williamson[84] expanded the allometric relation for body burden to include external factors such as day length, vapor pressure deficit, and rainfall.

ACCUMULATION MODELS
Overview

The simplest accumulation model describes the net effects of uptake and

Table 3
Summary of b and v Values by Element and Essentiality

Element	N	Median	Range	W	N	Median	Range	W
		All Regressions			Regression with Weight Range >50-fold			
As								
b	2	0.93	0.74-1.11					
v	2	0.99	0.82-1.15					
Cd								
b	38	0.98	0.65-2.05	0.81[a]	17	0.97	0.65-1.98	0.87[a]
v	33	1.03	0.83-2.13	0.80[a]	16	1.02	0.83-2.12	0.85
Co								
b	2	0.87	0.67-1.07		1	0.67		
v	2	0.88	0.68-1.08		1	0.68		
Cr								
b	2	0.77	0.64-0.89					
v	2	0.93	0.83-1.03					
^{137}Cs								
b	2	0.92	0.88-0.96					
v	2	1.09	1.04-1.13					
Cu								
b	35	0.81	0.57-1.13	0.94	19	0.81	0.57-1.10	0.94
v	27	0.89	0.66-1.37	0.94	14	0.83	0.66-1.17	0.88
Fe								
b	30	0.79	0.65-1.22	0.88[a]	18	0.77	0.65-1.04	0.91
v	22	0.83	0.72-1.29	0.87[a]	13	0.81	0.72-1.00	0.92
Mn								
b	23	0.77	0.45-1.19	0.96	14	0.75	0.45-1.12	0.94
v	16	1.00	0.73-1.80	0.84[a]	9	0.93	0.73-1.15	0.88
Ni								
b	15	0.75	0.63-1.05	0.81[a]	7	0.73	0.63-1.05	0.78[a]
v	13	0.82	0.69-1.13	0.87	6	0.89	0.69-1.13	0.81
Pb								
b	24	0.75	0.65-1.43	0.75[a]	11	0.73	0.65-0.78	0.95
v	20	0.87	0.69-1.28	0.92	11	0.80	0.69-1.09	0.86
Zn								
b	41	0.86	0.13-1.20	0.88[a]	20	0.86	0.71-1.20	0.94
v	31	0.95	0.20-1.62	0.88[a]	15	0.92	0.72-1.23	0.95
Essential (Co, Cr, Cu, Fe, Mn, Ni, Zn)								
b	147	0.81	0.13-1.22	0.96[a]	79	0.80	0.45-1.23	0.96[a]
v	111	0.93	0.20-1.80	0.94[a]	58	0.86	0.66-1.23	0.93[a]
Nonessential (As, Cd, Pb)								
b	64	0.96	0.65-2.05	0.78[a]	28	0.83	0.65-1.98	0.77[a]
v	55	0.99	0.69-2.13	0.79[a]	27	0.90	0.69-2.12	0.78[a]

[a] The H_0 of normality is rejected at an $\alpha = 0.05$.

elimination on the amount of metal within the organism. It is often expressed in the following form:

$$C_t = C_e(1 - e^{-k_e t}) \tag{26}$$

where k_e = elimination rate constant (1/time)
C_t and C_e = concentrations at time, t, and equilibrium, e, respectively

Concentration units are $\mu g/ml$ of tissue or $\mu g/g$ if uniform densities are assumed as in the following discussion. An uptake rate constant (k_u) can be incorporated into this model using the following relationship.

$$C_e = C_S(k_u/k_e) \tag{27}$$

where k_u = uptake rate constant (1/time)
C_s = concentration in the source ($\mu g/ml$ or $\mu g/g$ assuming equivalent densities)

Combining Equations (26) and (27), the simple model becomes

$$C_t = C_S(k_u/k_e)(1 - e^{-k_e t}) \tag{28}$$

Within the context of its application, this simple model assumes the following conditions: one constant source of metal, instantaneous and homogeneous distribution of metal atoms within the organism, one compartment for elimination, a constant K_e, and a constant K_u.*

If the process of uptake is simplified to involve only diffusion, the flux across the exchange surface ($\mu g/cm^2/sec$) will be a function of the diffusion coefficient (cm^2/sec) and the concentration gradient of the solute ($\mu g/ml$ or $\mu g/g$)[85] across the exchange surface (cm^2). The model described by Equation (28) assumes a constant area of exchange surface. As discussed previously, this assumption can be invalid in considerations of body burden scaling because the amount of gill or general body surface can change disproportionately with animal mass.[30-32] The incorporation of transport sites on surfaces of exchange lends an additional complication to the model as there are no compelling reasons to assume that the number of sites per unit surface for exchange will not change with animal size. The concentration gradient across gills will be influenced by respiratory processes such as ventilation volume.[21] These processes are related to size according to the allometric equation [Equation (1)]. If the route of uptake is associated with feeding, then scaling of related processes such as growth efficiency,[86] ingestion

---

* The model described here is the most parsimonious model available. As such, it is often inadequate for description of accumulation data. The reader is referred to this volume, Chapter 7 for a richer discussion of compartmental models of bioaccumulation and associated assumptions.

rate,[10] and particle size-conversion efficiency[87] can influence the size dependence of body burden.

The elimination rate may also be linked to size. Elimination associated with the alimentary tract is influenced by factors such as size-dependent ingestion rate.[10] Size-dependent changes in processes (e.g., renal clearance) or structures (e.g., kidney weight to body weight) will also influence elimination rate.[4]

In the form presented above, this model [Equation (28)] and related models assume a constant volume (or mass) for the compartment. As discussed previously, this may not be a valid assumption depending on the relative rates of accumulation and growth. Growth can significantly contribute to body burden allometry.[60]

It can be concluded from the above discussion that k_e and k_u are not constants in the context of modeling size-dependent body burdens. Equations such as Equation (28) are inadequate for describing body burden allometry unless they are modified to incorporate size-dependent changes in processes and structures. Selected attempts to do so are discussed below to highlight the advantages and disadvantages of such approaches. They will also be used to identify problems associated with using redescription models uncritically in predictive modeling.

Models Incorporating Allometry

Bioaccumulation of radionuclides has received attention primarily in the context of contaminated foodstuffs, health sciences, and radiotracers in ecological studies. Morgan[5] examined the accumulation of ^{134}Cs in finfish and shellfish near the Windscale nuclear facility in the United Kingdom and found that the biological half-life $[T_{1/2} = -(\ln 0.5/k_e)]$ increased slowly as weight increased (b values of 0.25 to 0.29 for plaice, eel, and lobster). He did not attribute this relationship to any single underlying mechanism; rather, he viewed the b values as multiple process statistics.

In an effort to extrapolate to humans, Fujita et al.[4] examined the interspecific, allometric relationship for equilibrium levels of Cs and K in mammals. Power relationships for both Cs and K had exponents of 0.45; however, the a values were 0.45 and 0.85 for Cs and K, respectively. The scatter in the K data was large;[4] therefore, a large model error term [Equation (12)] can be assumed. They used four relationships (with undefined model errors) to explain these findings: relationships between urinary excretion and total excretion, renal clearance and body size, kidney weight and body size, and total body concentration and plasma concentration. As judged by these four relationships, they predicted b values of 0.44 for both Cs and K, and a values of 0.39 and 0.67 for these same elements, respectively. This is an excellent fit of their observed data to a model involving four physiological and morphological relationships. However, it is suspected from the large model error term that this fit could be fortuitous. Similar interspecies studies with a primary focus on estimating radionuclide behavior in humans have demonstrated a size-dependent shift in pharmacokinetics. The basis for these shifts is most often linked to structural or physiological allometry.[88,89]

Anderson and Spear[16] examined Cu accumulation kinetics in gills of the pump-kinseed sunfish *(Lepomis gibbosus)* and fit the clearance of this metal to Equation (10) [see also Equation (22)]. The elimination rate constant relationship was $k_e = 0.29X^{-1.8}$. Total accumulation after 32 h of exposure was described by the relationship $\mu g\ Cu/g = 0.0077X^{-0.35}$. They suggested that physiological and morphological differences with size and perhaps changes in the number of binding sites on the gill with change in total gill surface likely accounted for these power relationships.

Newman and Mitz[55] used ^{65}Zn to measure the size-dependent accumulation kinetics of Zn in the mosquitofish, *Gambusia holbrooki* (formerly, *Gambusia affinis*). Although the model for elimination had significant error associated with it, the k_e and k_u for Zn were linked to size using the following relationships:

$$k_e = 0.001W^{-0.42} \tag{29}$$

$$k_u^* = 0.029W^{-0.90} \tag{30}*$$

As previously stated, no direct linkage of k_e or k_u can be made with mosquito-fish metabolic rate[90] or surface:volume based on similarity of exponents alone; however, the similarity between several of these relationships encourages further work. For example, the metabolic rate has a b value of 0.64 (b $-$ 1 = -0.36)[90] and the elimination rate constant for Zn had an mass exponent of -0.42. As suggested by Equation (24), the "b value" for k_e (= Θ) would be 0.58. This suggests, but certainly does not prove, that size-specific metabolic rate may play a significant role in determining the elimination rate for this element. Newman and Mitz[55] suggested that there was no direct connection between size-specific uptake (exponent = b $-$ 1 = -0.90) and gill surface:body mass allometry based on interpretation of b values for the gill surface:mass relationship. However, Murphy and Murphy[14] defined the b values for mosquitofish surfaces of exchange to mass as the following: whole body b value = 0.66; gill b value = 0.89. The amount of gill surface available for uptake per g of fish decreases disproportionately with fish mass (b $-$ 1 = 0.11). The uptake rate expressed in terms of $\mu g\ Zn/fish/day$ decreases disproportionately with fish weight (b = 0.10 assuming size-dependent elimination). This suggests a potentially significant role of surface:volume relationships in determining uptake rates for this fish.

The size-dependent relationships for k_e and k_u are shown in Figure 6 as is that estimated for time to reach 95% of equilibrium concentration (T_{95}). Clearly, Equation (28) must be modified to accommodate size-dependent accumulation kinetics. An allometric model of accumulation kinetics can be generated by substituting Equations (29) and (30) into Equation (28). By doing so and solving

---

* Note that the uptake rate (k_u^*) used here has the units of $\mu g/g/day$. It is identical to $C_s k_u$ in Equation (28) under the assumptions given for that equation.

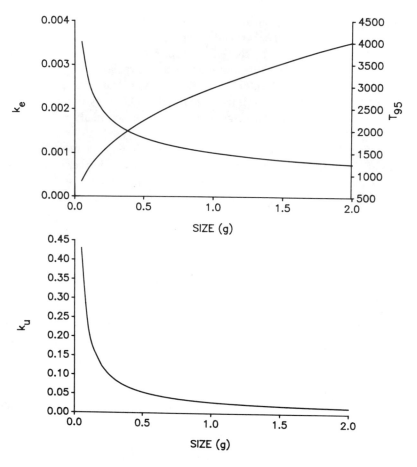

FIGURE 6. Size dependence of the elimination rate constant (k_e), time to 95% equilibrium concentration (T_{95}), and uptake rate constant (k_u) for zinc accumulation in mosquitofish *(Gambusia holbrooki)*.

this model at different times of exposure, the b values estimated from a range of simulated fish sizes were shown to increase with duration of exposure. It should also be noted that, considering the values for T_{95} (days) and the relatively short life-span of this species, any argument regarding these b values based on the assumption of equilibrium conditions is inappropriate. This was not the case, however, when Newman and Doubet[69] repeated this exercise with ionic Hg accumulation by mosquitofish. With this metal, concentrations rapidly approached equilibrium. Although uptake rates were higher for smaller than those for larger fish, there were no statistically significant effects of size on Hg elimination. These data are discussed more extensively in this volume, Chapter 7.

The model generated by combining Equations (29) and (30) with Equation

(28) remains unrealistic because no consideration is given to growth during the period of exposure. Growth will produce an apparent dilution of metal concentration and potentially shift the k_e and k_u as a fish becomes larger with time. A variety of growth models may be linked to these equations during simulation. In selecting the appropriate growth model, the availability of the necessary constants for the species, the model's ability to accurately describe growth of the particular species, and the potential for linkage to bioenergetics should be considered. Fagerström et al.[18] used the von Bertalanffy growth model [Equation (31)] combined with the relationship between fish length and fish size[50] [Equation (32)] to simulate fish growth in models of Hg accumulation.

$$l = l_\infty [1 - e^{\ k_g(u - u_0)}] \qquad (31)$$

$$W = \tau l^\phi \qquad (32)$$

where l = fish length
l_∞ = theoretical maximum fish length
k_g = growth rate constant
u   = age
u_0 = initial age of fish
W   = fish weight
τ, Φ = constants

The flexibility of the Richards model for description of growth[91] also makes it an attractive candidate during simulations. Depending on the value of the shape parameter (m), this relationship becomes the logistic (m = 2), von Bertalanffy (m = 0.67), Gompertz (m → 1.0), or single exponential (m = 0) models. Brisbin and co-workers recently reparameterized this model as discussed in Brisbin et al.[92] These reparameterized equations should be considered prior to use of the Richards model for growth.

Bioenergetics models that incorporate allometric relationships such as those discussed above have been developed for bioaccumulation of organic[84,86,93] and inorganic[18,20,22,94] contaminants. Fagerström et al.[18,63] incorporated growth, size-dependent uptake from food and water, and size-dependent clearance into a model describing methylmercury accumulation in northern pike *(Esox lucius)*. Boddington et al.[20] attempted to link oxygen uptake efficiency and pollutant (methylmercury) uptake efficiency of fish. Braune[94] included growth dilution, temperature, gill transfer efficiency, and size-dependent processes such as food conversion efficiency, metabolic rate, ventilation rate, and clearance to simulate Hg accumulation in herring *(Clupea harengus harengus)*. Recently, Rose et al.[22] examined radionuclide accumulation in shellfish using a similar bioenergetics approach. Bias in predicted values [Equations (16) and (17)] associated with the use of these redescription models for predictive purposes were not generally considered in development of these predictive models. The information necessary to perform the necessary bias corrections was usually not available.

ALLOMETRY AND TOXICITY
Overview

Allometric aspects of metal toxicity are often eliminated by use of a narrow range of size and/or age classes. This is done to enhance the precision of the toxicity test; however, most field populations are composed of individuals with a wide range of sizes. To describe or predict the toxic response of field populations to metals, an understanding of scaling is necessary. Table 3 in Anderson and Weber[23] attests to the frequent use of power equations [Equation (1) with $Y = LD_{50}$ or LC_{50}] to describe size effects on toxic end points. However, relationships between toxic response and size are often more complex than indicated by the allometric equation. Shepard[95] found that tolerance levels of small brook trout *(Salvelinus fontinalis)* to low concentrations of oxygen were similar to those of larger trout, but the small trout died more quickly when exposed to lethal concentrations. Threshold LC_{50} increased with fish size, but there were no size effects on the 96-h LC_{50} for goldfish *(Carassius auratus)*.[96] Developmental stage may further confound scaling effects.[1,97]

Models Incorporating Allometry

Dosage (amount given to an individual based on weight) can have different intensities of effect for different size, age, and sex classes.[98-100] In 1909, Moore, as cited in Bliss,[2] argued that dosage should be modified to be proportional to surface of absorptive tissue, e.g., $2/3$ power of mass. Early studies that modified dosage according to such power relationships incorrectly referred to this approach as "dose to body surface."[100] Bliss[2] formulated the following relationship for size effects on toxic response.

$$\text{Log(rate of toxic action)} = a + b\text{Log}(m/W^h) \qquad (33)$$

where rate of toxic action = 1/T (or 1000/min survival),
h = size coefficient
W = weight (cg)
m = mg As per individual × 1000
a,b = regression constants

This relationship was derived for silkworm larvae fed As. The h value was 1.5 under these test conditions. It took more As to get a toxic response from the larger than from the small larvae. Although he expressed caution regarding excessive speculation, Bliss suggested that some biochemical constituent to which As binds (such as glutathione) increases in concentration as animal size increases. This sequesters the As that otherwise would combine with and inactivate cell constituents present at relatively low concentrations.

When the Log (rate of toxic action) was plotted against Log of dose [10 (mg As/g body wt$^{1.50}$)], a straight line was generated. Bliss provided a time to response-concentration expression of the data set using the slope of this plot:

$$C^n t = \text{constant} \qquad (34)$$

where C = amount of As/unit mass
n = slope of Log-Log curve
t = time

Anderson and Weber[23] took the relationship developed by Bliss for dose [Equations (33) and (34)] and extended it to incorporate size-dependent responses of aquatic biota (% mortality as probit) to ambient concentrations of toxicants. Guppies *(Poecilia reticulata)* were exposed to a series of toxicants including Cu, Ni, and Zn. The mean daily toxicant concentration was used as m in Equation (33). This provided good fit to the data sets. Anderson and Weber[23] then linked these relationships to the following regression of weight effects on LC_{50}:

$$\text{Log}LC_{50} = \text{Log}a + b\text{Log}W \qquad (35)$$

or

$$LC_{50} = aW^b \qquad (36)$$

or

$$\text{Log}(LC_{50}/W^b) = \text{Log}a \qquad (37)$$

To show the similarity between Equation (37) and Bliss' linear time to response-concentration model, they substituted the 50% probit score into Bliss' relationship,

$$\text{Log}(m/W^h) = \text{Log}X \qquad (38)$$

where m is now the LC_{50} and X is the probit score at 50%.

Equation (36) was recommended by Anderson and Weber[23] "as a standard format for reporting LD_{50} and LC_{50} values for drugs, toxicants, and toxins for varying sizes of animals." Hedtke et al.[24] also began with Bliss' work[2] and developed a similar approach. As long as conditions regarding bias corrections are kept in mind and confounding biological factors[17] are clearly acknowledged, these relationships provide effective tools for simulation of size-dependent toxic response.

Other related approaches to modeling size-dependent mortality have been developed.[25,26] For example, proportional hazard model techniques were used by the authors and co-workers[101,102] to link time-to-death to a variety of factors, including size. No description of these techniques will be given here as a detailed description is presented in this volume, Chapter 8.

CONCLUSION

1. Allometry is the study of size and its consequences. Huxley established the power equation [Equation (1)] as the primary redescription model for the allometric relation. This model remains empirical, with no clear, underlying mechanism(s).

2. Concepts and techniques applied to ecotoxicological allometry were borrowed exclusively from those of physiological and morphological allometry. This linkage provided a rapid infusion of ideas and techniques. It also allowed linkage to bioenergetics during modeling efforts. However, many conceptual and technical errors were also transferred to ecotoxicological allometry.

3. The mass exponent (b) indicates the extent of disproportional change between some quality or structure and size. The mass coefficient (a) has no clear biological meaning.

4. Log-Log transformation, regression on transformed data, and backtransformation are the standard approaches for determining the constants, a and b. A variety of errors associated with this process are present. Methods for coping with these errors and correcting bias for predictive purposes are outlined in this chapter.

5. Significant measurement error in X and inherent variability in X are often ignored in allometric analyses. Often functional regression techniques should be used instead of predictive regression techniques.

6. Constants from allometric relationships of accumulation or toxicity should be considered multiple process statistics unless clearly shown to be otherwise. A complex array of factors, not simply metabolic rate or surface-to-volume ratio, can influence ecotoxicological allometry.

7. Boyden clearly defined testable hypotheses regarding allometry of body burden. Some have been rejected, whereas the validities of others are still unknown.

8. Contrary to the suggestions of Boyden, there is no clear evidence for the existence of three discrete types of allometric relationships for elemental body burden. Regardless, it is doubtful that such a classification scheme should have been based on b rather than v values.

9. Allometric models linked to bioenergetics provide excellent tools for modeling bioaccumulation. However, biases in predicted values from descriptive models can compromise the utility of predictive models which incorporate them.

10. Allometric aspects of metal toxicity are often eliminated by the use of a narrow range of animal sizes. This is done to enhance assay precision. However, this practice limits our ability to predict toxic impacts on field populations.

11. Allometric relationships for toxic effects are influenced by a variety of factors. The power relationship was used to describe the relationship between toxic end points and animal size.

12. Equations developed by Bliss were successfully modified to describe the relations among rate of toxic action, ambient toxicant concentrations, and animal size.

13. Despite the magnitude and prevalence of size-dependent effects, a variety of hypotheses remain untested or poorly tested in the area of ecotoxicological allometry. Necessary decisions are made and models are built despite this ambiguity. For example, many weak inferences about the body burdens of elements remain untested more than a decade after the original hypothesis was formulated. In our opinion, the field is progressing slowly because it lacks a tradition of strong inference.[103]

ACKNOWLEDGMENTS

Financial support was obtained from contract DE-AC09-76SR00819 between the U.S. Department of Energy and the University of Georgia. M. Heagler received support through the Savannah River Ecology Laboratory education program. The authors gratefully acknowledge excellent advice and criticism from Dr. P. Dixon, and two anonymous reviewers.

REFERENCES

1. Campbell, F. L. "Relative Susceptibility to Arsenic in Successive Instars of the Silkworm," *Gen. Physiol.* 9(6):727-733 (1926).

2. Bliss, C. I. "The Size Factor in the Action of Arsenic upon Silkworm Larvae," *Exp. Biol.* 13:95-110 (1936).

3. Eberhardt, L. L. "Relationship of Cesium-137 in Humans to Body Weight," *Health Phys.* 13:88-90 (1967).

4. Fujita, M., J. Iwamoto, and M. Kondo. "Comparative Metabolism of Cesium and Potassium in Mammals—Interspecies Correlation between Body Weight and Equilibrium Level," *Health Phys.* 12:1237-1247 (1966).

5. Morgan, F. "The Uptake of Radioactivity by Fish and Shellfish. I. ^{134}Cesium by Whole Animals," *J. Mar. Biol. Assoc. U.K.* 44:259-271 (1964).

6. Reichle, D. E. and R. I. Van Hook, Jr. "Radionuclide Dynamics in Insect Food Chains," *Manit. Entomol.* 4:22-32 (1970).

7. Gallegos, A. F. and F. W. Whicker. "Radiocesium Retention by Rainbow Trout as Affected by Temperature and Weight," in *Proceedings of the 3rd Natl. Symp. on Radiobiology* (Oak Ridge, TN: Dept. of Energy CONF-710501-P1, 1971), pp. 361-371.

8. Mishima, J. and E. P. Odum. "Excretion Rate of ^{65}Zn by *Littorina irrorata* in Relation to Temperature and Body Size," *Limnol. Oceanogr.* 8:39-44 (1963).

9. Pulliam, H. R., G. W. Barrett and E. P. Odum. "Bioelimination of Tracer ^{65}Zn in Relation to Metabolic Rates of Mice," in *Symposium on Radioecology*, D. J. Nelson and F. C. Evans, Eds. (Ann Arbor, MI: Dept. of Energy CONF-670503, 1967), pp. 725-730.

10. Williamson, P. "Use of ^{65}Zn to Determine the Field Metabolism of the Snail, *Cepaea nemoralis* L.," *Ecology* 56:1185-1192 (1975).

11. Boyden, C. R. "Trace Element Content and Body Size in Molluscs," *Nature* 251:311-314 (1974).

12. Boyden, C. R. "Effect of Size upon Metal Content of Shellfish," *J. Mar. Biol. Assoc. U.K.* 57:675-714 (1977).

13. Taylor, P. "Revising Models and Generating Theory," *Oikos* 54:121-126 (1989).

14. Murphy, P. G. and J. V. Murphy. "Correlations between Respiration and Direct Uptake of DDT in the Mosquitofish, *Gambusia affinis*," *Bull. Environ. Contam. Toxicol.* 6(6):581-588 (1971).

15. Fagerström, T. "Body Weight, Metabolic Rate and Trace Substance Turnover in Animals," *Oecologia (Berl.)* 29:99-104 (1977).

16. Anderson, P. D. and P. A. Spear. "Copper Pharmacokinetics in Fish Gills. I. Kinetics in Pumpkinseed Sunfish, *Lepomis gibbosus*, of Different Body Sizes," *Water Res.* 14:1101-1105 (1980).

17. Anderson, P. D. and P. A. Spear. "Copper Pharmacokinetics in Fish Gills. II. Body Size Relationships for Accumulation and Tolerance," *Water Res.* 14:1107-1111 (1980).

18. Fagerström, T., B. Åsell, and A. Jernelov. "Model for Accumulation of Methyl Mercury in Northern Pike, *Esox lucius*," *Oikos* 25:14-20 (1974).

19. Olsson, M. "Mercury Levels as a Function of Size and Age in Northern Pike, One and Five Years after the Mercury Ban in Sweden," *Ambio* 5:73-76 (1976).

20. Boddington, M. J., B. A. MacKenzie, and A. S. W. DeFreitas. "A Respirometer to Measure the Uptake Efficiency of Waterborne Contaminants in Fish," *Ecotoxicol. Environ. Safety* 3:383-393 (1979).

21. Neely, W. B. "Estimating Rate Constants for the Uptake and Clearance of Chemical by Fish," *Environ. Sci. Technol.* 13(12):1506-1510 (1979).

22. Rose, K. A., R. I. McLean, and J. Summers. "Development and Monte Carlo Analysis of an Oyster Bioaccumulation Model Applied to Biomonitoring Data," *Ecol. Model.* 45:111-132 (1989).

23. Anderson, P. D. and L. J. Weber. "Toxic Response as a Quantitative Function of Body Size," *Toxicol. Appl. Pharmacol.* 33:471-483 (1975).

24. Hedtke, J. L., E. Robinson-Wilson, and L. J. Weber. "Influence of Body Size and Developmental Stage of Coho Salmon *(Oncorhynchus kisutch)* on Lethality of Several Toxicants," *Fundam. Appl. Toxicol.* 2:67-72 (1982).

25. Logan, D. T. "Use of Size-Dependent Mortality Models to Estimate Reductions in Fish Populations Resulting from Toxicant Exposure," *Environ. Toxicol. Chem.* 5:769-775 (1986).

26. Hallam, T. G. and Y. Huang. "On The Dynamics of a Toxicant-Individual System," *Theoret. Biol.* 141:65-72 (1989).

27. Scholl, D. A. "The Theory of Differential Growth Analysis," *Proc. R. Soc. London B* 137:470-474 (1950).

28. Gould, S. J. "Allometry and Size in Ontogeny and Phylogeny," *Biol. Bull.* 41:587-640 (1966).

29. Huxley, J. S. "Relative Growth and Form Transformation," *Proc. R. Soc. London B* 137:465-470 (1950).

30. Muir, B. S. "Gill Dimensions as a Function of Fish Size," *J. Fish. Res. Bd. Can.* 26:165-170 (1969).

31. Niimi, A. J. "Relationship of Body Surface Area to Weight in Fishes," *Can. J. Zool.* 53:1192-194 (1975).

32. Hughes, G. M. "Measurement of Gill Area in Fishes: Practices and Problems," *J. Mar. Biol. Assoc. U.K.* 64:637-655 (1984).

33. Von Bertalanffy, L. "Metabolic Types and Growth Types, *Am. Natur.* 135:111-117 (1951).

34. Cammen, L. M. "Ingestion Rate: An Empirical Model for Aquatic Deposit Feeders and Detritivores," *Oecologia (Berl.)* 44:303-310 (1980).

35. Somero, G. N. and J. J. Childress. "A Violation of the Metabolism-Size Scaling Paradigm: Activities of Glycolytic Enzymes in Muscle Increase in Larger-Size Fish," *Physiol. Zool.* 53(3):322-337 (1980).

36. Emmett, B. and P. W. Hochachka. "Scaling of Oxidative and Glycolytic Enzymes in Mammals," *Resp. Physiol.* 45:261-272 (1981).

37. Magid, E. "Activity of Carbonic Anhydrase in Mammalian Blood in Relation to Body Size," *Comp. Biochem. Physiol.* 21:357-360 (1967).

38. Prothero, J. W. "Scaling of Blood Parameters in Mammals," *Comp. Biochem. Physiol.* 67A:649-657 (1980).

39. Huxley, J. S. "Constant Differential Growth-Ratios and Their Significance," *Nature* 114:895-896 (1924).

40. Heusner, A. A. "What Does the Power Function Reveal about Structure and Function in Animals of Different Size," *Annu. Rev. Physiol.* 49:121-133 (1987).

41. Heusner, A. A. "Biological Similitude: Statistical and Functional Relationships in Comparative Physiology," *Am. J. Physiol.* 246:R839-R845 (1984).

42. Butler, J. P., H. A. Feldman, and J. J. Fredberg. "Dimensional Analysis Does Not Determine a Mass Exponent for Metabolic Scaling," *Am. J. Physiol.* 253:R195-R199 (1987).

43. Schmidt-Nielsen, K. *Scaling. Why is Animal Size So Important?* (New York: Cambridge Univ. Press, 1984), p. 241.

44. Adolph, E. F. "Quantitative Relations in the Physiological Constitutions of Mammals," *Science* 109:579-585 (1949).

45. Smith, R. J. "Rethinking Allometry," *J. Theoret. Biol.* 87:97-111 (1980).

46. Smith, R. J. "Allometric Scaling in Comparative Biology: Problems of Concept and Method," *Am. J. Physiol.* 246:R152-160 (1984).

47. Koch, R. W. and G. M. Smillie. "Bias in Hydrologic Prediction Using Log-Transformed Regression Models," *Water Res.* 22(5):717-723 (1986).

48. Beauchamp, J. J. and J. S. Olson. "Correction for Bias in Regression Estimates after Logarithmic Transformation," *Ecology* 54(6):1403-1407 (1973).

49. Sprugel, D. G. "Correction for Bias in Log-Transformed Allometric Equations," *Ecology* 64(1):209-210 (1983).

50. Ricker, W. E. "Linear Regression in Fishery Research," *J. Fish. Res. Bd. Can.* 30:409-434 (1973).

51. Lobel, P. B. and D. A. Wright. "Relationship between Body Zinc Concentration and Allometric Growth Measurements in the Mussel, *Mytilus edulis,*" *Mar. Biol.* 66:145-150 (1982).

52. Williamson, P. "Opposite Effects of Age and Weight on Cadmium Concentration of a Gastropod Mollusc," *Ambio* 8(4):30-31 (1979).

53. Simkiss, K. "Calcium, Pyrophosphate and Cellular Pollution," *Trends in Biochemical Sciences* (April, 1985).

54. Cross, F. A., L. H. Hardy, N. Y. Jones, and R. T. Barber. "Relation Between Total Body Weight and Concentrations of Manganese, Iron, Copper, Zinc and Mercury in White Muscle of Bluefish *(Pomatonus saltatrix)* and a Bathyl-Demersal Fish *Antimora rostrata*," *J. Fish. Res. Bd. Can.* 30:1287-1291 (1973).

55. Newman, M. C. and S. V. Mitz. "Size Dependence of Zinc Elimination and Uptake from Water by Mosquitofish *Gambusia affinis* (Baird and Girard)," *Aquat. Toxicol.* 12:17-32 (1988).

56. Jeffree, R. A. "Patterns of Accumulation of Alkaline-Earth Metals in the Tissue of the Freshwater Mussel *Velesunio angasi*," *Arch. Hydrobiol.* 112(1):67-90 (1988).

57. Cossa, D., E. Bourget, and J. Piuze. "Sexual Maturation as a Source of Variation in the Relationship Between Cadmium Concentration and Body Weight of *Mytilus edulis* L.," *Mar. Pollut. Bull.* 10:174-176 (1979).

58. Simpson, R. D. "Uptake and Loss of Zinc and Lead by Mussels *(Mytilus edulis)* and Relationships with Body Weight and Reproductive Cycle," *Mar. Pollut. Bull.* 10:74-78 (1979).

59. Cossa, D., E. Bourget, D. Pouliot, J. Piuze, and J. Chanut. "Geographical and Seasonal Variations in the Relationship Between Trace Metal Content and Body Weight in *Mytilus edulis*," *Mar. Biol.* 58:7-14 (1980).

60. Strong, C. and S. Luoma. "Variations in the Correlation of Body Size with Concentrations of Cu and Ag in the Bivalve *Macoma balthica*," *Can. J. Fish Aquat. Sci.* 38:1059-1064 (1981).

61. Leggett, R. "Predicting the Retention of Cs in Individuals," *Health Phys.* 50(6):747-759 (1986).

62. Potter, L., D. Kidd, and D. Standiford. "Mercury Levels in Lake Powell. Bioamplification of Mercury in Man-Made Desert Reservoir," *Environ. Sci. Technol.* 9(1):41-46 (1975).

63. Fagerström, T., R. Kurten, and B. Åsell. "Statistical Parameters as Criteria in Model Evaluation: Kinetics of Mercury Accumulation in Pike, *Esox lucius*," *Oikos* 26:109-116 (1975).

64. Watling, R., T. McClurg, and R. Stanton. "Relation Between Mercury Concentration and Size in the Mako Shark," *Bull. Environ. Contam. Toxicol.* 26:352-358 (1981).

65. Rincón, F., G. Zurera, and R. Pozo-Lora. "Size and Mercury Concentration Relationship as Contamination Index," *Bull. Environ. Contam. Toxicol.* 38:515-522 (1987).

66. Roberts, J., A. DeFrietas, and M. Gidney. "Influence of Lipid Pool Size on Bioaccumulation of the Insecticide Chlordane by Northern Redhorse Suckers *(Moxostoma macrolepidotum)*," *J. Fish. Res. Bd. Can.* 34:89-97 (1977).

67. Newman, M. and I. L. Brisbin, Jr. "Variation of [137]Cs Levels Between Sexes, Body Sizes and Collection Localities of Mosquitofish, *Gambusia holbrooki* (Girard 1859), Inhabiting a Reactor Cooling Reservoir," *J. Environ. Radioactivity* 12:131-144 (1990).

68. Stebbing, A. "Hormesis—The Stimulation of Growth by Low Levels of Inhibitors," *Sci. Total Environ.* 22:213-234 (1982).

69. Newman, M. and D. Doubet. "Size-Dependence of Mercury (II) Accumulation Kinetics in the Mosquitofish, *Gambusia affinis* (Baird and Girard)," *Arch. Environ. Contam. Toxicol.* 18:819-825 (1989).

70. Newman, M. C. and J. J. Alberts, unpublished data, 1981.
71. Moriarty, F., H. Hanson, and P. Freeman. "Limitations of Body Burden as an Index of Environmental Contamination: Heavy Metals in Fish *Cottus gobio* L. From the River Ecclesbourne Derbyshire," *Environ. Pollut.* 34:297-320 (1984).
72. Romo, M. and M. Gnassia-Barelli. *"Donas trunculus* and *Venus verrucosa* as Bioindicators of Trace Metal Concentrations in Mauritanian Coastal Waters," *Mar. Biol.* 99:223-227 (1988).
73. Davies, I. and J. Pirie. "Evaluation of a 'Mussel Watch' Project for Heavy Metals in Scottish Coastal Waters" *Mar. Biol.* 57:87-93 (1980).
74. Amiard, J., C. Amiard-Triquet, B. Berthet, and C. Métayer. "Contribution to the Ecotoxicological Study of Cadmium, Lead, Copper and Zinc in the Mussel *Mytilus edulis* I. Field Study," *Mar. Biol.* 90:425-431 (1986).
75. Marks, P., D. Plaskett, I. Potter, and J. Bradley. "Relationship Between Concentration of Heavy Metals in Muscle Tissue and Body Weight of Fish from the Swan-Avon Estuary, Western Australia," *Aust. J. Mar. Freshwater Res.* 31:783-793 (1980).
76. Newman, M. C. and S. Mitz, unpublished data, 1989.
77. Parker, R. and P. Larkin. "A Concept of Growth in Fishes," *J. Fish. Res. Bd. Can.* 16(5) 1959.
78. Newman, M. and A. McIntosh. "Lead Elimination and Size Effects on Accumulation of Two Freshwater Gastrpods," *Arch. Environ. Contam. Toxicol.* 12:25-29 (1983).
79. Bergner, P-E. "On Relations Between Bioaccumulation and Weight of Organisms," *Ecol. Model.* 27:207-220 (1985).
80. Davis, J. and C. Boyd. "Concentrations of Selected Elements and Ash in Bluegill *(Lepomis marcochirus)* and Certain Other Freshwater Fish," *Trans. Am. Fish Soc.* 107(6):862-867 (1978).
81. Newman, M. C. unpublished data, 1990.
82. Ashraf, M. and M. Jaffar. "Weight Dependence of Arsenic Concentration in the Arabian Sea Tunafish," *Bull. Environ. Contam. Toxicol.* 40:219-225 (1988).
83. Kumagai, H. and K. Saeki. "The Variation with Growth in Heavy Metal Contents of Rock Shell," *Bull. Jpn. Soc. Sci. Fish.* 49(12):1917-1920 (1983).
84. Williamson, P. "Variables Affecting Body Burden of Lead, Zinc and Cadmium in a Roadside Population of the Snail, *Cepaea hortensis* Müller," *Oecologia (Berl.)* 44:213-220 (1980).
85. Barber, M., L. Suárez, and R. Lassiter. "Modeling Bioconcentration of Nonpolar Organic Pollutants by Fish," *Environ. Toxicol. Chem.* 7:545-558 (1988).
86. Niimi, A. "Gross Growth Efficiency of Fish (k_1) Based on Field Observations of Annual Growth and Kinetics of Persistent Environmental Contaminants," *Can. J. Fish. Aquat. Sci.* 38:250-253 (1981).
87. Borgmann, U. and D. Whittle. "Particle-Size Conversion and Contaminant Concentrations in Lake Ontario Biota," *Can. J. Fish. Aquat. Sci.* 40:328-336 (1983).
88. Richmond, C. R., J. E. Furchner, G. A. Trafton, and W. H. Langham. "Comparative Metabolism of Radionuclides in Mammal. I. Uptake and Retention of Orally Administered ^{65}Zn by Four Mammalian Species," *Health Phys.* 8:481-489 (1962).

89. Furchner, J. E. and C. R. Richmond. "Comparative Metabolism of Radioisotopes in Mammals. II. Retention of Iodine[131] by Four Mammalian Species," *Health Phys.* 9:277-282 (1963).

90. Mitz, S. and M. Newman. "Allometric Relationship Between Oxygen Consumption and Body Weight of Mosquitofish, *Gambusia affinis,*" *Environ. Biol. Fish.* 24(4):267-273 (1989).

91. Richards, F. "A Flexible Growth Function for Empirical Use," *J. Exp. Bot.* 10:290-300 (1959).

92. Brisbin, I., M. Newman, S. McDowell, and E. Peters. "Prediction of Contaminant Accumulation by Free-Living Organisms: Applications of a Sigmoidal Model," *Environ. Toxicol. Chem.* 9:141-149 (1990).

93. Griesbach, S. and R. Peters. "An Allometric Model for Pesticide Bioaccumulation," *Can. J. Fish. Aquat. Sci.* 39:727-735 (1982).

94. Braune, B. "Mercury Accumulation in Relation to Size and Age of Atlantic Herring *(Clupea harengus harengus)* from the Southwestern Bay of Fundy, Canada," *Arch. Environ. Contam. Toxicol.* 16:311-320 (1987).

95. Shepard, M. "Resistance and Tolerance of Young Speckled Trout *(Salvelinus fontinalis)* to Oxygen Lack, with Special Reference to Low Oxygen Acclimation," *J. Fish. Res. Bd. Can.* 12(3):387-449 (1955).

96. Adelman, I., L. Smith, Jr., and G. Siesennop. "Effect of Size or Age of Goldfish and Fathead Minnows on Use of Pentachlorophenol as a Reference Toxicant," *Water Res.* 10:685-687 (1976).

97. Chapman, G. "Toxicities of Cadmium, Copper and Zinc to Four Juvenile Stages of Chinook Salmon and Steelhead," *Trans. Am. Fish. Soc.* 107(6):841-847 (1978).

98. Pallotta, A., M. Kelly, D. Rall, and J. Ward. "Toxicology of Acetozycyclo-heximide as a Function of Sex and Body Weight," *J. Pharmacol. Exp. Therap.* 136:400-405 (1962).

99. Lamanna, C. and E. Hart. "Relationship of Lethal Toxic Dose to Body Weight of the Mouse," *Toxicol. Appl. Pharmacol.* 13:307-315 (1968).

100. Hogan, G., B. Cole, and J. Lovelace. "Sex and Age Mortality Responses in Zinc Acetate-Treated Mice," *Bull. Environ. Contam. Toxicol.* 39:156-161 (1987).

101. Diamond, S., M. Newman, M. Mulvey, P. Dixon, and D. Martinson. "Allozyme Genotype and Time to Death of Mosquitofish, *Gambusia affinis* (Baird and Girard), During Acute Exposure to Inorganic Mercury," *Environ. Toxicol. Chem.* 8:613-622 (1989).

102. Newman, M., S. Diamond, M. Mulvey, and P. Dixon. "Allozyme Genotype and Time to Death of Mosquitofish, *Gambusia affinis* (Baird and Girard) During Acute Toxicant Exposure: A Comparison of Arsenate and Inorganic Mercury," *Aquat. Toxicol.* 15:141-156 (1989).

103. Platt, J. "Strong Inference," *Science* 146(3642):347-353 (1964).

5

Metal Effects on Fish Behavior — Advances in Determining the Ecological Significance of Responses

Mary G. Henry[1] and Gary J. Atchison[2]

[1]U.S. Fish and Wildlife Service, Minnesota Cooperative Fisheries and Wildlife Research Unit, University of Minnesota, St. Paul, Minnesota 55108
and
[2]Department of Animal Ecology, Iowa State University, Ames, Iowa 50011

OVERVIEW

Sublethal metal toxicity has been shown to induce changes in many aspects of fish behavior. Studies have documented alterations in respiration, locomotion, social organization, reproduction, feeding, and predator avoidance. Atchison et al.[1] compiled an extensive review of this literature and concluded that metals, by inducing subtle behavioral changes, could mediate ecological death by disrupting the normal function and life history of exposed fish. Since this 1987 review, more studies have been conducted, some of which incorporate toxic mode of action data and field validation information. Consequently, the current literature, including some studies from Europe and the U.S.S.R., will be evaluated and attempts made to determine if the trends, elucidated by Atchison et

131

al.,[1] have been strengthened or weakened by recent research developments. The objective of this chapter is to carefully synthesize current literature from both laboratory and field studies to determine if sublethal changes in fish behavior due to sublethal metal toxicity can be linked to individual indirect mortality, community change, or population level alterations in natural systems or mesocosms.

INTRODUCTION

Determining the ecological significance of contaminant effects on aquatic systems, especially those effects predicted based on laboratory derived data, is difficult, at best. As a research community, we have struggled with this issue since the 1960s. At the 1977 Pellston Workshop, the issue of ecological significance was ranked most important in a list of criteria for a toxicity evaluation to be complete.[2] To what extent can laboratory derived data be used to predict contaminant effects at the population or community level in the wild? To what extent do changes in populations or communities, identified in the field, lead to significant changes in ecosystem structure and function? What are acceptable changes at the population, community, or ecosystem level? Complications arise because of (1) limitations in our understanding of ecological processes, (2) complexities involved in determining the source, chemical composition, fate, and flow of contaminants in aquatic ecosystems, and (3) experimental design limitations implicit in laboratory and field approaches.

It seems that a large part of the problem has to do with experimental approaches and selection of meaningful indices of change after metal exposure. Laboratory approaches are criticized because they are oversimplified and unrealistic.[3-6] However, in the laboratory, ecological realism is purposely sacrificed for control of exposure concentrations, in an attempt to elucidate a cause-effect relationship between contaminant and organism response. A well-designed laboratory experiment can address a hypothesis and test it under controlled and replicated conditions holding multiple, extraneous abiotic and biotic factors constant.

Field studies do not have to simulate elements of reality: they are reality. They trade knowledge of exposure and independent environmental factors for ecological realism. Factors such as competition, predation, seasonal and diel fluctuations in temperature, and dissolved oxygen are naturally incorporated into the study by the sheer fact that they cannot be controlled. The difficulty with a field approach is that results generated may be applicable only to the changes inherent in that unique system and the cause-effect relationship between contaminant and biological effect can only be inferred. Too many factors are operative at any one time for a cause-effect relationship between contaminant and biological response to be firmly established.

What can be done about this lab-field gap? When do we use one approach over another? How do we attack the issue of ecological significance and further the progress made in interpreting the role of contaminants, especially metals, in changing our aquatic ecosystems? Experimental approach and selection of an ecologically interpretable index of responses are critical. Therefore, we will approach these experimental design questions within a behavioral toxicology context, since there appears to be potential in pursuing answers to the above questions within this framework.

The objectives of this chapter are to (1) briefly review what has been done in behavioral toxicology with regard to determining the effects of metal exposure, (2) examine experimental options for determining the ecological effects of metals by using behavioral endpoints, and (3) in an attempt to generalize for application to other circumstances, examine the experimental design components in the studies where assessment of ecological significance of metal impacts was possible.

BEHAVIORAL TOXICOLOGY

Behavior is the organismal level manifestation of the motivational, biochemical, physiological, and environmentally influenced state of the organism. In the laboratory, fish behavior can be a sensitive indicator of toxicant-induced stress.[1,7-9] According to Rand,[8] in order for behavioral approaches to be utilized successfully in contaminant evaluations, the behavior studied should: (1) be easily observed in the laboratory or field; (2) be sensitive to the chemicals of interest; (3) be previously well-described; (4) be ecologically relevant to species survival; and (5) integrate several sensory and/or mechanical modalities. In addition, the methods should be routinely available and simple to employ.

Atchison et al.[1] reviewed the literature on the effects of metals on fish behavior. The behaviors evaluated after exposure to metals were avoidance/attractance, activity, critical swimming performance, respiratory behavior (ventilation and cough response), learning (change in conditioned response), intraspecific social interactions, reproductive behavior, feeding behavior, and predator avoidance. The metals evaluated from these behavioral perspectives were Al, Cd, Cr, Cu, Fe, inorganic/organic Hg, Ni, Pb, Se, and Zn.

Their review showed a paucity of field work (four studies, all on avoidance behavior) validating laboratory findings to determine ecological significance. It is clear that the balance of effort investigating behavioral responses to metals lies in the laboratory, emphasizing respiratory behavior and various aspects of swimming behavior including avoidance/attractance.

Studies conducted since the Atchison et al.[1] review are listed in Table 1. The focus on laboratory investigation is still apparent. Atchison et al.[1] concluded that "further research on behavioral toxicity tests is assuredly needed" and must possess ecological realism. They caution that "tests should be developed that can also be field validated; that is, laboratory results should be confirmed in the field." It is these two issues, incorporation of more ecological realism into laboratory testing and validation of results of these tests in the field, that lie at the heart of the experimental design problem in fish behavioral toxicology.

Table 1
Summary of Studies Dealing with Behavioral Toxicology of Metals and
Published Since 1986[a]

Behavior	Metals	Citation	Experimental Approach[a]
Locomotion			
Avoidance/attractance	Zn	11	L
	Cu	33	L
	Cr, Cu, As, Se	17	L-F
		18	L
	Cr, Cu, Cd, As, Se	19	L
Activity	A1	34	L
	Cu	35	L
Critical swimming			
performance	A1	36	L
	Cu, Cd, Cr	37	L
Respiratory behavior			
Ventilation/cough			
response	A1	38	
	Zn, Cd	39	L
Feeding behavior			
	Cu	10	L
	Cd	40	L
	A1	41	L
Reproductive behavior			
	Se	32	L

[a] Studies not included in Atchison et al.[1] review.
[b] L, laboratory; L-F, laboratory-field.

LIMITATIONS IN EXPERIMENTAL APPROACH

Most laboratory behavioral toxicity tests focus so narrowly on the behavior in question that the influence of many other motivational factors in nature are neglected. For instance, in feeding behavior studies, artificial food is often used, thus greatly simplifying the situation for the predator. In nature, feeding involves searching for prey within an array of living things, as well as specific prey selection, attack, and capture and handling, most of which is generally not addressed in laboratory toxicity tests.[10] Most laboratory avoidance tests do not account for motivational factors such as food, shelter, and reproduction.[11] Cairns[12] stressed the point that "the important issue in validation is the ability to predict the relationship between the response of the artificial laboratory system and the natural system." Field validation often attempts to duplicate a laboratory experiment in the field in the hopes of corroborating results from the two efforts. The practical obstacles of caging fish, multiple fluctuations in limnological parameters and contaminant levels, and natural influences, such as predation and seasonality, complicate such attempts. Direct observation of fish behavior is difficult in turbid freshwater systems receiving contaminants. Verification of toxicant exposure further complicates the picture. If direct transfer of laboratory experimental design to the field for validation is not possible, how then do we use the combination of behavior, laboratory data, and field data to get us closer to determining ecological significance of metal exposure?

Some experimental approach limitations need to be accepted. The laboratory was never meant to mimic the conditions of the field in their totality. The laboratory is oversimplified so that controlled conditions can be established and variables relating to cause-effect can be investigated systematically. However, experimental conditions in the laboratory can be made more ecologically realistic by giving more careful consideration to the natural life history habits of the test organisms.

Conversely, field studies cannot be "controlled" like laboratory studies. This is not to say, however, that aspects of good experimental design (i.e., treatment versus control, replication, and randomization) cannot be taken to the field. At best, results will be inferential and not cause-effect oriented. The cause-effect relationship will only be established in the laboratory and the realism of the effect will be obtained from field work inferring a relationship between contaminants and biological effect.

The combination of controlled laboratory experimentation, well-focused field studies and wise utilization of available historical/current information comes together to enable the investigator to determine some of the ecological significance of contaminant exposure. Once the limitations of laboratory and field work are understood and accepted, they can be reconciled and used to complement one another. If we stop trying to make them perform identically, we can use them in combination to our best advantage and to their fullest scientific potential.

IMPROVEMENTS IN EXPERIMENTAL DESIGN IN BEHAVIORAL TOXICOLOGY

An objective of this chapter is to critically evaluate the behavioral toxicology literature for experimental design advances that aid in addressing the issue of the ecological significance of metal exposure. We will focus on research done in two example areas, (1) avoidance of metals and (2) the effects of metals on fish foraging and energetics.

Example 1: Avoidance Behavior

Because of the emphasis on point-source pollution and because of intense interest in the commercial and sport value of anadromous fish, locomotion has been heavily investigated, especially as it relates to avoidance/attractance. The life history of anadromous fish which requires them to return to nursery streams to spawn offers an investigative opportunity in the field which is focused and geographically restricted, making the scope of addressing things experimentally more practical. The hypothesis that streams can become "chemosensorily blocked" due to the presence of contaminants has been tested and results indicate that, with many metals, avoidance occurs at concentrations equal to or less than the lowest observed effect concentration (LOEC) reported in standard chronic toxicity tests. Water quality and concentration gradient are two important factors mediating these locomotion-oriented responses.[1] Many avoidance/attractance studies have now been done, but no standard methodology yet exists. However, there have been more attempts to field validate these studies than with any other kind of behavioral toxicity test. We will review these field tests of metal effects and also some recent advances in developing more ecologically realistic approaches to laboratory testing.

The efforts of Sprague,[13] Sprague et al.,[14] Saunders and Sprague,[15] Geckler et al.,[16] and Hartwell et al.[17] demonstrated that the combination of laboratory and field data lends great credibility to an interpretation of sublethal, ecologically meaningful responses to the presence of metals in an aquatic system. Ecological significance of effect is most easily determined when the effect measured is directly related to the survival and reproductive success of the tested organisms. Disruption of the migration process (directly related to reproduction) through avoidance of natal streams is an example of selection of an ecologically meaningful behavioral parameter to examine after exposure to metals. In the case of resident species of fish, avoidance of a stream reach contaminated with metals is ecologically significant because production of stream fish is clearly impaired.

Sprague et al.[14] and Saunders and Sprague[15] reported that adult Atlantic salmon, *Salmo salar,* migrating upstream, avoided areas contaminated with a mixture of Cu and Zn. The threshold for avoidance was about 17 to 21 µg Cu/L and 210 to 258 µg Zn/L. Laboratory tests[13] demonstrated that Atlantic salmon parr avoided a concentration of 2.3 µg Cu/L and 53 µg Zn/L. Although the concentrations avoided in the laboratory were lower than those in the stream, the agreement between laboratory and field results was excellent if one considers the differences in the reproductive motivational states of adults versus parr. In another study, Geckler et al.[16] used stream water in a mobile laboratory to empirically determine a chronic LOEC for bluntnose minnows, *Pimephales notatus,* of 115 µg Cu/L. When 120 µg Cu/L was actually added to the stream itself (Shayler Run, Ohio, from which the mobile laboratory dilution water was taken), most fish moved out of the most heavily contaminated areas, a behavioral response not predicted by the standard laboratory studies. Spawning of bluntnose minnows occurred only in areas where Cu concentrations were below 35 to 77 µg/L. Overall stream fish production was reduced by the avoidance of the area by fish. Though not a field study attempting to validate laboratory avoidance tests, it did point out the ecological significance of avoidance behavior and that avoidance can be determined in the field for nonmigratory species.

More recently, Hartwell et al.[17-19] investigated the avoidance behavior of fish to a variety of metals. They found that, based on laboratory exposures, golden shiners, *Notemigonus crysoleucas,* avoided As, Cu, and Cr but not Cd and Se,[19] pointing out that avoidance is not necessarily a consistent response to all metal contaminants. However, avoidance can have very significant ecological effects when it occurs. Fathead minnows, *Pimephales promelas,* acclimated to a blend of four metals (As, Cu, Cr, and Se) prior to testing were much less sensitive to the presence of metals in an avoidance test than fish not previously exposed,[18] indicating that fish can behaviorally acclimate to metals in the laboratory and presumably in nature. Hartwell et al.[17] compared avoidance responses of fathead minnows to the same blend of metals as mentioned above in an artificial and a natural stream. They used a portable avoidance chamber for testing the response of fathead minnows to metal-blend solutions in the natural stream. They compared these results with their previously determined avoidance thresholds from the laboratory. Again, some fish were acclimated in the metal mixture, and some were acclimated in control water. The authors determined that the predicted results based upon unexposed laboratory fish may well overestimate the sensitivity of fish to metals in the wild. Laboratory tests indicated a lower avoidance threshold than found in the streams. In comparing the results of the artificial and the natural stream, considerable variability in results occurred with test system and season of testing. A large part of this variability was explained by water quality differences, especially for hardness. Laboratory tests with consistently high-quality dilution water and unexposed fish will likely overestimate the responsiveness of fish to metal contamination in the wild.

Korver and Sprague[11] have worked to develop an avoidance testing system that incorporates some relevant motivational factors. They provided reproductively mature fathead minnow males with shelters and allowed them to establish territories. The avoidance threshold was 0.284 mg Zn/L without shelters and 1.83 mg/L with them. Close attention should be paid to the species used and its life history requirements. Fathead minnow females are not territorial, and males are only territorial at certain times of the year, so perhaps single-factor avoidance tests, such as those traditionally run, are more appropriate for nonterritorial situations for that species.

Avoidance tests have ecological significance, as demonstrated by the field studies of Saunders and Sprague[15] and Geckler et al.[16] They can be field validated, and the laboratory testing can be made more ecologically relevant. These tests have been extensively used[1,20] and should continue to see much use in the future, especially if standardization of protocols is completed.

Example 2: Fish Foraging Behavior and Bioenergetics

Growth, along with survival and reproduction, is one of the most important end points currently used to evaluate the effects of toxicants on fish. Fish used in the laboratory toxicity tests are typically fed a high density and limited variety of nonliving food items (e.g., trout pellets, dried flakes, frozen brine shrimp) in an oversimplified structural and biological environment. Consequently, fish are not challenged to locate, pursue, and capture specific prey as they would be under natural conditions. The behavioral aspects of growth are seldom considered in these laboratory tests yet may be the most important components of toxicant effects on growth in the field. Of course, the indirect effects of food availability may also play a major role in fish growth in the field. Sandheinrich and Atchison[10] reviewed the literature on the sublethal effects of toxicants on fish foraging behavior and predator-prey interaction; the toxicants included Al, Cd, Cu, Hg, and Zn, among many others. Sandheinrich and Atchison[10] noted that most of the feeding and predator-prey contaminant literature has taken an empirical approach, essentially relying on observation or experimentation to derive a particular end point. Most of these tests have limited predictive capability and do not lend themselves to the development of readily testable hypotheses of fish behavior in nature. They suggested that by borrowing from both the empirical behavioral toxicology literature and from ecological theory (especially bioenergetics and optimal foraging), a mechanistic approach can be taken that is more predictive and more open to field validation than most empirical studies. Optimal foraging models assume that individuals most efficient in feeding will be most fit, i.e., optimally foraging fish will acquire more energy, grow faster, and produce more offspring than fish foraging suboptimally. These average-rate-maximizing models construct cost-benefit relationships that incorporate trade offs between searching for and handling prey. By using the models, the range of prey sizes eaten and the maximum net energy intake for a given sized fish can be determined. The optimal diet for a predator is that subset of available prey sizes that maximizes the predator's net, long-term rate of energetic intake. The components of the model that must be determined in the laboratory or from the literature are encounter rates with prey types and sizes, handling times for each prey size, costs of handling, and costs of searching. Optimal foraging theory has been extensively investigated.[21-23] Mittlebach[21] utilized the model to predict the diet and net energetic intake of bluegill foraging in the pelagic, littoral, and benthic zones of a lake. Laboratory feeding experiments were performed to determine prey encounter rates and prey handling times for bluegill foraging on different sizes of zooplankton, damselfly naiads, and midge larvae. Other parameters of the model, such as search costs for prey, were measured in the laboratory or obtained from the literature. Predictions of bluegill prey size selection and habitat use, based upon maximization of net energetic gain, were then examined for bluegill in a small Michigan lake. The ambient prey size-frequency distribution was compared with that of the model and the fishes' stomach contents. The model closely predicted the habitat use and subsequent diet of bluegill in the lake and confirmed the utility of this mechanistic approach to the study of fish foraging behavior. Sandheinrich and Atchison[24] applied to this model to the study of the effects of Cu on bluegill foraging. They found that prey handling times were the most sensitive component of the foraging sequence to Cu exposure; handling time increases of 50 to 100% were not unusual and were not equal across all prey size classes tested. Although the model itself has been field validated by Mittlebach,[21] its utility in behavioral toxicology has yet to be field validated. A replicated mesocosm study should be suitable for carrying out such a field validation study.

Sandheinrich and Atchison[10] also proposed the use of bioenergetics models to test the effects of contaminants on fish growth. Bioenergetics models complement foraging models by predicting changes in growth as a function of changes in physiological rates and amount of food consumed. Sandheinrich and Atchison[10] presented a brief review of fish bioenergetics modeling and provided an example of how such models might be used in behavioral toxicology. They constructed a model to simulate biomass (a function of growth) at different feeding rates. By incorporating the literature which describes fish foraging behavior and contaminant impacts on feeding and growth, a predictive model can be built to determine chemical effects on fish and prey in the wild from laboratory data. Similar models have been developed,[25-28] and the concept should have direct applications to toxicology; predictions should be open to field validation.

This is an interdisciplinary and creative use of available information expressed in an ecologically meaningful context. More attempts at developing similar predictive models for contaminant-induced changes in basic functions such as fecundity and year class strength, along with recruitment, should also be possible, utilizing a similar data base (fish management studies and ecological theory) and similar approach (modeling).

POTENTIAL FOR FUTURE APPROACHES

Historically, we have attempted to verify laboratory effects of metals in the field. Perhaps, by reversing the order of events, that is, by verifying field effects under controlled laboratory conditions, an even more realistic union of laboratory-field data could be formed. The following example illustrates such reversal of the order of events by observing changes in the field and then testing mechanisms mediating those changes under controlled, laboratory conditions.

Example 1: Examining the Significance of Field Exposure in the Laboratory

In 1986 Baumann and Gillespie[29] published a paper dealing with Se bioaccumulation in gonads of largemouth bass *(Micropterus salmoides)* and bluegills *(Lepomis machrochirus)* collected from power plant cooling reservoirs high in Se. The experimental approach consisted of the following: (1) field observation and (2) laboratory testing of field-collected fish. The field observations noted declines in population densities of largemouth bass and bluegill in cooling reservoirs receiving ash pond effluent. Also, the suggestion that Se preferentially accumulates in the gonads over the rest of the body was reported in the literature.[30] Building on these two pieces of information, Baumann and Gillespie[29] collected largemouth bass and bluegill from several representative Se-contaminated and uncontaminated cooling reservoirs and analyzed gonads versus whole carcasses for Se concentrations. Note that the experimental design of this field study includes "treatment versus control" as well as "replication" of experimental units (i.e., cooling reservoir). Results showed that whole carcass Se concentrations were not significantly different between sexes but were higher in all fish from cooling reservoirs receiving effluent from ash ponds. However, ovarian Se levels were higher than levels found in testicular tissue and were 1000 times higher in bluegill ovaries than in the water from which fish were taken. The next step taken was to determine the effects of Se bioaccumulation in gonads on reproduction. Gillespie and Baumann[31] did a series of cross-fertilization experiments in the laboratory. Artificial crosses of bluegill with all possible combinations of high and low Se concentrations showed that females high in Se produced larvae that formed edemas which prevented completion of swim-up and lead to death in the late stages of larval development. Controlled laboratory experiments, the design of which considered reports of recruitment failure in the literature and field data on bioaccumulation, enabled Gillespie and Baumnann[31] to elucidate a cause-effect relationship between Se and larval mortality while also providing a plausible explanation for declines in bluegill populations observed in cooling reservoirs receiving ash pond effluent.

Given these scientific developments, the stage was set for reproductive behavioral examination by Pyron and Beitinger,[32] who investigated more subtle Se effects on spawning behavior and fry production. They chose to use a species that can be easily spawned and observed under controlled laboratory conditions and has been well described reproductively, namely, the fathead minnow *(Pimephales promelas)*.

Building on available information, which strongly suggested that Se induced disruptions in reproduction, they examined seven aspects of territoriality, courtship displays, subsequent spawning, hatching success, and larval development. Although edema formation similar to that observed in bluegill larvae by Gillespie and Baumann[31] was observed also in fathead minnows larvae, it was observed at concentrations twice that inducing edemas in bluegill. Furthermore, reproductive behaviors, their form and frequency, were not affected by this level of Se. Perhaps behavioral disruptions would have been noted if the authors had examined spawning and development in bluegill already previously proven to be sensitive to Se.

Despite the insensitivity of fathead minnow reproductive behavior to Se, the reversed order of field to laboratory is still a noteworthy advance in approach. By observing changes in the field and then testing hypotheses in the laboratory, verification becomes more manageagble and rooted in the laboratory. Furthermore, by allowing the "field" to do the contaminant exposures for the investigator, the arguments about single compound, single water quality, fixed concentration, etc. laboratory oversimplifications become moot since the animal has integrated its exposure history under real field conditions.

CONCLUSIONS

Attempts to determine the ecological significance of metal-induced changes in fish have been difficult. After examining the behavioral toxicology literature, issues of experimental design approach and adequate selection of endpoints become increasingly important in the search for ways to determine the ecological significance of contaminant effects.

Empirical studies provide insight about cause-effect thresholds, differential species and behavior sensitivities, and ranges of response to metals. Field studies provide insight for hypothesis development and empirical testing. By preserving the laboratory's controlled environment and the field's realism, the combination is profound.

Interdisciplinary use of related available data bases from other fields can also produce worthwhile results. By employing ecological theory, existing models, and chronic and acute effects information from classical and behavioral toxicity tests, new models to predict contaminant impacts on fish in the wild can be constructed.

Reverse order, that is, field to laboratory verification, may also offer new insight to old questions. Caution must be used in choosing species, life stages, and behaviors observable under laboratory conditions.

More effort needs to be expended to utilize these approaches. By looking at things from a slightly different perspective, we may increase our understanding of the ecological significance of metal effects on fish, especially when mediated by behavioral adaptation.

REFERENCES

1. Atchison, G. J., M. G. Henry, and M. B. Sandheinrich. "Effects of Metals on Fish Behavior: A Review," *Environ. Biol. Fish.* 18:11-25 (1987).
2. Brungs, W. A. and D. I. Mount. "Introduction to a Discussion of the Use of Aquatic Toxicity Tests for Evaluation of the Effects of Toxic Substances," in *Estimating the Hazard of Chemical Substances to Aquatic Life,* J. Cairns, Jr., K. L. Dickson, and A. W. Maki, Eds., ASTM STP 657, (Philadelphia, PA: American Society for Testing and Materials, 1987), pp. 15-26.
3. Cairns, J., Jr. "The Case for Simultaneous Toxicity Testing at Different Levels of Biological Organization," in *Aquatic Toxicology and Hazard Assessment: 6th Symp.* W. E. Bishop, R. D. Cardwell, and B. B. Heidolph, Eds., ASTM STP 802, (Philadelphia, PA: American Society for Testing and Materials, 1983), pp. 111-127.
4. Hendrix, P. F. "Ecological Toxicology: Experimental Analysis of Toxic Substances in Ecosystems," *Environ. Toxicol. Chem.* 1:193-199 (1982).
5. National Research Council. "Testing for Effects of Chemicals on Ecosystems," (Washington, D. C.: National Academy Press, 1981), p. 103.
6. Kimball, K. D. and S. A. Levin. "Limitations of Laboratory Bioassays: The Need for Ecosystem-Level Testing," *BioScience* 35:165-171 (1985).
7. Little, E. E., B. A. Flerov, and N. N. Ruzhinskaya. "Behavioral Approaches in Aquatic Toxicity Investigations: A Review," in *Toxic Substances in the Aquatic Environment: An International Aspect,* P. M. Mehrle, R. H. Gray, and R. L. Kendall, Eds. (Bethesda, MD: Water Quality Section, American Fisheries Society, 1985), pp. 72-98.
8. Rand, G. M. "Behavior," in *Fundamentals of Aquatic Toxicology: Methods and Applications,* G. M. Rand and S. R. Petrocelli, Eds. (New York: Hemisphere Publishing Corp., 1985), pp. 221-263.

9. Westlake, G. F. "Behavioral Effects of Industrial Chemicals on Aquatic Animals," in *Hazard Assessment of Chemicals: Current Developments, Vol. 3,* J. Saxena, Ed. (San Diego: Academic Press, 1984), pp. 233-250.

10. Sandheinrich, M. B. and G. J. Atchison. "Sublethal Toxicant Effects on Fish Foraging Behavior: Empirical vs. Mechanistic Approaches," *Environ. Toxicol. Chem.* 9:107-119 (1990).

11. Korver, R. M. and J. B. Sprague. "Zinc Avoidance by Fathead Minnows *(Pimephales promelas):* Computerized Tracking and Greater Ecological Relevance," *Can. J. Fish. Aquat. Sci.* 46:494-502 (1989).

12. Cairns, J., Jr. "What is Meant by Validation of Predictions Based on Laboratory Toxicity Tests?" *Hydrobiologia* 137:271-278 (1986).

13. Sprague, J. B. "Avoidance of Copper-Zinc Solutions by Young Salmon in the Laboratory," *J. Water Pollut. Control Fed.* 36:990-1004 (1964).

14. Sprague, J. B., P. F. Elson, and R. L. Saunders. "Sublethal Copper-Zinc Pollution in a Salmon River—A Field and Laboratory Study," *Int. J. Air Water Pollut.* 9:531-543 (1965).

15. Saunders, R. L. and J. B. Sprague. "Effects of Copper-Zinc Mining Pollution on a Spawning Migration of Atlantic Salmon," *Water Res.* 1:419-432 (1967).

16. Geckler, J. R., W. B. Horning, T. M. Neiheisel, Q. H. Pickering, E. L. Robinson, and C. E. Stephen. "Validity of Laboratory Tests for Predicting Copper Toxicity in Streams," U.S. Environmental Protection Agency, Ecol. Res. Ser. EPA-600/3-76-116 (1986).

17. Hartwell, S. I., D. S. Cherry, and J. Cairns, Jr. "Field Validation of Avoidance of Elevated Metals by Fathead Minnows *(Pimephales promelas)* Following *In Situ* Acclimation," *Environ. Toxicol. Chem.* 6:189-200 (1987).

18. Hartwell, S. I., D. S. Cherry, and J. Cairns, Jr. "Avoidance Responses of Schooling Fathead Minnows *(Pimephales promelas)* to a Blend of Metals During a 9-Month Exposure," *Environ. Toxicol. Chem.* 6:177-187 (1987b).

19. Hartwell, S. I., D. S. Cherry, and J. Cairns, Jr. "Toxicity Versus Avoidance Response of Golden Shiner, *Notemigonus crysoleucas,* to Five Metals," *J. Fish Biol.* 35:447-456 (1989).

20. Belitginer, T. L. and L. Freeman. "Behavioral Avoidance and Selection Responses of Fishes to Chemicals," *Residue Rev.* 90:35-55 (1983).

21. Mittelbach, G. G. "Foraging Efficiency and Body Size: A Study of Optimal Diet and Habitat Use by Bluegills," *Ecology* 62:1370-1386 (1981).

22. Stephens, D. W. and J. R. Krebs. *Foraging Theory* (Princeton, NJ: Princeton University Press, 1986).

23. Werner, E. E. and D. J. Hall. "Optimal Foraging and the Size Selection of Prey by the Bluegill Sunfish *(Lepomis machrochirus),"* *Ecology* 55:1042-1052 (1974).

24. Sandheinrich, M. B. and G. J. Atchison. "Sublethal Copper Effects on Bluegill *Lepomis machrochirus,* Foraging Behavior," *Can. J. Fish. Aquat. Sci.* 46:1977-1985 (1989).

25. Kitchell, J. F., D. J. Stewart, and D. Weininger. "Applications of a Bioenergetics Model to Yellow Perch *(Perca flavescens)* and Walleye *(Stizostedion vitreum vitreum),"* *J. Fish. Res. Bd. Can.* 34:1922-1935 (1977).

26. Stewart, D. J., D. Weininger, D. V. Rottiers, and T. A. Edsall. "An Energetics Model for Lake Trout, *Salvelinus namaycush:* Application to the Lake Michigan Population," *Can. J. Fish. Aquat. Sci.* 40:681-698 (1983).

27. Rice, J. A., J. E. Breck, S. M. Bartell, and J. F. Kitchell. "Evaluating the Constraints of Temperature, Activity and Consumption on Growth of Largemouth Bass," *Environ. Biol. Fish.* 9:263-275 (1983).

28. Stewart, D. J. and F. P. Binkowski. "Dynamics of Consumption and Food Conversion by Lake Michigan Alewives: An Energetics-Modeling Synthesis," *Trans. Am. Fish. Soc.* 115:643-661 (1986).

29. Baumann, P. C. and R. B. Gillespie. "Selenium Bioaccumulation in Gonads of Largemouth Bass and Bluegills from Three Power Plant Cooling Reservoirs," *Environ. Toxicol. Chem.* 5:695-701 (1986).

30. Cumbie, P. M. and S. L. VanHorn. "Selenium accumulation associated with fish mortality and reproductive failure," in *Proceedings of Annual Conference Southeastern Association of Fish and Wildlife Agencies* 32:612-624 (1978).

31. Gillespie, R. B. and P. C. Baumann. "Effects of High Tissue Concentrations of Selenium on Reproduction by Bluegills," *Trans. Am. Fish. Soc.* 115:208-213 (1986).

32. Pyron, M. and T. L. Beitinger. "Effect of Selenium on Reproductive Behavior and Fry of Fathead Minnows," *Bull. Environ. Contam. Toxicol.* 42:609-613 (1989).

33. Welch, T. J., J. R. Stauffer, Jr., and R. P. Morgan, II. "Temperature Preference After Exposure to Copper in *Tilapia,*" *Bull. Environ. Contam. Toxicol.* 43:761-768 (1989).

34. Kane, D. and C. Rabeni. "Effect of Aluminum and pH on Early Life Stages of Smallmouth Bass," *Water Res.* 21(b):633-639 (1987).

35. Ellgaard, E. G. and J. L. Guillot. "Kinetic Analysis of the Swimming Behaviour of Bluegill Sunfish, *Lepomis machrochirus* Rafinesque, Exposed to Copper: Hypoactivity Induced by Sublethal Concentrations," *J. Fish Biol.* 33:601-608 (1988).

36. Cleveland, L., E. E. Little, S. J. Hamilton, D. R. Buckler, and J. B. Hunn. "Interactive Toxicity of Aluminum and Acidity to Early Life Stages of Brook Trout," *Trans. Am. Fish. Soc.* 115:610-620 (1986).

37. Little, E. E. and S. E. Finger. "Swimming Behavior as an Indicator of Sublethal Toxicity in Fish," *Environ. Toxicol. Chem.* 9:13-19 (1990).

38. Walker, R. L., C. M. Wood, and H. L. Bergman. "Effects of Low pH and Aluminum on Ventilation in the Brook Trout *(Salvelinus fontinalis),*" *Can. J. Fish. Aquat. Sci.* 45:1614-1622 (1988).

39. Diamond, J. M., M. J. Parson, and D. Gruber. "Rapid Detection of Sublethal Toxicity Using Fish Ventilator Behavior," *Environ. Toxicol. Chem.* 9:3-11 (1990).

40. Borgmann, U. and K. M. Ralph. "Effects of Cadmium, 2,4, Dichlorophenol, and Pentachlorophenol on Feeding, Growth, and Particle-Size Conversion Efficiency of White Sucker Larvae and Young Common Shiners," *Arch. Environ. Contam. Toxicol.* 15:473-480 (1986).

41. Gunn, J. M. and D. L. G. Noakes. "Latent Effects of Pulse Exposure to Aluminum and Low pH on Size, Ionic Composition, and Feeding Efficiency of Lake Trout *(Salvelinus namaycush)* Alevins," *Can. J. Fish. Aquat. Sci.* 44:1418-1424 (1987).

6

The Developmental Toxicity of Metals and Metalloids in Fish

Peddrick Weis and Judith S. Weis

Department of Anatomy, New Jersey Medical School, University of Medicine and Dentistry of New Jersey, Newark, New Jersey 07103
and
Department of Biological Sciences, Rutgers University, Newark, New Jersey 07102

OVERVIEW

The early life stages of an animal's life cycle are the most sensitive to environmental contaminants. Responses of fish embryos and larvae to many metals and metalloids include mortality, as well as morphological, physiological, and biochemical alterations. As teratogens, metals are fairly nonspecific, since the same abnormalities can be elicited by a number of different chemicals. Common morphological abnormalities include optic, skeletal, and cardiovascular malformations. Mechanisms of teratogenicity include disruption of mitosis, interference with transcription and translation, and disturbances in energy utilization. These can interfere with cell interactions, migration, and growth. Physiological responses include decreased motility and precocious or delayed hatching. Biochemical effects are reflected in reduced or inhibited enzymatic activity and reduced growth. Larval studies have usually been restricted to observations of mortality. Salmonids tend to be among the most sensitive species, responding to some metals at, or less than, 1 µg/L. The chorion protects the fish embryo from some metals but enhances toxicity to others by acting as an ion exchanger. In fish, metal interactions with salinity, pH, and temperature have been noted, as well as interactions between toxicants. Interactions between metals and/or metalloids can be additive or antagonistic.

Most reports of teratogenesis have been qualitative; quantitation has usually been in the form of percentages of affected embryos. If severity of effects is indexed, more critical analyses can be achieved.

INTRODUCTION

Developmental toxicity is a broad term, encompassing abnormal development (teratogenicity), mortality, growth inhibition, and functional deficiencies during early life stages. Embryonic and larval stages are often the most sensitive periods in the life cycle of aquatic organisms, responding to relatively subtle chemical and physical changes in the environment. Therefore, environmental toxicologists have found it expedient to replace whole life cycle tests or tests using adults with early life stage tests.[1] Typical data gathered in such tests include such parameters as total hatching success and survival and growth of the larvae.[2] These are useful parameters because they are easy to screen in large-scale operations and do not require special expertise. The U.S. Environmental Protection Agency (EPA) includes two such fish embryo-larval survival/teratogenicity tests among their protocols.[3] However, these tests, while utilizing embryos, do not provide insight into specific teratogenic or biochemical effects of the chemicals tested. Early life stage tests were designed not as tests for developmental toxicants in particular, but as general toxicity screening methods.

There are, however, toxicants that exert their effects particularly on developing systems, and regulatory agencies are concerned about pollutants that might cause birth defects. The EPA has issued guidelines for risk assessment for suspected developmental toxicants.[4] The term "developmental toxicity" is somewhat broader than "teratology," encompassing embryotoxicity, altered growth, and functional deficiency in the offspring (which may not be apparent until after birth), in addition to structural abnormalities. The use of fish embryos for screening potential developmental toxicants has been recommended[5] because of several unique advantages: the transparent chorions allow continuous monitoring, large numbers of specimens can be processed at minimal cost, the developmental time of some species is only a few days, and the use of "lower" vertebrates is less unpalatable to the animal welfare movement. On the other hand, the absence of a placenta makes correlations to human fetal health more remote.

There are many studies of reproductive and developmental effects of environmental pollutants on aquatic biota. In addition to studies of the nature and degree of abnormalities produced, there are attempts to analyze mechanisms of action of the teratogens. The effects studied include not only embryonic malformations but also effects on developmental processes that occur at later life stages, such as hatching and larval growth.

When considering the developmental toxicity of metals and metalloids, attention must be given to the fact that some of these elements are essential micronutrients, and their toxicity thresholds tend to be two or three orders of magnitude greater than those of nonessential elements. That is, while embryos of a given species may be sensitive to nanomolar concentrations of such nonessential metals as Cd and Hg, micromolar concentrations are necessary to elicit a toxic response to Cu and Zn. Another facet of metal toxicity to be considered is the speciation of the metal. For example, Cr(VI) is substantially more toxic than Cr(III) and is the more prevalent species in seawater.[6] A third facet in metal toxicity is alkylation. Organometals are frequently much more toxic than their inorganic counterparts. This is generally because the alkyl group substantially reduces solubility and the resultant hydrophobic (and thus lipophilic) molecule penetrates cell membranes (lipid bilayers), and thus gill and gastrointestinal epithelia, more readily.

Typically, experimental embryos are exposed to metals after fertilization. However, embryos in the natural environment can be exposed to pollutants in two additional ways: via yolk, which is synthesized during oogenesis by exposed females, and during the brief period (at most, a few minutes[7]) between shedding of the gametes and elevation of the chorion. A few studies have shown that metals incorporated into the egg during oogenesis can produce malformations in the embryos that subsequently develop from those eggs (Zn [ref 8], Hg [refs. 9, 10]). Once the chorion is elevated and hardened, it can act as a barrier to partially protect the developing embryo from the toxic effects of the pollutant.[11] However, there have been a few reports which have shown that dechorionated embryos were actually less susceptible to the toxic effects than embryos with an intact chorion.[12,13] This has been explained for metal ions by Rombough[14] in relation to the Donnan equilibrium: those cations with negative standard electrode potentials (e.g., Zn^{2+}, Cd^{2+}, and Pb^{2+}) can readily penetrate the chorion (which acts as an ion exchanger) and are concentrated in the perivitelline fluid. Thus, embryos with a chorion are more susceptible to these ions than those without. The reverse is true of cations with positive standard electrode potentials (e.g., Hg^{2+}, Cu^{2+}, Ag^{2+}). These bind to the chorion, which thus acts as a barrier. This is reinforced by the findings of Shazili and Pascoe,[15] who found larvae to be more susceptible to Cu than embryos of the rainbow trout, while Cd and Zn were more toxic to early embryonic stages than to larvae. There was a particularly sensitive stage in early development. However, during later stages, the chorion did offer substantial protection against Cd and Zn.

MECHANISMS OF EFFECTS OF DEVELOPMENTAL TOXICANTS
Teratogenic Effects

It is an axiom of teratology that a defect in morphogenesis of a structure can occur if a teratogenic insult is present during the "critical stage" of development of that structure, the critical stage being the sensitive period of tissue differentiation and rapid growth. Many different substances can produce similar deformities if present at the appropriate time, even though the actual modes of action of the different chemicals may differ. Teratogens thus tend to be fairly nonspecific in the nature of the defects that they cause.[16] General developmental mechanisms that can lead to abnormal development include abnormal cell or tissue differentiation, excessive or inadequate cell death during development, inadequate cell migration, improper cellular communication, and disrupted metabolism (respiration, absorption, excretion, or secretion).

Fish embryos, in general, tend to become abnormal in certain ways. The most sensitive system appears to be the developing skeletal system, and flexures (scoliosis, lordosis), as well as stunting, are seen in many species treated with a variety of teratogens. Another common set of abnormalities involves the developing circulatory system. Circulatory stasis, a failure of the heart tube to bend, and edema of the pericardial cavity are also commonly observed defects. The developing optic system is also very sensitive, and many investigators have observed optic malformations, such as microphthalmia and anophthalmia, as well as cyclopia and intermediate conditions of fusion of the two optic vesicles (synophthalmia). Again, these tend to be nonspecific responses; the appearance of a certain syndrome in a certain species of fish cannot, in general, be used to diagnose the presence of a specific teratogen.

Although not strictly a developmental anomaly, another phenomenon often observed in teratogen-exposed embryos is a general retardation of development. This decrease in developmental rate, sometimes seen as an arrest of development, may permit teratogens to act for longer than normal times during critical stages, and thus intensify the severity of the anomalies produced. Developmental arrest in itself will not cause abnormalities if recovery occurs and normal development resumes. Laale and McCallion[17] found that homogenates of zebrafish embryos caused intact embryos to arrest their development at stage 17-18. These embryos showed no mitotic figures. Upon return to freshwater, the embryos resumed normal development. Therefore, it is not the developmental arrest itself, but the continuous exposure to a teratogen at that critical time that causes the anomalies once development resumes. Conversely, environmental or inherent factors that enable embryos to develop more rapidly, spending less time at critical stages, can make them less susceptible to teratogenic influences.[18,19]

Physiological and Biochemical Effects

Many heavy metals have been shown to act specifically by inhibiting certain enzymes, thus interfering with metabolic processes in development. Divalent

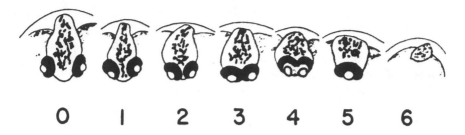

$$0 \quad | \quad 2 \quad 3 \quad 4 \quad 5 \quad 6$$

FIGURE 1. Diagram of heads of 1-week-old mummichog *(Fundulus heteroclitus)* embryos, showing the range of responses when treated with methylmercuric chloride at 50 μg/L; the various degrees of craniofacial abnormality are arranged in order of severity to demonstrate the craniofacial index (CFI). 0, normal; 3, synophthalmia; 5, cyclopia; 6, anophthalmia. Reproduced from Weis et al.[35] with permission of the American Society for Testing and Materials.

cations can cross-link sulfhydryl groups and thereby disrupt (denature) the tertiary structure of enzymes, rendering them less active or inactive. Such activities may result in teratogenic responses, but sometimes an organism exhibits less dramatic but, nevertheless, serious consequences, especially if the exposure occurs outside the critical stages. Such sequellae include diminished or retarded growth, inhibited, delayed, or precocious hatching, or outright mortality.

DEVELOPMENTAL TOXICITY OF SELECTED METALS
A. Mercury

Abnormalities in the development of the optic cups have been observed by many investigators. Dial[20] observed disorganized retinas, abnormal pigment distribution, and invasive blood sinuses in eyes of medakas *(Oryzias latipes)* treated with 80 μg/L methylmercury. There was also evidence of mitotic arrest. Many investigators, starting with Stockard,[21] have noted a defect in forebrain development in which the eye rudiments converge, sometimes to the point of cyclopia. The cyclopic eye noted by Weis and Weis[22,23] in $HgCl_2$ and methylmercury-treated mummichog *(Fundulus heteroclitus)* embryos exhibits an unusually large optic cup and lens (Figure 1), since it is the result of the fusion of the two separate rudiments. The mechanism underlying the fusion of the optic vesicles is believed to be an inadequate induction of the forebrain, which then allows the convergence of the two vesicles in the anterior midline of the embryo. This anomaly, therefore, is not strictly an optic one — rather, its genesis involves a defect in craniofacial development. The organic Hg was effective at one third the equivalently effective concentration of inorganic Hg ($HgCl_2$) in this species.

Fundulus is also susceptible to cardiovascular anomalies, ranging from a slight inhibition of morphogenesis of the heart, through a contracting but undifferentiated simple tube, to complete failure of cardiovascular development. Both organic and inorganic Hg compounds will bring about these cardiovascular anomalies in *F. heteroclitus*,[22,23] as does mercuric chloride in *O. latipes*.[24] In the latter species, Dial[25] noted failure of the incorporation of the heart into the body after 5 days exposure to 40 μg/L methylmercury. Other abnormalities included stunting, reduced pectoral fins, and reduced pigmentation. Those embryos that hatched exhibited erratic swimming, indicating neurological effects.

Skeletal anomalies, namely, flexures and stunting, have been observed in *F. heteroclitus* treated with Hg compounds,[22,23,30] (Figure 2) and *O. latipes* treated with methylmercury.[25] Birge et al.[9] noted immobile jaws, partial twinning, and rigid coiling in treated rainbow trout (formerly *Salmo gairdneri*, now *Oncorhynchus mykiss*). They noted that terata are a much more common response in some species than others, being commonly seen in treated rainbow trout, but much less often seen in treated channel catfish *(Ictalurus punctatus)*, goldfish *(Carassius auratus)*, or largemouth bass *(Micropterus salmoides)*. These latter species more commonly responded with mortality rather than with abnormalities. In their 1983 study of six metals and metalloids, Birge et al.[26] exposed rainbow trout embryos and found that Hg (both organic and inorganic) was the most toxic and teratogenic. The LC_{50} was 5 μg/L, and both mortality and terata were elevated at as little as 1 μg/L of either organic or inorganic Hg. This is an example, sometimes seen in freshwater fish, of the inorganic form of Hg being as toxic as the organic form. Also, this report demonstrates the extreme sensitivity of salmonid species to teratogens. Furthermore, the partial twinning seen in rainbow trout embryos was elicited only by Hg — an example of a metal-specific teratogenic response.

Weis et al.[27] found differential responses to mercuric chloride by embryos in two populations of *Fundulus*. One population responded to 50 μg/L with a moderate degree of teratogenic responses, primarily the craniofacial, cardiovascular, and skeletal abnormalities described above. The other population (which was from a highly polluted environment and had acquired increased resistance to methylmercury-induced teratogenesis) responded to mercuric chloride exposure with high mortality but not a high incidence of abnormalities.

Exposure of adults to Hg has been investigated in relation to the developmental effects on offspring. Brook trout *(Salvelinus fontinalis)* transferred Hg from adult tissues to developing oocytes, resulting in some malformations.[10] Birge et al.[9] treated adult rainbow trout with 0.2 and 0.7 μg/L Hg and raised fertilized eggs from these chronically exposed fish in clean water. Larval survival at 4 days posthatching was less than 50%, and many of the larvae were abnormal. Mortality rates were higher for the embryonic stages than for the larvae. A significant reduction in reproductive success (i.e., increased mortality and abnormalities) was also noted for adults chronically exposed to 0.2 μg/L Hg. When exposed fish were crossed with control fish to see whether sperm or eggs were selectively affected, it was found that both sperm and eggs were affected in exposed fish, with the sperm being affected to a greater degree. The authors felt their data indicated that the freshwater criterion of 0.05 μg/L Hg may not adequately protect reproduction in salmonid species. Embryos produced by adult zebrafish exposed to and spawned in water containing 0.1 μg/L phenylmercuric acetate had decreased hatching.[28] The mummichog *Fundulus heteroclitus* appears to be able to protect its embryos from Hg — the egg Hg was only a small fraction of the liver Hg in each female analyzed.[19] Apparently, transport of vitellogenin, a

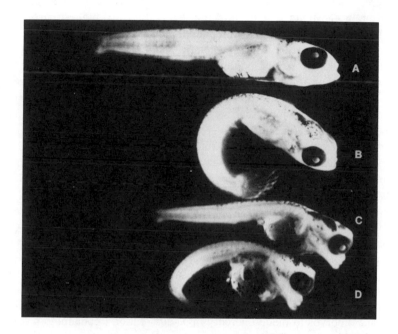

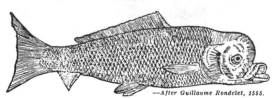

—*After Guillaume Rondelet, 1555.*
RONDELET'S PUG-NOSED CARP.
THE FIRST PUBLISHED FIGURE OF THIS ABNORMALITY.

FIGURE 2. One-day posthatch mummichog larvae, at approximately 13x. (A), control; (B-D), experimental. Such inability to uncurl after hatching has been elicited by Pb and by methylmercury. Specimens (C) and (D) also demonstrate synophthalmia and a tubular snout; these are reminiscent of (E), the earliest known publication of a fish abnormality.[92] Reproduced from Weis and Weis[23] with permission of Wiley-Liss, Inc. 2E is a 1555 illustration reproduced in Gudger.[92]

lipoprotein, from the liver to the ovary does not carry a significant amount of the lipid-soluble methylmercury with it.

Critical periods in development were considered in several studies with Hg. Weis and Weis[23] reported that gastrulation is the critical period in *Fundulus* for the genesis of craniofacial defects by methylmercury. This corresponds to the time of induction of the forebrain, defects in which are believed responsible for the convergence of the optic cups. Akiyama[29] identified early cleavage and the time of brain and optic vesicle formation as the times of highest susceptibility of *O. latipes* to two organic Hg compounds, phenylmercuric acetate and methoxyethylmercuric chloride. Sharp and Neff[30] found that the duration of exposure to mercuric chloride was important in the genesis of spinal curvature in *F. heteroclitus*. Using 96-h survival, total hatching success, and percentage incidence of spinal curvature as parameters of effect, these investigators reported no critical period. Hatching success of chronically exposed embryos was significantly reduced at concentrations greater than 10 μg/L Hg.

Perry et al.[31] observed decreased mitotic index and increased percentage of abnormal mitoses in methylmercury-treated embryos of *F. heteroclitus*. Embryos with more severe teratogenic responses also exhibited more severe mutagenic responses.

In a study of biochemical effects of methylmercury exposure in *S. fontinalis* embryos, Christensen[32] noted a decrease in glutamic-oxalacetic transaminase after exposure to 1.03 μg/L. Alevins showed decreased weight and increased enzyme level.

Varying environmental conditions such as salinity after the susceptibility of embryos to toxic effects of pollutants. Decreased salinity increased the toxicity of methylmercury to *F. heteroclitus*.[18] However, Sakaizumi[33] found that exposure of medaka embryos to methylmercury in a salt solution (Yamamoto's solution) reduced hatching to a greater extent than comparable exposure in deionized water. When various inorganic salts were tested, only halides were found to reduce hatching, while nitrate and sulfate had no effect. At the same time, these halides (Cl and Br) were found to reduce the uptake of mercuric chloride and methylmercuric chloride into the yolk of the embryos, while uptake into the embryo proper was increased, thus accounting for the intensified effects.

Quantification of Effects

Most reports of teratogenic effects present the data in a qualitative way, describe the defects, and give a percentage of embryos affected. However, in comparing effects of different chemicals or responses of different populations of the same species, it is useful to have a more quantitative approach to the effects. Since many malformations can be more or less severe, it is possible to devise indices to rank embryos in terms of how severely they are affected. Anderson and Battle,[34] for example, rated chloramphenicol-treated zebra fish embryos on a three-part scale. More refined indices of craniofacial, cardiovas-

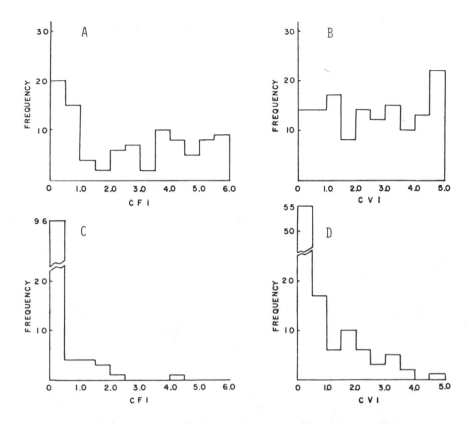

FIGURE 3. Craniofacial and cardiovascular indices for methylmercury-treated mummichog embryos. Frequency (Y-axis) represents the number of females whose eggs have that mean response. The distributions of malformations in a reference population are shown in (A) and (B) and for a polluted (and apparently resistant) population in (C) and (D). Reproduced from Weis et al.[36] with the permission of Springer-Verlag (Heidelberg).

cular, and skeletal anomalies produced by methylmercury have been devised[23,25] which can give a quantitative estimate of the severity of the defect (Figure 1). These values allow for much more detailed and precise analysis than simply reporting the percentages of embryos affected.

Using indices described above, we have noted striking population differences in susceptibility of *Fundulus* embryos to teratogenic effects of methylmercury. Embryos from a polluted environment were much less affected than those from a more pristine environment (Figure 3).[36] Possible mechanisms for the increased resistance are a more rapid development time and a less permeabile chorion.[19]

Another possible mechanism that has been investigated is the possibility that metal-binding proteins, e.g., metallothionein, which protect against toxic effects of metals by sequestering them, might be found in higher amounts in the embryos from the polluted population. However, significant amounts of this protein were not found until late developmental stages, well past the period of genesis of the malformations.[37]

B. Cadmium

Cd, like Hg, is nonessential, and its effects on fish development have been widely reported. Effects of Cd on embryos and larvae of seven freshwater fish species (*Salvelinus fontinalis, Salmo trutta, Salvelinus namaycush, Oncorynchus kisutch, Esox lucius, Catostomus commersoni,* and *Micropterus dolomieui*) were studied by Eaton et al.[38] Larvae were consistently more sensitive than the embryos and showed diminished growth. Of the embryos, only those of the white sucker *(C. commersoni)* were not able to tolerate the highest test concentration (100 μg/L). The reduced growth which is often observed in larvae at hatching may be due to decreased enzyme activity, which has been noted in Cd-exposed embryos. Mounib et al.[39] found that Pacific herring eggs exposed to 10 mg/L Cd showed decreases in NAD and NADP-malic enzymes, propionyl-CoA carboxylase, and phosphoenolpyruvate carboxykinase, enzymes that are involved in biosynthetic processes.

Cd at 239 μg/L produced pericardial edema, delayed yolk resorption, microcephalia, and skeletal malformations in the bluegill.[40] Skeletal malformations were seen in medaka embryos treated with 10 and 50 μg/L Cd.[41]

Birge et al.[42] developed a short-term embryo-larval test in which exposure began immediately after fertilization and continued for 4 days after hatching. For the fathead minnow, the LC_{50} for Cd was about 80 μg/L. Newly hatched larvae frequently showed developmental defects, primarily skeletal. Most of these larvae died during the 4-day larval exposure period, supporting the authors' defining of gross malformations as mortalities in probit calculations of LC_{50} and LC_1 (threshold) values. The LC_1's calculated for Cd were fairly similar to maximum acceptable toxicant concentrations (MATC) from the published literature for the particular species tested. In addition to the fathead minnow, largemouth bass, channel catfish, carp, and goldfish were studied.

Cd treatment of the rainbow trout, *O. mykiss,*[43] at 10 and 100 μg/L produced classical cardiac malformations, premature hatching, retarded growth, and increased rates of mortality. In this study, exposures started at 19 days after fertilization, so early developmental effects could not have been detected. *Oncorhynchus mykiss*[43] and the minnow, *Phoxinus phoxinus,*[44] treated with Cd also developed typical skeletal deformities. Rombough and Garside[45] studied effects in the Atlantic salmon, *Salmo salar,* and found that, although the eggs take up Cd rapidly, the alevin growth was much more sensitive, being affected at 0.47 μg/L, while the LC_{50} for embryos was 300 to 800 μg/L. Gastrulation, axiation, the time of development of vitelline circulation, and the period just before hatch were the most sensitive embryonic stages. High concentrations produced skeletal malformations. The embryonic toxicity appeared to be greater at 5° than 10°C.

This may have been due to a prolongation of sensitive stages by the reduced developmental rate at the lower temperature. However, increased temperature increased the toxicity of Cd to embryos of the fathead minnow[42] in a short-term embryo-larval test. Cd was also more toxic at 10° than at 5°C to winter flounder embryos *(Pseudopleuronectes americanus.)*[46] These latter reports suggest a metabolic effect that differs from the greater teratogenic effects usually seen in other systems at reduced temperatures.

Rosenthal and Sperling[47] found accelerated hatching of herring *(Clupea harengus)* eggs at Cd concentrations of 100 µg/L, which they attributed to reduced embryonic movement preventing the normal distribution of hatching enzyme, causing it to concentrate in the head region where the egg capsule then ruptures. They also found that short-term Cd exposure changed the properties of the chorion. Westernhagen et al.[48] reported that pectoral fin movements, responsible for circulating the perivitelline fluid, were reduced in Cd-exposed (5 mg/L) embryos of the garpike, *Belone belone*. Pectoral fin abnormalities were also noted.

Dethlefsen et al.[49] noted that larvae of the garpike hatched after incubation in Cd exhibited vertebral flexure, and reduced activity and swimming ability. Muramoto[50] attributed Cd-induced vertebral column damage in the carp, *Cyprinus carpio,* to a decrease in Ca and P in the bones which weakened them and made them susceptible to curvature by muscle action. Westernhagen et al.[51] observed rapid "shivering" muscular movements in Cd-treated embryos of herring.

Some studies have noted that a very small percentage of the Cd actually reaches the embryo. Beattie and Pascoe[52] found that, in rainbow trout embryos, 98% of the Cd was associated with the chorion, rather than with the embryo itself or the yolk. Peterson et al.[53] found that 90% of the Cd uptake by salmon embryos was associated with the chorion. Similarly, Michibata[54] found that over 94% of the Cd in medakas was found in the chorion, and Rosenthal and Sperling[47] found over 60% of the Cd in herring eggs was in the chorion. These observations are difficult to resolve with the chorionic function described by Rombough and Garside,[45] who reported that dechorionated *Salmo* embryos were more resistant to Cd and took up less of the metal than did embryos with intact chorions; they concluded that the chorion acted to concentrate the Cd in the embryonic system.

Decreased salinity increased the toxicity of Cd to winter flounder embryos[46] as measured by viable hatch. The enhanced toxicity at lower salinity may be due to greater uptake of water, and, therefore, of the toxicants. The optimum salinity for resisting Cd effects was 20 to 25 parts per thousand (‰), a salinity typically found in this species' spawning grounds. Similarly, reduced salinity intensified the effects of Cd on herring embryos *(Clupea harengus).*[51] Prehatching mortality and malformation rates (skeletal bending, stunting, reduction in otic capsule and otolith size, optic size), and behavioral abnormalities after hatching were highest at low salinities. The chorion was softened in the Cd-treated embryos. The optimal salinity was 25‰ and Cd uptake was greater at lower salinities. At these lower salinities, more of the Cd is available as free ions (the toxic form) and less would be complexed. A comparable situation in freshwater systems occurs with hardness, whereby increased water hardness reduces the uptake of the Cd and thus reduces its toxicity to *O. latipes* embryos.[55] As noted earlier, this investigator also found that relatively little Cd penetrated the chorion. More recently, Michibata et al.[56] analyzed the effect of water hardness by studying the effects of Ca and Mg separately in Cd-exposed medaka eggs. Ca, but not Mg, was able to reduce the acute embryotoxicity by 80%. However, the reduction in Cd uptake brought about by Ca and Mg was similar. The authors concluded that, since 94% of the bound Cd was associated with the chorion, very small changes in total uptake could account for the significant changes in toxicity to the embryo proper.

Similarly, pH had a major influence on Cd uptake and toxicity in Atlantic salmon *(Salmo salar)*.[53] At lower pH levels, more Cd was available for embryonic uptake due to lowered chorionic binding. On the other hand, Dave[57] studied effects of Cd and pH on embryos and larvae of *B. rerio* and found that Cd was more toxic at higher pH levels. Survival was not impaired for the first 3 days of Cd exposure up to 4.1 mg/L at any pH, but, after hatching, survival was reduced at concentrations as low as 30 µg/L, with the most toxic pH being 6 to 7 and the least toxic being 9.0. It should be noted that, in this study, embryonic exposures did not start until the optic cup stage, past the most sensitive embryonic stages.

C. Tin

As the metal, Sn is relatively nonreactive and nontoxic; the chloride is a strong reducing agent which readily oxidizes to an insoluble material. Organotin compounds, however, are extremely toxic and are used as biocides. The trialkyltins are the most toxic. Tributyltin (TBT) has been studied for effects on aquatic organisms because it has been incorporated into antifouling paints (see Chapter 12).

TBT was used in larval toxicity tests by Bushong et al.[58] The inland silversides *(Menidia beryllina)* larval 96-h LC_{50} value was 3.0 µg/L. Treatment of adult sheepshead minnows *(Cyprinodon variegatus)* with as little as 0.24 µg/L caused their F_1 posthatch larvae to have excess mortality.[59] Rainbow trout fry had reduced growth and hemoglobin at as little as 0.2 µg/L.[60] Embryos of the mummichog *(F. heteroclitus)* exhibited batch-to-batch variability in response to TBT. One batch had teratological effects, including eye and skeletal defects, at 30 µg/L (Figure 4); others had no terata but high mortality in as little as 3 µg/L, while some batches exhibited only delayed hatching in 3 to 30 µg/L.[61] Medaka embryos were exposed to TBT by direct topical application to the egg.[62] This involved subnanogram amounts in the solvent dimethylsulfoxide. Delayed hatching and vertebral abnormalities were observed at doses as low as 0.02 ng/egg;

this was calculated to be equivalent to 25.4 ng/L. This is the lowest reported effective dose of any metal compound studied to date.

D. Copper

Cu is a required micronutrient; nevertheless, many species of fish are sensitive to its toxic effects at relatively low concentrations. Herring *(Clupea harengus)* eggs had high mortality when incubated in 30 μg/L Cu.[63] Those larvae that hatched in 30 μg/L were deformed. If exposure was initiated 4 days after fertilization, deformities were not apparent, indicating that the early embryonic stages are the most sensitive. Cu exposure tended to accelerate the hatching rate. Larvae were 30 times less susceptible than the embryos.

McKim et al.[64] studied effects of Cu on the embryos and larvae of eight species

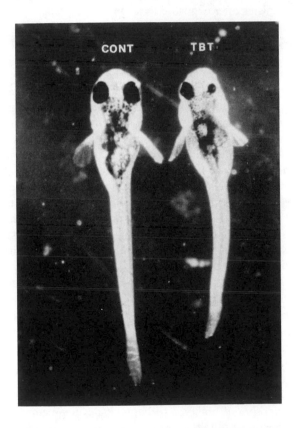

FIGURE 4. Mummichog larvae, 1 day posthatch, of which one had been treated with 30 μg/L tributyltin (TBT) oxide, showing unilateral microphthalmia, reduced growth, and slight skeletal malformation.

of freshwater fish (*Catostomus commersoni, Coregonus artedi, Salvelinus fontinalis, S. namaycush, Salmo trutta, Oncorhynchus mykiss, Esox lucius,* and *Micropterus dolomieui.*) Larvae and juveniles were all more sensitive than embryos. Embryo survival was affected only at the higher concentrations tested (northern pike at 500 μg/L, rainbow trout at 37 μg/L, white sucker at 333 μg/L, brook trout at 555 μg/L, lake trout at 555 μg/L, brown trout at 111 μg/L, and herring at 555 μg/L). No effect was seen on smallmouth bass embryos at any concentration tested. However, embryonic exposure reduced the larval resistance to Cu; brown trout that were exposed for the longest embryonic period were the most sensitive as larvae. Benoit[65] found that in the bluegill, larvae were also more sensitive than embryos; 77 μg/L had no effect on the embryos but did reduce larval survival. He also observed that larvae at 12 μg/L survived better than the controls, which may be an example of hormesis (overcompensation by regulating mechanisms after exposure to a low concentration of a toxicant).

Cosson and Martin[66] investigated the effects of Cu on embryos and larvae of *Dicentrarchus labrax.* Hatching rate was reduced at levels of Cu as low as 5 μg/L. Unlike some of the other species, embryos were more sensitive than larvae, alevins, or juveniles. Rice et al.[67] also found that embryos of the northern anchovy, *Engraulis mordax,* were more sensitive than larvae and indicated that, in general, larvae of freshwater fish are more sensitive than embryos, while embryos of marine fish are more sensitive than their larvae. Salmon, living in the ocean but spawning in freshwater, behave like saltwater fish in this regard.

Engel and Sunda[68] compared the sensitivities of embryos of the spot, *Leiostomus xanthurus,* which has planktonic eggs, and the silversides, *Menidia menidia,* which has demersal eggs, to Cu concentration, measured as the free ion (pCu) in an ion-buffer system. They found the silversides were more sensitive in terms of hatching success, which was surprising, since they had considered planktonic eggs to be more delicate than demersal eggs. For the silversides, the greatest toxicity occurred at the time of hatch; the greatest mortality of spot took place prior to hatch.

Cu concentrations of 10 μg/L and higher increased mortality of embryos, decreased hatching frequency, prolonged incubation time, and produced abnormalities in developing cod *(Gadus morrhua)* embryos.[69] Among the abnormalities noted at 10 μg/L were bent axis and cyclopia. Cu toxicity was reduced by the surfactant linear alkylbenzene sulfonate (LAS), a chemical that, when used alone, had about the same level of toxicity as the Cu.

Perhaps the least sensitive fish embryos studied to date are those of the common carp. Kaur and Virk[70] found reduced hatching and various skeletal abnormalities resulted from exposure of embryos throughout development. The 50% effect and incipient effect concentrations were 4 and 1.5 mg/L, respectively. Those concentrations are nominal, however, since no indications are given for type of application, rate of renewal of test solution, or of water quality. These data are especially relevant for Cu toxicity studies because Cu is more likely than most metal ions to form complexes that render it less biologically available; thus, a metal-buffering system in which the free ion concentration is controlled[68] is particularly useful for Cu.

E. Zinc

Zn is another required micronutrient which can be embryotoxic above naturally occurring concentrations. Bengtsson[71] found that Zn exposure caused vertebral damage in the minnow, *Phoxinus phoxinus*. In the zebra fish, *Brachydanio rerio*, Dave et al.[72] reported that the no-effect for hatching was 500 μg/L. Exposure to 1 and 2 mg/L Zn caused damage to the vertebral centra and deflection of the caudal peduncle in medaka.[42] Exposure to as little as 0.05 mg/L caused various craniofacial and skeletal malformations in herring embryos.[73]

Holcombe et al.[74] found that exposure to 1.4 mg/L Zn decreased embryo survival and chorionic strength in the brook trout. The latter effect may be ecologically important in this species since the eggs are normally buried in gravel. The effect was greater in eggs that were transferred to Zn prior to chorionic hardening. Similarly, Benoit and Holcombe[75] found that 145 μg/L Zn exposure of newly fertilized fathead minnow eggs caused a decrease in egg adhesiveness and an increase in their fragility. At 295 μg/L, skeletal deformities were observed, and hatchability and larval survival were diminished.

No deleterious effects of Zn exposure of 240 μg/L were seen in egg to smolt stages of the sockeye salmon, *Oncorynchus nerka*. Rather, a slight stimulation of growth was noted.[76] This may be another example of hormesis. Speranza et al.[8] observed malformations in offspring of *B. rerio* adults previously exposed to Zn. Thus, Zn accumulated in the adult tissues could be transferred to the eggs during oogenesis, causing the developmental defects.

Zn concentrations of 10 mg/L reduced survival, and 5 mg/L increased the frequency of bent larvae in the cod, *G. morrhus*.[69] The relatively minor effects of Zn were attributed to its being primarily bound to the chorion, thus not reaching the embryo. At 0.5 mg/L, survival of eggs was slightly better than the control. This may have been due to a reduction in fungi and bacteria or could be an example of hormesis.

Embryonic exposure to zinc increased larval Zn resistance in the flagfish, *Jordanella floridae*.[77] Preexposure to a low concentration of the metal thus resulted in enhanced tolerance. Induction of metal-binding proteins, such as metallothionein, is one mechanism by which this can occur.

DEVELOPMENTAL TOXICITY OF SELECTED METALLOIDS
A. Boron

Boron is not known to have any nutritive value. It does not exist in nature as the free element but as borates (i.e., borax). Elemental B is relatively nontoxic, but its compounds do bioaccumulate to toxic levels.

Boron, as either boric acid or borax, was found to be teratogenic to the rainbow trout at levels as low as 1 μg/L, although the LC_{50} was 79 mg/L. This is a much greater difference than is usually reported. The deformities produced were in the skeletal system and consisted of moderate vertebral bends and extreme rigid coiling.[26]

B. Arsenic

Arsenic has micronutrient value but is toxic above trace levels. It is also carcinogenic, mutagenic, and teratogenic. It exists in four oxidation states, of which two, As(III) and As(V), are prevalent. Of these, As(V) is the more prevalent and less toxic. Arsenic is associated with certain metallic ores, so it can be an environmental problem near smelters. Fortunately, bioconcentration factors are relatively low.[78]

The studies of As effects on fish embryos that have come to our attention include two on rainbow trout embryos. In one, a 28-day LC_{50} of 0.54 mg/L and no teratogenic response were reported.[79] In the other, Birge et al.[26] found skeletal malformations at concentrations as low as 0.25 mg/L. In the latter study, the use of As(III) was specified.

C. Selenium

Selenium, a required micronutrient, is of special interest to toxicologists because of its antagonistic interactions with some toxic heavy metals, especially Hg, as already discussed. It is toxic by itself. When bluegills *(Lepomis macrochirus)* were exposed to 30 ppm dietary Se (as Se-methionine or as selenite, the predominant organic and inorganic forms), the F_1 larvae were edematous and had excess skeletal anomalies and subsequent mortality. The organic Se was much more toxic than the inorganic Se, producing 100% mortality and 100% terata vs 25 and 15%, respectively.[80] Bluegills taken from a Se-contaminated reservoir had offspring with similar defects.[81] Fathead minnows did not show reproductive effects at concentration (20 and 30 ppm Se as selenite) in the diet that decreased growth of juveniles of that species.[82]

In a study in which Se (as selenate) was added to rearing water, rainbow trout embryos showed mortality at concentrations as low as 10 μg/L. Teratogenesis was not obvious below 1 mg/L.[26]

METAL-METAL AND METAL-METALLOID INTERACTIONS

In most polluted environments, multiple pollutants would be expected to be found, rather than one single chemical. Thus, interactions between and among metals-metalloids would be expected. Some of these possible interactions have been studied in developing fish embryos. Many of such studies with metals and metalloids have shown antagonisms, in that the toxicity of one metal was reduced by the presence of the other. Zn and Cd reduced the teratogenic effects of methylmercury on *F. heteroclitus,*[34] and Se reduced the toxicity of Hg to medaka embryos.[83] In the latter study, anomalies produced by mercuric chloride included lack of hemoglobin and blood circulation, abnormal heart development, and malformations in the tail region. The Se, which had no effect by itself, did not reduce the Hg effects until after the embryos' livers had developed, suggesting an important role of that organ in Se protection against Hg toxicity. Klaverkamp et al.[84] studied the interactions of mercuric chloride and Se on salmonid (*Oncorhynchus mykiss* and *Salvelinus namaycush*) eggs and found that up to 10 mg/L Se had no effect on Hg toxicity, but that 100 mg/L Se did have a protective effect in lake trout eggs. Conversely, Huckabee and Griffith[85] found a marked synergistic effect between Hg and Se in carp eggs. The failure of Se to protect here may have been due to the early stage of the embryos (pre-liver development) at the time of treatment. Birge et al.[9] found that Hg-Cu and Hg-Se combinations were additive to antagonistic at lower exposure levels to eggs of channel catfish, goldfish, largemouth bass, and rainbow trout, while higher concentrations were synergistic. These results may explain some of the discrepancies in the literature. Another explanation may relate to the form of Hg. Baatrup and Danscher[86] demonstrated that Se will complex with methylmercury but not $HgCl_2$ in trout liver. Nearly all (> 90%) of Hg in fish is methylated because of the relative mobility of the alkylated form, as well as an *in vivo* methylating process,[87] although the *in vivo* methylating process is not universal.[88]

Interactions of metal ions can also reduce the toxicity of Cd to fish embryos by reducing the bioavailability of the free Cd ion. The presence of another divalent cation can trigger competition, thus lowering the amount of the Cd affecting the embryo. Silver reduced the toxicity of Cd to winter flounder embryos.[89] Similarly, the presence of Zn increased the viable hatch of herring eggs in 5.0 mg/L Cd from 0 to 11.4%.[48]

DISCUSSION

There is great variability in the papers in this field, partly because different investigators are using fish embryos for different purposes. Some are using them in screening tests for aquatic toxicity assessment in order to arrive at maximum acceptable toxicant concentrations and the like. These studies generally use end points of hatching success and may include early larval survival and growth as components of an "embryo-larval test." In some cases, the LC_{50} is the only end point of interest. Analysis of the developmental abnormalities is generally not included in such studies, although the fact that abnormal embryos were present is sometimes mentioned in passing. It is true, from the perspective of the population, that hatching success and larval growth and survival are important end points. It is also true that, for a number of toxicants, the larvae are more sensitive than the embryos.

Other investigators are looking for nonmammalian "alternatives" in testing for teratogenic effects of chemicals.[90] These studies generally involve greater analysis of the developmental abnormalities, which are "validated" when parallels can be drawn between the data from the fish embryos and responses of mammalian embryos. There is a general, though not absolute, correlation between substances that are teratogenic in fishes and those that are teratogenic in humans. Even among fish, substances can vary greatly in their effects on different species. Fish embryo-larval stages can serve as effective models for the investigation of teratogenesis in the screening of chemicals that might pose a threat to human development.

Data from the many studies reported here show that these are excellent systems for the study of developmental and reproductive toxicology. Large numbers of embryos can be obtained from many fish species, enabling better data analysis. The transparency of the fish embryo enables the investigator to examine it at frequent intervals, make detailed examinations on the structure and functioning of the living embryo, and follow the progress of a developing abnormality.

Among the most commonly observed end points in toxicity testing for effects of pollutants on fish embryos is successful hatch. Many factors could account for an embryo failing to hatch, embryonic death being a common cause, but grossly abnormal embryos also fail to hatch. In addition, there are embryos that often appear morphologically normal but do not hatch. These three conditions represent a progressive decrease in severity of effect of the toxicant. Studies that differentiate among these different states are more informative than those that merely report percentage hatch.

The sensitive periods in embryonic development are often around the time of gastrulation when the primary inductive events occur. Another important period is the time in which the liver develops. This organ can synthesize metallothioneins to protect the organism against deleterious effects of certain metals and can be responsible for enhanced tolerance seen after preexposure to low levels of metals.

It is sometimes useful to have a more quantitative approach to teratogenic effects than simply reporting the percentage of affected embryos. Since many malformations can vary in severity, it is possible to devise indices to rank embryos in terms of how severely they are affected. This allows for much more detailed and precise analysis. Using such indices, striking population differences in susceptibility of *Fundulus* embryos to teratogenic effects of methylmercury were noted.[37] Possible mechanisms for the increased resistance are a more rapid development time and a less permeable chorion.[29] Danzmann et al.[91] have found that heterozygotes of a strain of rainbow trout tend to develop more rapidly than homozygotes. This provides a way for heterozygotes, which also have greater developmental stability as measured by bilateral symmetry of meristic traits, to be more resistant to teratogens. Faster developmental rates are expected to "decrease the probability of accidents during critical periods of development, resulting in a more stable or uniform phenotype."

In mammalian teratology, a field much more intensively studied than fish teratology, it has become clear that functional deficiencies are a more subtle response to toxicants than the production of gross anatomical abnormalities. A considerable amount of attention is now being devoted to the study of "behavioral teratology," i.e., abnormal behavior studied after birth of anatomically normal animals which were exposed to teratogens during their embryonic development. Other than occasional reports of lethargic larvae, there has been little or no attention paid to this aspect of developmental responses of fish embryos to environmental pollutants. It would seem to be an area ripe for study, particularly since there have been numerous studies on direct effects of toxicants on behavior of fishes, including feeding, learning, schooling, and breeding behavior. It is likely that embryonic exposure to low concentrations of certain pollutants will result in effects on behavior that would be seen in later life history stages.

REFERENCES

1. McKim, J. M. "Evaluation of Tests with the Early Life Stages of Fish for Predicting Long Term Toxicity," *J. Fish. Res. Bd. Can.* 36:1148-1154 (1977).
2. Birge, W. J., J. A. Black, J. E. Hudson, and D. M. Bruser. "Embryo-larval Toxicity Tests with Organic Compounds," in *Aquatic Toxicology*, L. Marking and R. Kimmerle, Eds., ASTM S.T.P. 667 (Philadelphia, PA: American Society for Testing and Materials, 1979), pp. 131-147.
3. U.S. Environmental Protection Agency. "Short Term Methods for Estimating the Chronic Toxicity of Effluents and Receiving Waters to Marine and Estuarine Organisms," EPA/600/4-87/028, (Cincinnati OH 1987), 417 pp.
4. U.S. Environmental Protection Agency. "Proposed Guidelines for Health Assessment of Suspect Developmental Toxicants," *Fed. Reg.* 49(227):46324-46331 (1984).
5. Weis, J. S. and P. Weis. "Pollutants as Developmental Toxicants in Aquatic Organisms," *Environ. Health Perspect.* 1:77-85 (1987).
6. Towill, L. E., C. R. Shriner, J. S. Drury, A. S. Hammons, and J. W. Holleman. "Reviews of the Environmental Effects of Pollutants: III. Chromium," U.S. Environmental Protection Agency. Rep. 600/1-78-023 (Washington DC, 1978), 287 pp.
7. Brummett, A. R. and J. N. Dumont. "Cortical Vesicle Breakdown in Fertilized Eggs of *Fundulus heteroclitus*," *J. Exp. Zool.* 216:63-79 (1981).
8. Speranza, A. E., R. J. Seeley, V. A. Secley, and A. Perlmutter. "The Effects of Sublethal Concentrations of Zinc on Reproduction in the Zebrafish, *Brachydanio rerio* (Hamilton-Buchanan)," *Environ. Pollut.* 12:217-222 (1977).
9. Birge, W. J., J. A. Black, A. G. Westerman, and J. E. Hudson. "The Effects of Mercury on Reproduction of Fish and Amphibians," in *The Biogeochemistry of Mercury in the Environment*, J. Nriagu, Ed. (North Holland, Amsterdam: Elsevier/North Holland Biomedical Press, 1979), pp. 629-655.
10. McKim, J. M., G. F. Olson, G. W. Holcombe, and E. P. Hunt. "Long Term Effects of Methylmercuric Chloride on Three Generations of Brook Trout *(Salvelinus fontinalis)*: Toxicity, Accumulation, Distribution and Elimination," *J. Fish. Res. Bd. Can.* 33:2726-2739 (1976).
11. Ozoh, P. T. "Effects of Reversible Incubations of Zebrafish Eggs in Copper and Lead Ions with or without Shell Membranes," *Bull. Environ. Contam. Toxicol.* 24:270-275 (1980).

12. Skidmore, J. F. "Resistance to Zinc Sulphate on Zebrafish, *Brachydanio rerio*, Embryos after Removal or Rupture of the Outer Egg Membrane," *J. Fish. Res. Bd. Can.* 23:1037-1041 (1966).

13. Rombough, P. J. and E. T. Garside. "The Influence of the Zona Radiata on the Toxicity and Uptake of Cadmium in Embryos of the Atlantic Salmon *(Salmo salar)*," *Can. J. Zool.* 61:2338-2343 (1983).

14. Rombough, P. J. "The Influence of the Zona Radiata on the Toxicities of Zinc, Lead, Mercury, Copper, and Silver Ions to Embryos of Steelhead Trout *Salmo gairdneri*," *Comp. Biochem. Physiol.* 82C:115-117 (1985).

15. Shazili, N. A. M. and D. Pascoe. "Variable Sensitivity of Rainbow Trout *(Salmo gairdneri)* Eggs and Alevins to Heavy Metals," *Bull. Environ. Contam. Toxicol.* 36:468-474 (1986).

16. Laale, H. W. "Teratology and Early Fish Development," *Am. Zool.* 21:517-533 (1981).

17. Laale, H. W. and D. J. McCallion. "Reversible Developmental Arrest in the Embryo of the Zebrafish, *Brachydanio rerio*," *J. Exp. Zool.* 167:117-128 (1968).

18. Weis, J. S., P. Weis, and J. Ricci. "Effects of Cadmium, Zinc, Salinity, and Temperature on the Teratogenicity of Methylmercury to the Killifish, *Fundulus heteroclitus*," *Rapp. Proces. Verb. Reunions Cons. Perman. Int. Explor. Mer.* 178:64-70 (1981).

19. Toppin, S. V., M. Heber, J. S. Weis, and P. Weis. "Changes in Reproductive Biology and Life History of *Fundulus heteroclitus* in a Polluted Environment," in *Pollution Physiology of Marine Organisms*, W. Vernberg, A. Calabrese, F. Thurberg, and F. J. Vernberg, Eds. (Columbia, SC: University of South Carolina Press, 1985), pp. 171-184.

20. Dial, N. A. "Some Effects of Methylmercury on Development of the Eye in Medaka Fish," *Growth* 42:309-318 (1978).

21. Stockard, C. R. "The Artificial Production of a Single Median Cyclopean Eye in the Fish Embryo by Means of Sea Water Solutions of Magnesium Chloride," *Wilhelm Roux Arch. Entwickslungsmech. Organismen* 23:249-258 (1907).

22. Weis, J. S. and P. Weis. "The Effects of Heavy Metals on Embryonic Development of the Killifish, *Fundulus heteroclitus*," *J. Fish Biol.* 11:49-54 (1977).

23. Weis, P. and J. S. Weis. "Methylmercury Teratogenesis in the Killifish, *Fundulus heteroclitus*," *Teratology* 16:317-327 (1977).

24. Heisinger, J. F. and W. Green. "Mercuric Chloride Uptake by Eggs of the Ricefish and Resulting Teratogenic Effects," *Bull. Environ. Contam. Toxicol.* 14:665-673 (1975).

25. Dial, N. A. "Methylmercury: Some Effects on Embryogenesis in the Japanese Medaka, *Oryzias latipes*," *Teratology* 17:83-92 (1978).

26. Birge, W. J., J. A. Black, A. G. Westerman, and B. A. Ramey. "Fish and Amphibian Embryos - A Model System for Evaluating Teratogenicity," *Fundam. Appl. Toxicol.* 3:237-242 (1983).

27. Weis, J. S., P. Weis, M. Heber, and S. Vaidya. "Investigations into Mechanisms of Heavy Metal Tolerance in Killifish *(Fundulus heteroclitus)* Embryos," in *Physiological Mechanisms of Marine Pollutant Toxicity*, W. B. Vernberg, A. Calabrese, F. P. Thurberg, and F. J. Vernberg, Eds. (San Diego: Academic Press, 1982), pp. 311-330.

28. Kihlstrom, J. E., C. Lundberg, and L. Hulth. "Number of Eggs and Young Produced by Zebrafish *(Brachydanio rerio* Hamilton-Buchanan) Spawning in Water Containing Small Amounts of Phenylmercuric Acetate," *Environ. Res.* 4:355-359 (1971).

29. Akiyama, A. "Acute toxicity of Two Organic Mercury Compounds to the Teleost, *Oryzias latipes,* in Different Stages of Development," *Bull. Jpn. Soc. Sci. Fish.* 36:563 (1970).

30. Sharp, J. R. and J. M. Neff. "Effects of the Duration of Exposure to Mercuric Chloride on the Embryogenesis of the Estuarine Teleost, *Fundulus heteroclitus,*" *Mar. Environ. Res.* 3:195-213 (1980).

31. Perry, D., J. S. Weis, and P. Weis. "Cytogenic Effects of Methylmercury in Embryos of the Killifish, *Fundulus heteroclitus,*" *Arch. Environ. Contam. Toxicol.* 17:569-574 (1988).

32. Christensen, G. M. "Biochemical Effects of Methylmercury Chloride, Cadmium Chloride, and Lead Nitrate on Embryos and Alevins of the Brook Trout, *Salvelinus fontinalis,*" *Toxicol. Appl. Pharmacol.* 32:191-197 (1975).

33. Sakaizumi, M. "Effect of Inorganic Salts on Mercury-compound Toxicity to the Embryos of the Medaka, *Oryzias latipes,*" *J. Fac. Sci. Univ. Tokyo* 14:369-384 (1980).

34. Anderson, P. D. and H. I. Battle. "Effects of Chloramphenicol on the Development of the Zebrafish, *Brachydanio rerio,*" *Can. J. Zool.* 45:191-204 (1967).

35. Weis, J. S., P. Weis, and M. Heber. "Variation in Response to Methylmercury by Killifish *(Fundulus heteroclitus)* Embryos," in *Aquatic Toxicology and Hazard Assessment:* 5th Conf., J. G. Pearson, R. Foster, and W. Bishop, Eds., ASTM, S.T.P. 766. (Philadelphia, PA: American Society for Testing and Materials, 1982), pp. 109-119.

36. Weis, J. S., P. Weis, M. Heber, and S. Vaidya. "Methylmercury Tolerance of Killifish *(Fundulus heteroclitus)* Embryos from a Polluted vs Non-polluted Environment," *Mar. Biol.* 65:283-287 (1981).

37. Weis, P. "Metallothioneins and Methylmercury Tolerance in *Fundulus heteroclitus,*" *Mar. Environ. Res.* 14:153-166 (1984).

38. Eaton, J. G., J. M. McKim, and G. W. Holcombe. "Metal Toxicity to Embryos and Larvae of Seven Freshwater Fish Species. I. Cadmium," *Bull. Environ. Contam. Toxicol.* 19:95-103 (1978).

39. Mounib, M. S., H. Rosenthal, and J. S. Eisan. "Some Effects of Cadmium on the Metabolism of Developing Eggs of Pacific Herring," I.C.E.S. Paper 1975/E:20, *Rapp. Proces. Verb. Reunions Cons. Perma. Int. Explor. Mer.* Copenhagen (1975), 6 pp.

40. Eaton, J. G. "Chronic Cadmium Toxicity to the Bluegill *(Lepomis macrochirus* Rafinesque)," *Trans. Am. Fish. Soc.* 103:729-735 (1974).

41. Hiraoka, Y. and H. Okuda. "Characteristics of Vertebral Abnormalities of *Medaka* as a Water Pollution Indicator," *Hiroshima J. Med. Sci.* 32:261-266 (1983).

42. Birge, W. J., J. A. Black, and A. G. Westerman. "Short-term Fish and Amphibian Embryo-Larval Tests for Determining the Effects of Toxicant Stress on Early Life Stages and Estimating Chronic Values for Single Compounds and Complex Effluents," *Environ. Toxicol. Chem.* 4:807-821 (1985).

43. Woodworth, J. and D. Pascoe. "Cadmium Toxicity to Rainbow Trout, *Salmo gairdneri* Richardson: a Study of Eggs and Alevins," *J. Fish Biol.* 21:47-57 (1982).

44. Bengtsson, B.-E., C. H. Carlin, A. Larsson, and O. Svanberg. "Vertebral Damage in the Minnow *Phoxinus phoxinus* L. Exposed to Cadmium," *Ambio* 4:166-168 (1975).

45. Rombough, P. J. and E. T. Garside. "Cadmium Toxicity and Accumulation in Eggs and Alevins of the Atlantic Salmon *Salmo salar,*" *Can. J. Zool.* 60:2006-2014 (1982).

46. Voyer, R. A., C. E. Wentworth, E. P. Barry, and R. J. Hennekey. "Viability of Embryos of the Winter Flounder *Pseudopleuronectes americanus* Exposed to Combinations of Cadmium and Salinity at Selected Temperatures," *Mar. Biol.* 44:117-124 (1977).

47. Rosenthal, H. and K.-R. Sperling. "Effects of Cadmium on Development and Survival of Herring Eggs," in *Early Life History of Fish,* J. H. Blaxter, Ed. (Berlin: Springer-Verlag, 1974), pp. 383-396.

48. Westernhagen, H. von, V. Dethlefsen, and H. Rosenthal. "Combined Effects of Cadmium and Salinity on Development and Survival of Garpike Eggs," *Helgolaender Wiss. Meeresunters.* 27:268-282 (1975).

49. Dethlefsen, V., H. von Westernhagen, and H. Rosenthal. "Cadmium Uptake by Marine Fish Larvae," *Helgolaender Wiss. Meeresunters.* 27:396-407 (1975).

50. Muramoto, S. "Vertebral Column Damage and Decrease of Calcium Concentration in Fish Exposed Experimentally to Cadmium," *Environ. Pollut.* (A)24:125-133 (1981).

51. Westernhagen, H. von, H. Rosenthal, and K.-R. Sperling. "Combined Effects of Cadmium and Salinity on Development and Survival of Herring Eggs," *Helgolaender Wiss. Meeresunters.* 26:416-433 (1974).

52. Beattie, J. H. and D. Pascoe. "Cadmium Uptake by Rainbow Trout, *Salmo gairdneri,* Eggs and Alevins," *J. Fish Biol.* 13:631-637 (1978).

53. Peterson, R. H., J. L. Metcalfe, and S. Ray. "Uptake of Cadmium by Eggs and Alevins of Atlantic Salmon *(Salmo salar)* as Influenced By Acidic Conditions," *Bull. Environ. Contam. Toxicol.* 34:359-368 (1985).

54. Michibata, H. "Uptake and Distribution of Cadmium in the Egg of the Teleost *Oryzias latipes,*" *J. Fish Biol.* 19:691-696 (1981).

55. Michibata, H. "Effects of Water Hardness on the Toxicity of Cadmium to the Egg of the Teleost *Oryzias latipes,*" *Bull. Environ. Contam. Toxicol.* 27:187-192 (1981).

56. Michibata, H., S. Sahara, and M. K. Kojima. "Effects of Calcium and Magnesium Ions on the Toxicity of Cadmium to the Egg of the Teleost, *Oryzias latipes,*" *Environ. Res.* 40:110-144 (1986).

57. Dave, G. "The Influence of pH on the Toxicity of Aluminum, Cadmium, and Iron to Eggs and Larvae of the Zebrafish, *Brachydanio rerio,*" *Ecotoxicol. Environ. Safety* 10:253-257 (1985).

58. Bushong, S. J., W. S. Hall, W. E. Johnson, and L. W. Hall, Jr. "Toxicity of Tributyltin to Selected Chesapeake Bay Biota," *Oceans '87 Proc. Int. Organotin Symp.* 4:1499-1503 (1987).

59. Ward, G. S., G. C. Cramm, P. R. Parrish, H. Trachman, and A. Slesinger. "Bioaccumulation and Chronic Toxicity of Bis(tributyltin) Oxide (TBTO): Tests With A Saltwater Fish," in *Aquatic Toxicology and Hazard Assessment: 4th Conf.,* D. R. Bronson and K. L. Dickson, Eds., ASTM S.T.P. 737 (Philadelphia, PA: American Society of Testing and Materials, 1981), pp. 183-200.

60. Seinen, W., R. Helder, H. Vernij, A. Penninks, and P. Leeuwangh. "Short-term Toxicity of Tributyltin Chloride in Rainbow Trout (*Salmo gairdneri,* Richardson) Yolk Sac Fry," *Sci. Total Environ.* 19:155-166 (1981).

61. Weis, J. S., P. Weis, and F. Wang. "Developmental Effects of Tributyltin on the Fiddler Crab, *Uca pugilator,* and the killifish, *Fundulus heteroclitus,*" Oceans '87 Proc. Int. Organotin Symp. 4:1456-1460 (1987).

62. Helmstetter, M. F. and R. W. Alden, III. "The Effects of Tributyltin on the Eggs of the Japanese Medaka *(Oryzias latipes),*" Paper presented at Atlantic Estuarine Research Federation, Gloucester Point, VA, May 1990.

63. Blaxter, J. H. S. "The Effect of Copper on the Eggs and Larvae of Plaice and Herring," *J. Mar. Biol. Assoc. U.K.* 57:849-858 (1977).

64. McKim, J. M., J. G. Eaton, and G. W. Holcombe. "Metal Toxicity to Embryos and Larvae of Eight Species of Freshwater Fish. II. Copper," *Bull. Environ. Contam. Toxicol.* 19:608-616 (1978).

65. Benoit, D. A. "Chronic Effects of Copper on Survival, Growth, and Reproduction of the Bluegill *(Lepomis macrochirus),*" *Trans. Am. Fish. Soc.* 104:353-358 (1975).

66. Cosson, R. and J. L. Martin. "The Effects of Copper on Embryonic Development, Larvae, Alevins, and Juveniles of *Dicentrarchus labrax (L),*" *Rapp. Proces. Verb. Reunions Cons. Perman. Int. Explor. Mer.* 178:71-75 (1981).

67. Rice, D. W., F. L. Harrison, and J. Jearla. "Effects of Copper on Early Life History Stages in Northern Anchovy, *Engraulis mordax,*" *Fish. Bull.* 78:675-683 (1980).

68. Engel, D. W. and W. G. Sunda. "Toxicity of Cupric Ion to Eggs of the Spot *Leiostomus xanthurus* and Atlantic Silverside, *Menidia menidia,*" *Mar. Biol.* 50:121-126 (1979).

69. Swedmark, M. and A. Granmo. "Effects of Mixtures of Heavy Metals and a Surfactant on the Development of Cod *(Gadus morhua),*" *Rapp. Proces. Verb. Reunions Cons. Perman. Int. Explor. Mer.* 178:95-103 (1981).

70. Kaur, K. and S. Virk. "Toxic Effects of Copper Sulphate Residues in Water on the Development of the Eggs of the Common Carp: *Cyprinus carpio* Linn.," *Ind. J. Ecol.* 7:294-297 (1980).

71. Bengtsson, B.-E. "Vertebral Damage in the Minnow *Phoxinus phoxinus* Exposed to Zinc," *Oikos* 25:134-139 (1974).

72. Dave, G., B. Damgaard, M. Grande, J. E. Martelin, B. Rosander, and T. Viktor. "Ring Test of an Embryo-larval Toxicity Test with Zebrafish *(Brachydanio rerio)* Using Chromium and Zinc as Toxicants," *Environ. Toxicol. Chem.* 6:61-71 (1987).

73. Somasundaram, B., P. E. King, and S. E. Shackley. "Some Morphological Effects of Zinc upon the Yolk Sac Larvae of *Clupea harengus* L.," *J. Fish Biol.* 25:333-343 (1984).

74. Holcombe, G., D. Benoit, and E. Leonard. "Long Term Effects of Zinc Exposure on Brook Trout *(Salvelinus fontinalis),*" *Trans. Am. Fish. Soc.* 108:76-87 (1979).

75. Benoit, D. A. and G. W. Holcombe. "Toxic Effects of Zinc on Fathead Minnows, *Pimephales promelas,* in Soft Water," *J. Fish Biol.* 13:701-708 (1978).

76. Chapman, G. "Effects of Continuous Zinc Exposure on Sockeye Salmon during Adult-to-Smolt Freshwater Residency," *Trans. Am. Fish. Soc.* 107:828-836 (1978).

77. Spehar, R. L., E. N. Leonard, and D. L. DeFoe. "Chronic Effects of Cadmium and Zinc Mixtures on Flagfish *(Jordanella floridae),*" *Trans. Am. Fish. Soc.* 107:354-360 (1978).

78. Eisler, R. "Arsenic Hazards to Fish, Wildlife, and Invertebrates: A Synoptic Review." U.S. Fish and Wildlife Service Biol. Rep. 85 (1.12) (1988), 92 pp.

79. U.S. Environmental Protection Agency. "Ambient Water Quality Criteria for Arsenic." U.S.EPA Rep. 440/5-80-021 (1980), 205 pp.

80. Woock, S. E., W. R. Garrett, W. E. Partin, and W. T. Bryson. "Decreased Survival and Teratogenesis During Laboratory Selenium Exposure to Bluegill, *Lepomis macrochirus,*" *Bull. Environ. Contam. Toxicol.* 39:998-1005 (1987).

81. Gillespie, R. B. and P. Bauman. "Effects of High Tissue Concentrations of Selenium on Reproduction by Bluegills," *Trans. Am. Fish. Soc.* 115:208 (1986).

82. Ogle, R. S. and A. W. Knight. "Effects of Elevated Foodborne Selenium on Growth and Reproduction of the Fathead Minnow *(Pimephales promelas),*" *Arch. Environ. Contam. Toxicol.* 18:795-803 (1989).

83. Bowers, M. A., D. Dostal, and J. F. Heisinger. "Failure of Selenite to Protect Against Mercuric Chloride in Early Developmental Stages of the Japanese Ricefish *(Oryzias latipes),*" *Comp. Biochem. Physiol.* 66C:175-178 (1980).

84. Klaverkamp, J. F., W. A. MacDonald, W. R. Lillie, and A. Lutz. "Joint Toxicity of Mercury and Selenium in Salmonid Eggs," *Arch. Environ. Contam. Toxicol.* 12:415-419 (1983).

85. Huckabee, J. W. and N. A. Griffith. "Toxicity of Mercury and Selenium to the Eggs of Carp *(Cyprinus carpio),*" *Trans. Am. Fish. Soc.* 103:822-825 (1974).

86. Baatrup, E. and G. Danscher. "Cytochemical Demonstration of Mercury Deposits in Trout Liver and Kidney Following Methylmercury Intoxication: Differentiation of Two Mercury Pools by Selenium," *Ecotoxicol. Environ. Safety* 14:129-141 (1987).

87. Westöö, G. "Determination of Methylmercury Compounds in Foodstuffs. II. Determination of Methylmercury in Fish, Egg, Meat, and Liver," *Acta Chem. Scand.* 21:1790-1806 (1967).

88. Huckabee, J. W., S. A. Janzen, B. G. Blaylock, Y. Talmi, and J. J. Beauchamp. "Methylated Mercury in Brook Trout *(Salvelinus fontinalis):* Absence of an *In Vivo* Methylating Process," *Trans. Am. Fish. Soc.* 107:848-852 (1985).

89. Voyer, R. A., J. A. Cardin, J. F. Heltshe, and G. L. Hoffman. "Viability of Embryos of the Winter Flounder, *Pseudopleuronectes americanus,* Exposed to Mixtures of Cadmium and Silver in Combination with Selected Fixed Salinities," *Aquat. Toxicol.* 2:223-233 (1982).

90. Goss, L. B. and T. D. Sabourin. "Utilization of Alternative Species for Toxicity Testing: an Overview," *J. Appl. Toxicol.* 5:193-219 (1985).

91. Danzmann, R. G., M. M. Ferguson, F. W. Allendorf, and K. Knudson. "Evolution and Developmental Rate in a Strain of Rainbow Trout *(Salmo gairdneri)*," *Evolution* 40:86-93 (1986).

92. Gudger, E. W. "Beginnings of Fish Teratology, 1555-1642. Belon, Rondelet, Gesner and Aldrovandi, the Fathers of Ichthyology, the First to Figure Abnormal Fishes," *Sci. Monthly* 43:252-261 (1936).

7

Stochastic Models of Bioaccumulation

James H. Matis,[1] Thomas H. Miller,[1] and David M. Allen[2]

[1]Laboratory for Biological Systems Modelling, Department of Statistics,
Texas A&M University, College Station, Texas 77843
and
[2]Department of Statistics, University of Kentucky,
Lexington, Kentucky 40506

OVERVIEW

Many useful models for the uptake and depuration of pollutants are based on the concept of homogeneous compartments with deterministic flows. Such compartmental models may be described mathematically by systems of differential equations. Recently corresponding stochastic compartmental models based on Markov process theory have also been proposed, and new tools such as mean residence times derived. This chapter reviews the theoretical basis and practical application of both the deterministic and the Markov process models. The stochastic model is also generalized to include a semi-Markov process formulation, which introduces time-varying transit rates. The statistical analysis of the stochastic models is addressed, and issues relating to the parameter estimation and hypothesis testing are illustrated with data from the literature on the bioaccumulation of pollutants.

INTRODUCTION

This chapter develops a family of mathematical models to be used for describing the bioaccumulation of metals in animals, and illustrates the use of such models with data on Hg accumulation in fish. The chapter shares the philosophy of Box that "all models are wrong, but some are useful;"[1] however the restatement that "all models are incomplete, yet some are indispensable" seems more appropriate for present purposes. The overall objective is to provide succinct, rigorous mathematical structures which may be instrumental in summarizing current knowledge and which may be helpful in testing pertinent scientific hypotheses related to bioaccumulation.

Models are sometimes broadly classified as either empirical or mechanistic. Most of the models previously reported for describing the bioaccumulation of metals in small animals have been somewhere between the two extremes; however, they are usually mostly empirical, with only a limited mechanistic basis. A particular objective of this chapter is to provide alternative models which are more mechanistic than most previously used, and which, therefore, would help to better explain the accumulation process itself rather than merely predicting the response. This more mechanistic structure is introduced through the linear compartmental model framework, which is being successfully utilized for many other kinetic modeling problems.[2,3]

The data used to illustrate these models come from a paper by Newman and Doubet on Hg accumulation in mosquitofish.[4] The article outlines a basic mechanistic structure relating to uptake and elimination, and their assumptions yield a mathematical model which fits the data adequately. However, these data will be reanalyzed with the primary objective of identifying more mechanistic structure rather than just finding better fitting models. In the context of Hg kinetics in fish, the identification of such structure would be helpful in developing more detailed physiological models, analogous to the toxicokinetic models of Hg in humans.[5]

The next section reviews the basic one-compartment model with uptake and elimination. The basic assumptions for the model are presented, after which the model is derived and fitted to the Hg accumulation data. The stochastic one-compartment model is then introduced, and the results are reinterpreted in the context of this new model formulation. A new parameter, the mean residence time, is developed from the stochastic model. Several multicompartment models are investigated, using first the deterministic and then the stochastic formulations. A family of stochastic models based on generalized compartments with gamma distribution retention times is also outlined. All of these models are fitted to the Hg accumulation data, and four particular compartment models are found to be useful for the present data. The results from these four different models are then compared in order to help identify the mechanistic structure. Some final comments address the utility of compartmental models for describing and explaining bioaccumulation data from other experimental designs.

ONE-COMPARTMENT DETERMINISTIC MODEL
Model Derivation
The specific objective of the bioaccumulation models for the given experiments is to describe the change in Hg concentration in mosquito fish of various sizes exposed to a Hg radionuclide in water over a 6-day period. The experimental details are given in Newman and Doubet[4] and discussed in Chapter 4. The following notation is helpful:

1. $X(t)$ = amount (μg) of Hg in a particular fish at (elapsed time) t
2. V = size (gm dry wt) of fish
3. $C(t)$ = $X(t)/V$ = concentration (μg/gm dry wt) of Hg in a fish at t
4. $U(t)$ = rate of Hg uptake (μg/day) by fish at t
5. $E(t)$ = rate of Hg elimination (μg/day) by fish at t
6. $\dot{X}(t)$, $\dot{C}(t)$ = derivatives, or rates of change, of $X(t)$ and $C(t)$
7. $X(\infty)$, $C(\infty)$ = equilibrium values of $X(t)$ and $C(t)$

One can write a model for the change in Hg accumulation using the "mass balance" equation

$$\dot{X}(t) = U(t) - E(t) \tag{1}$$

Consider now the following two simplifying assumptions:

A1. The uptake rate is time-invariant, i.e., $U(t) = U$
A2. The elimination rate is a constant fraction of the amount of Hg in the fish, i.e., $E(t) = cX(t)$

This suggests the following additional notation:

8. k_e = $E(t)/X(t)$ = (fractional) elimination rate constant (day^{-1}) of Hg from fish
9. k_u = U/V = proportional uptake rate (μg/gm dry wt/day) of Hg by fish

These assumptions and definitions may be used to rewrite Equation (1) as the following differential equation:

$$\dot{X}(t) = U - k_e X(t) \tag{2}$$

Upon dividing both sides by V, one has

$$\dot{C}(t) = k_u - k_e C(t) \tag{3}$$

FIGURE 1.   Schematic of Model
1, a one-compart-
ment model.

The solution to Equation (3), assuming no initial contamination, i.e. $C(0) = 0$,
is

$$C(t) = (k_u/k_e)(1 - e^{-k_e t}) \tag{4}$$

Clearly, the equilibrium concentration under this model is

$$C(\infty) = k_u/k_e \tag{5}$$

This constant uptake/constant fractional elimination model has been widely used,
and indeed is discussed and fitted to data by Newman and Doubet.[4] The model
has only two parameters, k_u and k_e, and hence is relatively easy to fit to data.
The present assumptions (A1 and A2) of linear time-invariant kinetics are represented by the one-compartment schematic in Figure 1.

Data Analysis

The predicted concentration curve in Equation (4) was fitted separately to the
observed Hg accumulation data for each of the 21 fish in the experiment. The
following additional notation is useful:

10. $c(t)$ = observed concentration of Hg in fish at t
11. $\epsilon(t)$ = random error of observed concentration
12.   RSS = residual (observed-fitted value) sum of squares
13.   MSE = residual (or "error") mean square

The statistical regression model for the observed concentration is

$$c(t) = C(t) + \epsilon(t)$$
$$= (k_u/k_e)(1 - e^{-k_e t}) + \epsilon(t) \tag{6}$$

where the random errors, $\epsilon(t)$, are assumed to be independent and identically distributed normal random variables.

The regression model is nonlinear in the parameters, and may be fitted to the data using standard nonlinear least-squares programs such as PROC NLIN in SAS.[6] An alternative software package called KINETICA was used for the present study.[7] KINETICA does not require analytical solutions, such as Equation (6), but rather only symbolic representations of the differential equations, and hence it is usually much easier to use, particularly for the subsequent, more involved models. The residual mean square, MSE, is used as a measure of model goodness-of-fit.

Three of the 21 observed data sets in the analysis had an initial convex uptake curve, in which the observed concentration at elapsed time 1, $c(1)$, was substantially less than one half the observed concentration at elapsed time 2, $c(2)$. Such a convex pattern is not consistent qualitatively with the model in Equation (4) which predicts an "exponential" (or concave) initial accumulation. Therefore the data for the three fish with the convex accumulation will be analyzed separately subsequently.

The fish are ranked, and also numbered, in decreasing order of size, V, in Table 1. The fish with the initial convex accumulation tended to be rather small (Fish 7, 17, and 21).

Some results, including the estimated parameters [$\hat{k}_u$, $\hat{k}_e$, and $\hat{C}(\infty)$], and the residual mean square, are also given for the remaining 18 fish in Table 1. There is significant correlation between the size (V) and uptake rate $(\hat{k}_u)$, but not between V and the elimination rate $(\hat{k}_e)$. Newman and Doubet discuss this correlation and also plot all of the fitted curves.[4]

For subsequent purposes, the 18 fish are subdivided into the following three groups:

- Group 1: Fish 2, 3, 4, 11, 12, 14
- Group 2: Fish 1, 5, 6, 8, 13, 15, 16
- Group 3: Fish 9, 10, 18, 19, 20

The observed data, the fitted curves, and the residuals are given for each fish separately, by group, in Figures 2 to 4. It is apparent that the residuals from most observed accumulation curves for this fitted model have noticeable patterns. In Figure 2, for the fish in Group 1, the residuals are negative for $t = 3$ and positive for $t = 4$, but they are apparently unbiased at $t = 6$. In Figure 3, for Group 2, the residuals are negative for $t = 3$ and positive at $t = 6$, and they also have a distinct cubic pattern over time. In Figure 4, for Group 3, the residuals also have a distinct cubic pattern, but of opposite sign, with positive residuals at $t = 4$ and negative ones at $t = 6$. Obviously, systematic lack-of-fit could lead to poor estimates of model parameters; for example, it is clear that a predicted curve which has a biased estimated concentration at the last observed value ($t = 6$) might yield a badly biased estimate of the extrapolated equilibrium concentration, $\hat{C}(\infty)$. Alternative compartmental models which reduce the bias at $t = 6$ for the Group 2 and the Group 3 fish and also incorporate more mechanistic structure will be developed subsequently.

ONE-COMPARTMENT STOCHASTIC MODEL
Model Derivation

An analogous stochastic formulation which parallels the previous deterministic model formulation may be developed. Two justifications are frequently cited for this alternative formulation. One is that it adds more realism to the model by broadening the basic conceptualization of the underlying mechanism of accumulation. The other is more pragmatic and relates to additional variables which are inherent in the stochastic model and which facilitate the subsequent data analysis. This chapter focuses on the latter justification.

One well-known stochastic formulation is based on assumptions concerning the kinetics of individual Hg molecules. The variable $X(t)$ is considered to be the number of such molecules in the fish at time t, and it changes stochastically with random unit additions and deletions. Formally, the probability assumptions

Table 1
Estimated Accumulation Parameters from Model 1 for the Individual Fish

Fish #	V	k_u	k_e	$C(\infty)$	MRT	MSE($\times 10^4$)
1	0.391	0.310	0.619	0.501	1.62	1.74
2	0.303	0.166	0.459	0.362	2.18	1.97
3	0.280	0.244	0.638	0.382	1.57	0.47
4	0.275	0.133	0.400	0.334	2.50	0.49
5	0.246	0.322	0.387	0.832	2.58	2.08
6	0.129	0.373	0.805	0.463	1.24	1.75
7	0.122			Initially convex		
8	0.095	0.522	0.691	0.756	1.45	5.81
9	0.089	0.173	0.399	0.434	2.50	1.40
10	0.088	0.113	0.518	0.218	1.93	0.15
11	0.084	0.541	0.834	0.648	1.20	2.64
12	0.083	0.325	0.428	0.760	2.34	3.46
13	0.081	0.271	0.621	0.436	1.61	0.71
14	0.078	0.385	0.620	0.621	1.61	0.29
15	0.068	0.452	0.545	0.830	1.83	8.48
16	0.041	0.601	0.622	0.967	1.61	5.63
17	0.034			Initially convex		
18	0.024	0.445	0.446	0.998	2.24	2.15
19	0.024	0.488	0.496	0.982	2.02	10.58
20	0.016	0.379	0.477	0.796	2.10	4.40
21	0.007			Initially convex		

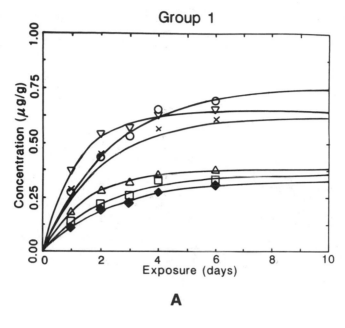

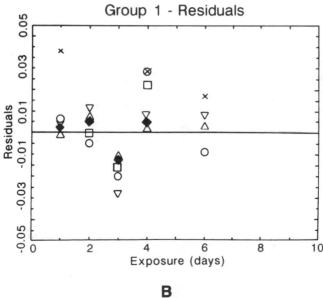

FIGURE 2. Results from fitting Model 1 to data for Group 1 fish. (A) Observed data and fitted values vs time; (B) residuals vs time. (Symbols with fish numbers: □, 2; △, 3; ◆, 4; ▽, 11; ○, 12; x, 14.)

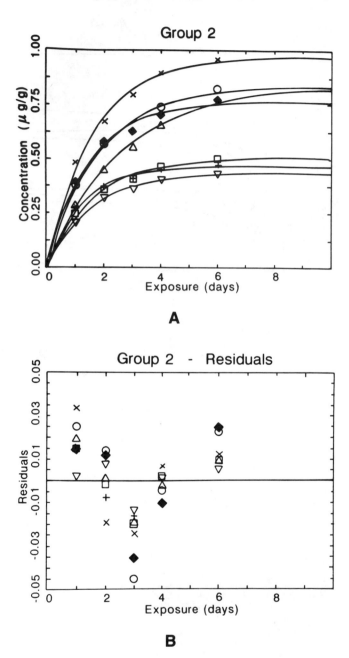

FIGURE 3. Results from fitting Model 1 to data for Group 2 fish. (A) Observed data and fitted values vs time; (B) residuals vs time. (Symbols with fish numbers: □, 1; △, 5; +, 6; ◆, 8; ▽, 13; ○, 15; x, 16.)

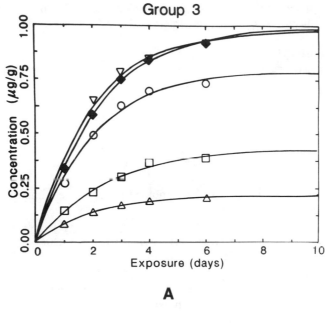

A

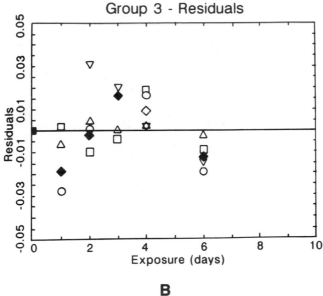

B

FIGURE 4. Results from fitting Model 1 to data for Group 3 fish. (A) Observed data and fitted values vs time; (B) residuals vs time. (Symbols with fish numbers: □, 9; △, 10; ◆, 18; ▽, 19; ○, 20.)

for molecule uptake and elimination for this stochastic model in a suitably small time interval from t to $t + \Delta t$ (i.e., interval size Δt) are

A3. Prob {new Hg molecule is absorbed into fish} = $U\Delta t$
A4. Prob {any one of the $X(t)$ molecules is eliminated from fish} = $E(t)\Delta t$

which are analogous to the previous definitions of U and $E(t)$.
   Two additional assumptions in interval Δt are necessary:

A5. All molecules are independent with respect to their kinetic behavior
A6. Prob {a specific molecule in the fish is eliminated} = $k_e\Delta t$

These last two assumptions may be combined to replace assumption A4 with the following:

A7. Prob {any one of the $X(t)$ molecules is eliminated} = $k_e X(t)\Delta t$

In this stochastic formulation, the variables $X(t)$ and $C(t)$ are considered to be random variables, which change accordingly to the probability laws given previously; however, for convenience, the size V is considered to be nonrandom (i.e., fixed). Because $X(t)$ and $C(t)$ are random variables, they are not constants at given elapsed times, $t > 0$, but rather they have probability distributions. For example, one could show that $X(t)$ is a Poisson random variable. The complete probability solution is not of present interest; however, the means of $X(t)$ and $C(t)$ are required. This motivates the following additional notation:

14. $\mu_x t, \mu_c t$ = means, or expected values of $X(t)$ amd $C(t)$
15. $\dot{\mu}_x \dot{\mu}t, \dot{\mu}_c t$ = derivatives of mean values
16. $\mu_x(\infty), \mu_c(\infty)$ = mean values for fish in equilibrium

   It has been shown that the differential equations for these mean values are

$$\dot{\mu}_x(t) = U - k_e\mu_x(t) \tag{7}$$

$$\dot{\mu}_c(t) = k_u - k_e\mu_c(t) \tag{8}$$

which are direct analogs of Equations (2) and (3).[8]
   The solution for the mean concentration, with $\mu_c(0) = 0$, is

$$\mu_c(t) = (k_u/k_e)(1 - e^{-k_e t}) \tag{9}$$

with

$$\mu_c(\infty) = k_u/k_e \tag{10}$$

In comparing these solutions of the stochastic model, for example, Equations (9) and (10), to previous ones of the deterministic model, for example, Equations (4) and (5), the corresponding equations have the same mathematical form. However, the key philosophical difference is that the stochastic model allows random changes (i.e., uncertainty) in the exact accumulation at any specified time and purports only to model the expected accumulation, whereas the deterministic model rules out any such uncertainty (or process error).

One practical contribution of the stochastic model lies in its inherent ability to describe the distribution of the "retention time" (or length of stay) of molecules in fish. Assumption A6 implies that there is a constant, time-invariant probability of the elimination of a specific particle, independent of how long the molecule may have already been in the fish. Such an elimination process, which does not discriminate on the basis of the past "age" of the molecule in the fish, is said to "lack a memory" of its age. In turn, it is well known that the retention time for molecules which satisfy this property in A6 follows an exponential distribution.[9] This leads to the following additional notation:

17. R = retention time variable for random molecule in fish
18. $f_R(t)$ = probability density function of random variable R
19. μ_R, σ_R^2 = mean and variance of R
20. Exp (Θ) = exponential distribution with parameter Θ

Assumption A6 may now be restated as

A8. R is an exponentially distributed random variable

which is denoted as

$$R \sim \text{Exp}(k_e) \tag{11}$$

The exponential distribution has the following properties of interest:[10]

$$f_R(t) = k_e e^{-k_e t}, \ t \geq 0 \tag{12}$$

$$\mu_R = k_e^{-1} \tag{13}$$

$$\sigma_R^2 = k_e^{-2} \tag{14}$$

The mean in Equation (13) is often called the mean residence time, MRT, and its utility is illustrated and discussed subsequently.

Data Analysis

The predicted mean value function in Equation (9) may now be fitted to the data. The statistical regression model is

$$c(t) = \mu_c(t) + \epsilon(t) \tag{15}$$

which is similar in form to Equation (6). One fundamental difference is that the random errors, $\epsilon(t)$, now include "process error" which is inherent in the stochastic process. If such error is substantial compared to the latent, nonstochastic "measurement error," it should be included in the estimation.[11] For simplicity, this chapter will assume that the process error is not large, and, therefore, the model in Equation (15) would give identical parameter estimates as the one for the deterministic model in Equation (6). However, the interpretation of the parameters changes, and the stochastic model also gives estimated MRT. The MRT values from the stochastic model are also listed in Table 1.

MULTICOMPARTMENT DETERMINISTIC MODELS
Model Derivation

There are two fundamental approaches to searching for better, alternative models to either reduce or eliminate the systematic lack of fit which is apparent in fitting the one-compartment model to some observed accumulation curves. One approach is to search for better empirical (also called correlative) models. The Richards model is an excellent example and this very flexible family of curves was used by Brisbin et al. to describe these same Hg accumulation data.[12]

The other approach is to search for better mechanistic models, and a natural extension along such lines is to postulate that the fish might have more than one "compartment." In the present context, this would imply that there are several subdivisions of the fish. Each subdivision (or compartment) is assumed to be relatively uniform within itself, but different from the others with respect to Hg kinetics. The literature contains a wealth of examples of such multicompartment models used to describe physiology and pharmacokinetics in other species.[2,3] The four-compartment model for the elimination of Hg in humans is an excellent example.[5]

It is possible in principle to fit rather sophisticated multicompartment models to aggregate accumulation data of the present type. However, the present data sets consist of, at most, six observed data points which limits the number of parameters that may be estimated with suitable precision. Therefore, the present study will be limited to, at most, two compartments with no more than three parameters. Figures 5 and 6 contain schematics for two particular models which satisfy these criteria and which will be investigated.

Additional notation is required for the two-compartment models. Definitions are given first for a deterministic model and parallel those given previously. In particular, the variables have the same units of measurement as before. The expanded notation follows:

21. $X_i(t)$ = amount of Hg in (compartment) i at (elapsed time) t
22. $X.(t) = X_1(t) + X_2(t)$ = total amount in fish at t
23. V_i = size of compartment i
24. $V. = V_1 + V_2$ = total size of fish
25. $C_i(t) = X_i(t)/V_i$ = concentration in i at t
26. $C.(t) = X.(t)/V.$ = mean concentration in fish at t
27. $\dot{X}_i(t), \dot{C}_i(t)$ = derivatives of $X_i(t)$ and $C_i(t)$
28. $X.(\infty), C.(\infty)$ = equilibrium values of $X.(t)$ and $C.(t)$
29. $U_i(t)$ = rate of uptake into i at t
30. $E_i(t)$ = rate of elimination from i at t
31. $T_{ij}(t)$ = rate of transfer of Hg from i to j at t

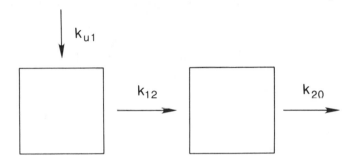

FIGURE 5. Schematic of a two-sequential-compartment model.

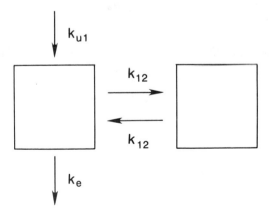

FIGURE 6. Schematic of Model 2, a two-compartment model with storage.

The model in Figure 5 assumes that all Hg uptake into the fish occurs through compartment 1 and all elimination from the fish occurs through compartment 2. The mathematical model for this "sequential" model is

$$\dot{X}_1(t) = U_1(t) - T_{12}(t)$$

$$\dot{X}_2(t) = T_{12}(t) - E_2(t) \tag{16}$$

On the other hand, the model in Figure 6 assumes that all uptake and elimination occur through compartment 1 and that compartment 2 is a "storage" compartment, perhaps consisting of tissue with more permanent Hg binding. This model has also been proposed by Spacie and Hamelink.[13] Its set of defining equations is

$$\dot{X}_1(t) = U_1(t) - E_1(t) - T_{12}(t) + T_{21}(t)$$

$$\dot{X}_2(t) = T_{12}(t) - T_{21}(t) \tag{17}$$

The following simplifying assumptions, similar to previous ones, are now made:

A9. The uptake rate is time-invariant, i.e., $U_1(t) = U_1$
A10. Each elimination rate is some constant fraction of the amount in the respective compartment, i.e., $E_i(t) = f_i X_i(t)$
A11. Each transfer rate is also some constant fraction of the amount in the respective compartment, i.e., $T_{ij}(t) = f_{ij} X_i(t)$

These latter assumptions define the notion of "donor-controlled" kinetics. Useful additional notation for the rate constants is:

32. $k_{io} = E_i(t)/X_i(t) =$ fractional elimination rate, from i to the exterior (which may be denoted as compartment 0)
33. $k_{ij} = T_{ij}(t)/X_i(t) =$ fractional transfer rate from i to j
34. $k_{ui} = U_i/V. =$ proportional uptake rate into i

With this notation, the sequential compartment model in Equations (16) may be rewritten as

$$\dot{X}_1(t) = U_1 - k_{12}X_1(t)$$

$$\dot{X}_2(t) = k_{12}X_1(t) - k_{20}X_2(t) \tag{18}$$

The solutions to this set of equations, assuming $k_{12} \neq k_{20}$ and $X_1(0) = X_2(0) = 0$, are[2]

$$X_1(t) = (U_1/k_{12})(1 - e^{-k_{12}t})$$

$$X_2(t) = U_1[(1 - e^{-k_{20}t})/k_{20} - (e^{-k_{12}t} - e^{-k_{20}t})/(k_{20} - k_{12})] \qquad (19)$$

It is easy to show that the model for the mean concentration is

$$C.(t) = k_{u1}[(1 - e^{-k_{12}t})/k_{12} + (1 - e^{-k_{20}t})/k_{20} -$$

$$(e^{\lambda_{2u}} - e^{\lambda_{20}t})/(k_{20} - k_{12})] \qquad (20)$$

which may be fitted to the observed data. It is apparent that this model has three parameters, namely, k_{12}, k_{20}, and k_{u1}. The equilibrium concentration for this sequential model is

$$C.(\infty) = k_{u1}(k_{12}^{-1} + k_{20}^{-1}) \qquad (21)$$

The general model with the tissue storage compartment has four parameters, rendering it intractable for present purposes. Therefore, the following assumption is made:

A12. The fractional transfer coefficients in the model in Figure 6 are equal, i.e.,
$$k_{12} = k_{21}$$

With this assumption, the Equations in (17) may be rewritten as

$$\dot{X}_1(t) = U_1 - (k_{10} + k_{12})X_1(t) + k_{12}X_2(t)$$

$$\dot{X}_2(t) = k_{12}X_1(t) - k_{12}X_2(t) \qquad (22)$$

The solutions to these equations, assuming $X_1(0) = X_2(0) = 0$, are given in Jacquez as[2]

$$X_1(t) = U_1[(k_{12} - \lambda_1)(1 - e^{-\lambda_1 t})/\lambda_1 - (k_{12} - \lambda_2)$$

$$(1 - e^{-\lambda_2 t})/\lambda_2]/(\lambda_2 - \lambda_1)$$

$$X_2(t) = U_1 k_{12}[(1 - e^{-\lambda_1 t})/\lambda_1 - (1 - e^{-\lambda_2 t})/\lambda_2]/(\lambda_2 - \lambda_1) \qquad (23)$$

where

$$\lambda_2, \lambda_1 = [(2k_{12} + k_{10}) \pm \sqrt{4k_{12}^2 + k_{10}^2}]/2$$

The model for mean concentration, which would be fitted to data, is

$$C.(t) = [X_1(t) + X_2(t)]/(V_1 + V_2) \tag{24}$$

and the equilibrium concentration may be obtained using some algebra in Equation (23) as

$$C.(\infty) = 2k_{u1}/k_{10} \tag{25}$$

Data Analysis

Both two-compartment predicted bioaccumulation curves, Equations (20) and (24), have a so-called "sums of exponentials" form, and both have an initial concave accumulation. Therefore, these models also were fitted only to the 18 fish with such initial shape characteristics. The statistical regression models for the observed data were

$$c(t) = C.(t) + \epsilon(t)$$

where the $\epsilon(t)$ were again assumed to be independent normals with constant variance.

Regression equation (20) gave a lower MSE, and hence yielded a better fit of the data than Equation (4) for six of the 18 fish (Fish 9, 10, 14, 18, 19, and 20). This finding suggests that, for these fish, the two-sequential compartment model provides a better description of Hg accumulation kinetics than the previous one-compartment model. However, five of the six curves (all except Fish 14) had degenerate solutions where the $\hat{k}_{12}$ and $\hat{k}_{21}$ parameter estimates were equal. This limiting case of the model in Equations (20) is more readily interpreted through a stochastic formulation which will be developed subsequently. Therefore, a discussion of the best model of Hg accumulation for this set of five fish with the degenerate solution is deferred until later. It is apparent from the fish rank numbers that this group of five fish tended to be smaller than the others.

The regression Equations (23) and (24) for the two-compartment model with storage were also fitted to each of the 18 fish using KINETICA. These equations provided a better fit of the data than any previous equation for seven fish (Fish 1, 5, 6, 13, 15, and 16), which previously were classified as Group 2. The fitted curves and residuals for each of the fish in this group are presented in Figure 7. The residuals are much smaller with the new model. There is still some evidence of model lack-of-fit with negative residuals at $t = 3$ and positive residuals at $t = 4$; however, the predicted curve has no apparent bias at $t = 6$ and hence is expected to give far superior estimates of some parameters, particularly $C(\infty)$.

Table 2 gives the size, V, the estimated parameter values [$\hat{k}_{u1}$, $\hat{k}_{10}$, $\hat{k}_{12}$, and $\hat{C}(\infty)$] and the MSE for the seven fish in this group. Although there is no question that the present model provides a better fit for the data, the parameter estimates

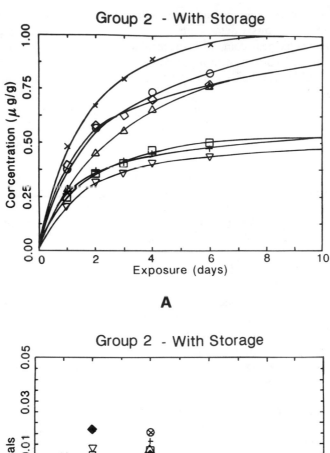

A

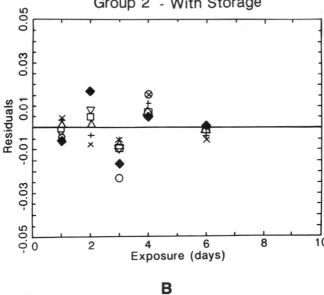

B

FIGURE 7. Results from fitting Model 2 to data for Group 2 fish. (A) Observed data and fitted values vs time; (B) residuals vs time. (Symbols with fish numbers: □, 1; △, 5; +, 6; ◆, 8; ▽, 13; ○, 15; x, 16.)

have sizable standard errors as one faces multicollinearity problems in estimating three parameters from six data points.

One striking change in comparing the results from this model to the comparable results for the one-compartment model is the substantial increase ($>60\%$) in the estimated equilibrium concentration, $C(\infty)$, for some fish (FIsh 6, 8, and 13). The new estimates are also a large increase over the estimates reported by Brisbin et al.[12] These large increases are due to storage compartments with low estimated transfer rates ($k_{12} < 0.100$). The effect of such storage compartments would be minimal by $t = 6$; however, the compartments would become "saturated" and have a large impact over time. An important principle is that extrapolations, in this case from $t = 6$ to $t = \infty$, based on mechanistic laws are generally preferable to those based on empirical laws.[14] However, in the present case, the mechanistic model is restricted by the constraint $k_{12} = k_{21}$ which might bias the result somewhat even though the curves fit very well. Other comparisons of results in Table 1 with those in Table 2 follow subsequently.

MULTICOMPARTMENT STOCHASTIC MODEL
Model Derivation

The previous stochastic one-compartment model may be generalized to give a stochastic multicompartment model. In this formulation, the variables $X_1(t)$ and $X_2(t)$ are interpreted as the number of molecules in the respective compartments, and they are stochastic. The following assumptions describe the fluctuations in these counts by specifying probabilities for molecule uptake, elimination, and transfer in a suitably small time interval from t to $t + \Delta t$:

A13. Prob {new Hg molecule is absorbed into fish} $= U_1 \Delta t$
A14. Prob {any one of the $X_1(t)$ molecules is eliminated} $= k_{10} X_1(t) \Delta t$
A15. Prob {any one of the $X_1(t)$ molecules in 1 is transferred to 2} $= k_{12} X_1(t) \Delta t$
A16. Prob {any one of the $X_2(t)$ molecules in 2 is transferred to 1} $= k_{12} X_2(t) \Delta t$

Table 2
Estimated Accumulation Parameters from Model 2 for the Fish in Group 2

Fish #	V	k_{u1}	k_{10}	k_{12}	$C(\infty)$	MRT	MSE($\times 10^4$)
1	0.391	0.460	1.719	0.602	0.535	2.27	0.57
5	0.246	0.408	0.875	0.432	0.933	2.29	0.43
6	0.129	0.436	1.090	0.051	0.800	1.83	0.71
8	0.095	0.624	1.001	0.071	1.246	2.00	2.12
13	0.081	0.290	0.723	0.029	0.801	2.76	3.58
15	0.068	0.583	0.986	0.146	1.181	2.03	3.58
16	0.041	0.886	1.732	0.646	1.023	1.16	1.29

The assumptions are analogous to the assumed rate structure in the corresponding deterministic model. It could also be shown that the corresponding set of differential equations, with notation similar to definitions A14 to A16, is

$$\dot{\mu}_{x1}(t) = U_1 - (k_{10} + k_{12})\mu_{x1}(t) + k_{12}\mu_{x2}(t)$$

$$\dot{\mu}_{x2}(t) = k_{12}\mu_{x1}(t) - k_{12}\mu_{x2}(t) \tag{26}$$

The solution to these equations would be identical to that given in Equation (22), with $X_i(t)$ being replaced by $\mu_{xi}(t)$. Therefore, the model for the expected concentration in the whole fish is

$$\mu_c(t) = [\mu_{x1}(t) + \mu_{x1}(t)]/(V_1 + V_2) \tag{27}$$

and the stochastic equilibrium concentration is

$$\mu_c(\infty) = 2k_{u1}/k_{10} \tag{28}$$

The stochastic model may again be used to derive new variables based on the concept of MRT, and these new variables will be used as practical tools in the subsequent analysis. The fundamental conceptualization of the stochastic model is that molecules move randomly, coming into the fish and moving within and out of the fish according to the probability laws given in assumptions A13 to A16. It is now useful to focus on the amounts of time molecules spend in the compartments, which are new random variables. The following notation is helpful:

35. R_{ij} = retention time during a single ''visit'' in (compartment) i of a random molecule whose next transfer will be to (compartment) j
36. R_i = retention time during a single visit in i of a random molecule prior to its next transfer out of i, whether to some other compartment or to the fish's exterior
37. S_{ij} = total residence time that a molecule which initially enters the fish in i will accumulate in j during all of its visits prior to exiting the fish
38. S_i = total residence time that a molecule which initially enters the fish in i will accumulate in all of the compartments prior to exiting the fish
39. $\mu_{R_{ij}}, \mu_{R_i}, \mu_{S_{ij}}, \mu_{S_i}$, = means of R_{ij}, R_i, S_{ij}, and S_i, respectively

Assumptions A14 to A16 may now be restated, as in the one-compartment model A7 was restated to A8, to specify that R_{10}, R_{12}, and R_{21} are exponential random variables. These equivalent assumptions are:

A17. $R_{10} \sim \text{Exp}(k_{10})$
A18. $R_{12} \sim \text{Exp}(k_{12})$
A19. $R_{21} \sim \text{Exp}(k_{12})$

which then are also stochastic analogs to the deterministic linear kinetic assumptions A11 and A12.

A number of results concerning MRT follow directly from properties of the exponential. For example, it can be shown that

$$\mu_{R10} = k_{10}^{-1}$$

$$\mu_{R12} = k_{12}^{-1}$$

$$\mu_{R1} = (k_{10} + k_{12})^{-1}$$

$$\mu_{R21} = \mu_{R2} = (k_{12})^{-1}$$

Other results relating to MRT may be obtained from general theorems for stochastic multicompartment models.[15] The chief result of current interest for the present model is[16]

$$\mu_{S1} = 2(k_{10})^{-1} \tag{29}$$

Results for other variables, such as the expected number of particle transfers, are also available, but they do not appear to have immediate application in the present problem.

Data Analysis

The estimated rate coefficients, $\hat{k}_{ij}$'s, are obviously very model-specific, and, hence, the comparison of such rate coefficients between models is usually non-informative. On the other hand, the mean residence time concept has universal application across models and, therefore, provides a measure which is interpretable within each model and comparable between models. The MRT for the seven fish in Group 2 are given in Table 2. It is obvious in comparing the results in Tables 1 and 2 for these seven fish that the MRT tended to increase substantially for this two-compartment model as compared to the corresponding one-compartment model. Some comparisons between the groups of fish with different preferred models will be given subsequently.

ONE-COMPARTMENT STOCHASTIC MODEL WITH GAMMA TRANSIT TIMES
Model Derivation

A previous section outlined a two-sequential-compartment model which provided an improved fit for 6 of the 18 fish. However, it became apparent that, for 5 of these 6 fish, the best fitting curve was the degenerate case of Equation (20), where $k_{12} = k_{20}$. The special case of the sequential model with equal rates could be solved directly by introducing this constraint into Equations (18). The solutions would then be

$$X_1(t) = (U_1/k_{12})(1 - e^{-k_{12}t})$$

$$X_2(t) = U_1[(1 - e^{-k_{12}t})/k_{12} - te^{-k_{12}t}] \qquad (30)$$

which are also the limiting solutions of Equations (19) as k_{20} approaches k_{12}. The concentration curve is

$$C.(t) = k_{u1}[2(1 - e^{-k_{12}t})/k_{12} - te^{-k_{12}t}] \qquad (31)$$

with equilibrium concentration

$$C.(\infty) = 2k_{u1}/k_{12}$$

Equations such as (30) and (31) could be fitted to data directly, yet they are seldom used in practice in a deterministic context. Perhaps one reason for its generally being overlooked is that it does not have the characteristic "sum of exponentials" form usually associated with compartmental modeling. Another reason might be the mechanistic interpretation of two exactly equal rate coefficients, which seems implausible and which might also lead to such subsequent inferences as equal compartment volumes. Whatever the reasons, the lack of such models would significantly limit many kinetic studies, including the present one.

Fortunately, the stochastic model may be used to provide a relatively simple interpretation of the observed result. In a stochastic context, the assumptions for the two sequential compartment model with equal coefficients could be written as

A20. A random molecule in the fish has two sequential retention time variables, R_1 and R_2, both of which have independent and identical exponential distributions with parameter k_{12}, i.e., $R_1 \sim \text{Exp}(k_{12})$ and $R_2 \sim \text{Exp}(k_{12})$

It has often been pointed out that this assumption is equivalent to the following:

A21. A random molecule in the fish has a retention time variable R_1 which has a gamma distribution with "shape" parameter $n = 2$ and "scale" parameter $\lambda = k_{12}$

This is denoted as

$$R_1 \sim G(2, \lambda)$$

A schematic for this model is given in Figure 8.

It is important to note that, although assumption A20 of two sequential exponential retention times and assumption A21 of a single gamma retention time

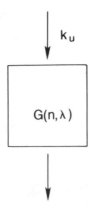

FIGURE 8. Schematic of Model 3, a one-compartment model with gamma, $G(n, \lambda)$, retention times.

both lead to a regression model of the form in Equation (31), the latter assumption is less restrictive mechanistically. The point is that the gamma variable in question might indeed have been produced as a sum of two identical exponential variables; however, there are also many other mechanistic models which would lead to gamma retention times. A number of these alternative mechanisms are discussed in Matis and Wehrly.[17]

The gamma retention time assumption in A21 is useful from another perspective. As noted, the previous assumptions relating to molecule transfer in A6, A7, and A14 to A16 imply that the transfer/elimination process "lacks a memory" of the "age" of the molecule in the compartment. This leads to a time-invariant conditional transfer probability, also called a constant hazard rate in the chapter by Dixon and Newman (this volume, Chapter 8).[18] The assumption of a gamma retention time leads to an increasing hazard rate function within the combined compartment. This extension, which links classical statistical survival analysis with compartmental modeling, is called generalized compartmental modeling and is discussed elsewhere.[17] In summary, the reformulation in A21 opens up modeling possibilities beyond classical linear compartmental analysis, which has largely overlooked the model in Equation (31). One practical advantage of the $G(2, k_{12})$ model in (31) is that it has only two parameters, U_1 and k_{12}. The model could be expanded to include gamma distributions with shape parameters $n > 2$. For example, a $G(3, \lambda)$ model would give

$$X_3(t) = U_1[(1 - e^{-\lambda t})/\lambda - e^{-\lambda t}(t + t^2/2)] \tag{32}$$

with

$$C.(t) = k_{u1}[3(1 - e^{-\lambda t})/\lambda - e^{-\lambda t}(2t + t^2/2)] \tag{33}$$

and

$$C.(\infty) = 3k_{u1}/\lambda$$

In general, a $G(n, \lambda)$ model would have solution

$$C.(t) = k_{u1} \left[n(1 - e^{-\lambda t})/\lambda - e^{-\lambda t} \sum_{i=1}^{n-1} (n - i)t^i/i! \right] \qquad (34)$$

and

$$C.(\infty) = nk_{u1}/\lambda \qquad (35)$$

The mean and variance of the residence time of a gamma distribution are easy to find from the results in Equations (15) and (16). A gamma (n, λ) variable has mean and variance

$$\mu_{R_1} = n/\lambda \qquad (36)$$

$$\sigma_{R_1}^2 = n/(\lambda)^2 \qquad (37)$$

Data Analysis

Gamma models of the form in Equation (34) with increasing n were fitted to the observed curves of each of the six fish (Fish 9, 10, 14, 18, 19, and 20) previously identified with the two-sequential-compartment model. The residual mean square would decrease initially with n but then reach a minimum and increase sharply thereafter. Fish 14 is unique in that none of the gamma models could improve on the two-sequential model with *unequal* rate coefficients in Equation (2). The fitted curves and residuals for the other five fish, which together constitute Group 3, are presented in Figure 9. The residuals are somewhat smaller on average, and more importantly, there is no apparent pattern nor bias.

The results for the parameter estimates for this group are given in Table 3. It is immediately obvious that these fish tend to be relatively small in size. In comparing the corresponding results in Tables 1 and 3 for this set of fish, it is apparent that the gamma model has lead consistently to a larger MRT than that given by the corresponding one-compartment model. The predicted equilibrium concentrations, $C(\infty)$, are consistently lower due to the elimination of the bias at $t = 6$.

ONE-COMPARTMENT STOCHASTIC MODEL FOR INITIAL CONVEX ACCUMULATION
Model Derivation

One common property of the three previous compartmental accumulation curves [in Equations (4), (24), and (34)] is the concave shape of the curves. It

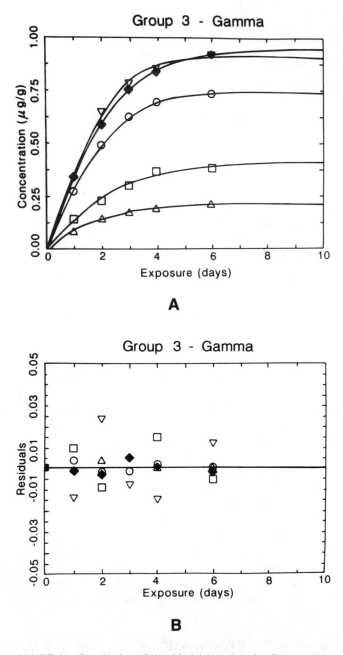

FIGURE 9. Results from fitting Model 3 to data for Group 3 fish. (A) Observed data and fitted values vs time; (B) residuals vs time. (Symbols with fish numbers: □, 9; △, 10; ◆, 18; ▽, 19; ○, 20.)

could be shown, for example, that one necessary condition for such curves is that the ratio of the concentrations at times 1 and 2, i.e., $c(1)$ and $c(2)$, exceed 0.50, for example,

$$r = c(1)/c(2) > 0.50$$

However, it was previously noted that three of the observed accumulation curves, namely, those for Fish 7, 17, and 21, were initially convex, with the above ratio ranging from 0.25 to 0.40 for the three fish.

One general explanation of such an event is to attribute it to mere measurement error of $c(1)$ or $c(2)$ or both. That simple explanation does not seem plausible (nor consistent with empirical analysis) with the present data sets. An alternative explanation is that these observed curves represent some phenomenon, such as an initial "resistance" to Hg uptake by the fish.[4,12] However, the mechanistic basis for such a phenomenon was not discussed in the previous investigation and is not clear to the present authors.

Still, although a mechanistic foundation is lacking, one could construct empirical models which might predict the accumulation well. Previously, the Richards' curves were used for such predictions; however, compartmental analysis also provides such empirical curves. For example, Figure 10 depicts a compartmental model with a sequence of compartments but with the data being observed only in the terminal compartment. Such a model would provide a plausible mechanistic interpretation in the case where the observed concentration was some chemically altered form (or metabolite) of the uptake metal, but no such interpretation is intended for the present application. The equations for the amount in the terminal compartment have been previously derived for several models in Equations (30) and (32), and they could be used on a purely empirical basis. In general, the amount in the terminal compartment of a $G(n, \lambda)$ model is

$$X_n(t) = U_1 \left[(1 - e^{-\lambda t})/\lambda - e^{-\lambda t} \sum_{i=1}^{n-1} t^i/i! \right] \tag{38}$$

Table 3
Estimated Accumulation Parameters from Model 3 for the Fish in Group 3

Fish #	V	k_{u1}	n	λ	$C(\infty)$	MRT	MSE($\times 10^4$)
9	0.089	0.144	2	0.701	0.409	2.85	1.30
10	0.088	0.092	2	0.877	0.209	2.28	0.053
18	0.024	0.366	2	0.772	0.949	2.59	0.093
19	0.024	0.354	4	1.556	0.910	2.57	2.98
20	0.016	0.275	4	1.492	0.736	2.68	0.059

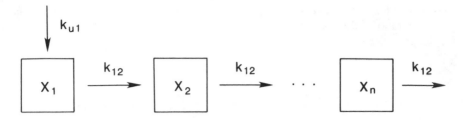

FIGURE 10. Schematic of Model 4, an *n*-sequential compartment model with observed concentration in the terminal (n^{th}) compartment.

from whence the model for the concentration curve is

$$C(t) = k_{u1}\left[(1 - e^{-\lambda t})/\lambda - e^{-\lambda t}\sum_{i=1}^{n-1} t^i/i!\right] \qquad (39)$$

with

$$C(\infty) = k_{u1}/\lambda$$

The mean retention time of a molecule in this observed compartment is

$$\mu_R = (\lambda)^{-1}$$

However, the notion of a MRT for this model is not directly comparable to the MRT concept for the previous mechanistic models without this initial resistance phenomenon and, therefore, is not pursued at present.

Data Analysis

Models of the form in Equation (39) with increasing *n* were fitted to the observed curves of each of the three fish. The residual mean square would again initially decrease as *n* increased but then reach a minimum and increase sharply thereafter. The results for this set of three fish are given in Table 4, and the three fitted curves are graphed in Figure 11. The comparable results for fitting the other models derived in this chapter to these three fish are not given, yet it is obvious that the present curves in Equation (39) provide vastly superior predicted values, including the predicted equilibrium value $C(\infty)$, as compared to any of the previous models in this chapter.

Table 4
Estimated Accumulation Parameters from Model 4 for the Fish in Group 4

Fish #	V	k_{u1}	n	λ	$C(\infty)$	MSE($\times 10^4$)
7	0.122	0.345	4	2.717	0.127	0.030
17	0.034	1.222	3	1.536	0.796	14.010
21	0.007	2.828	7	2.850	0.992	13.080

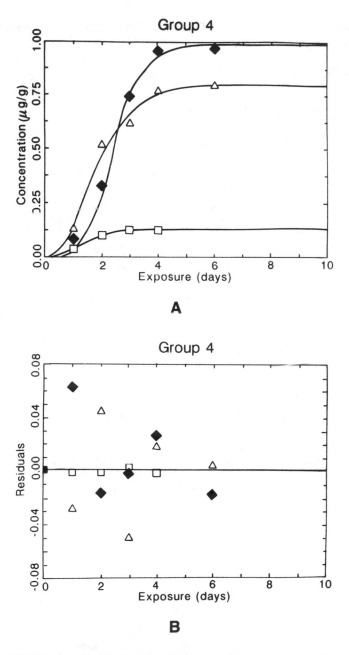

FIGURE 11. Results for fitting Model 4 to data for Group 4 fish. (A) Observed data and fitted values vs time; (B) residuals vs time. (Symbols with fish numbers: □, 7; △, 17; ♦, 21.)

COMPARISON OF RESULTS BETWEEN ACCUMULATION MODELS
Comparison of Mean Values

Table 5 lists the results of merging Tables 1 to 4 to find for each fish the best estimates of the parameters across the accumulation models derived in the previous sections. The four parameters of interest are size (V), uptake rate (for simplicity denoted k_u), equilibrium concentration [$C(\infty)$], and mean residence time (MRT).

The three mechanistic, compartmental models for Hg accumulation with a concave pattern will be denoted Models I to III. Model I denotes the one-compartment model with simple uptake and elimination, Model II the two-compartment model with a storage compartment, and Model III the one-compartment model with the initial age-dependent Hg elimination. For the sake of simplicity, the two-compartment sequential model with unequal rate coefficients, with Equation (20), was not used in the subsequent considerations. This model provided the best fit for only one of the 21 fish (Fish 14), but even in this case the one-compartment model provided an acceptable alternative on the basis of MSE (MSE $= 2.85 \times 10^{-5}$ vs MSE $= 2.93 \times 10^{-5}$). The compartment model with the initial resistance in uptake, which gives the initial convex accumulation, will be denoted Model IV. Only the size, V, and the equilibrium concentration, $C(\infty)$, parameters will be compared for the fish with this preferred model.

As previously noted, the standard errors of some of the estimates are substantial due to multicollinearity problems in estimating two or three parameters from

Table 5
Estimated Accumulation Parameters for the Individual Fish from the Best Fitting Individual Accumulation Model

Fish #	Best Model	V	k_u	$C(\infty)$	MRT
1	II	0.391	0.460	0.535	2.27
2	I	0.303	0.166	0.362	2.18
3	I	0.280	0.244	0.382	1.57
4	I	0.275	0.133	0.334	2.50
5	II	0.246	0.408	0.933	2.29
6	II	0.129	0.436	0.800	1.83
7	IV	0.122		0.127	
8	II	0.095	0.624	1.246	2.00
9	III	0.089	0.144	0.409	2.85
10	III	0.088	0.092	0.209	2.28
11	I	0.084	0.541	0.648	1.20
12	I	0.083	0.325	0.760	2.34
13	II	0.081	0.290	0.801	2.76
14	I	0.078	0.385	0.621	1.61
15	II	0.068	0.583	1.181	2.03
16	II	0.041	0.886	1.023	1.16
17	IV	0.034		0.796	
18	III	0.024	0.366	0.948	2.59
19	III	0.024	0.354	0.910	2.57
20	III	0.016	0.275	0.736	2.68
21	IV	0.007		0.992	

Table 6
Estimated Accumulation Parameters for
the Four Groups of Fish

Group	V	k_u	MRT	$C(\infty)$
1	0.184	0.290	1.94	0.517
2	0.150	0.527	2.05	0.931
3	0.048	0.246	2.59	0.642
4	0.054			0.638

only five or six data points. Also, the number of fish within each group is small, for example, only three fish in Group 4. Notwithstanding these problems, there are distinct differences among the group means for each of the four accumulation parameters. The estimated means for each parameter within each group are listed in Table 6.

Previously it was observed that the fish in Groups 3 and 4 had high rank numbers, indicating that the fish in these groups tended to be small. This fact is also apparent when comparing mean sizes. The mean sizes for the four groups are 0.184, 0.150, 0.048, and 0.054, respectively. These obviously vary substantially but are not statistically significantly different due to the large variability within Group 4.

The mean estimated uptake rates are 0.290, 0.527, and 0.246, respectively. The mean rate for Group 2 is significantly larger ($p < 0.05$) than the other means.

The mean estimated MRT's are 1.94, 2.05, and 2.59 days, respectively. The mean MRT of Group 3 is significantly larger than the other means.

The mean estimated equilibrium concentrations are 0.517, 0.931, 0.642, and 0.638, respectively. The mean concentration for Group 2 is substantially larger but not significantly different from the other means.

Effect of Size, V, on Prediction of Accumulation Parameters

The size of the fish is an important variable in determining its accumulation parameters. One manifestation of this is the significant linear correlation ($r = -0.460$, $p = 0.036$) between size, V, and equilibrium concentration, $C(\infty)$, in the data in Table 5 for all four groups of fish. The effect of size is even more clearly manifested through its strong association with some accumulation parameter *within* the four groups of fish. The following figures illustrate these associations.

Figure 12 illustrates the relationship between V and $C(\infty)$ within each of the four groups of fish. The following results are evident in the figure:

1. The larger mean $C(\infty)$ for Group 2, noted previously.
2. The significant correlation between $C(\infty)$ and V over all groups, also noted previously.

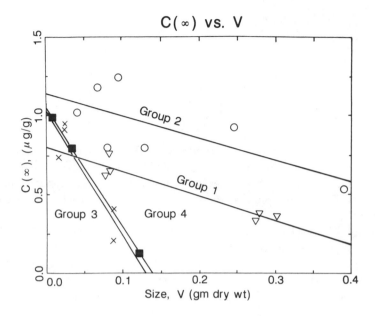

FIGURE 12.   Scatterplot of estimated equilibrium concentration, $C(\infty)$, vs size, V, for all fish by group, with estimated regression line for each group. (Symbols with fish numbers: $\triangledown$, 1; $\bigcirc$, 2; x, 3; $\blacksquare$, 4.)

3.   The significant linear regression relationship between V and $C(\infty)$ within the groups. There are significant differences ($p < 0.01$) between the four slopes, but it is clear that slopes for Groups 1 and 2 are equal, as are the slopes for Groups 3 and 4.

Figure 13 illustrates the regression relationships between V and the estimated intake rate, k_u. The following results are apparent in the graphs:

1.   The significantly faster mean intake for Group 2, which, on average, is roughly twice the rate of the other groups.
2.   A linear regression relationship between V and k_u within the groups. There is no statistical evidence to conclude that the slopes within the groups differ ($p = 0.44$). The best estimate of the common slope within the groups is $b = -0.91$, which is significant ($p < 0.05$).
3.   Differences between mean k_u after adjusting for V used as a covariate. The adjusted mean of Group 2 is also significantly faster than the others.

Figure 14 illustrates the relationship between V and the estimated MRT. The significantly longer mean MRT for Group 3 is apparent, but there is no statistical evidence that the MRT is a function of V.

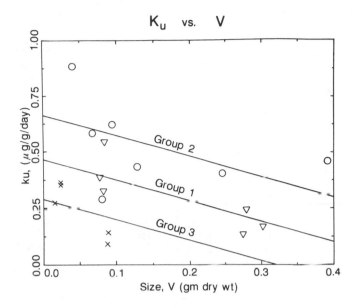

FIGURE 13. Scatterplot of estimated intake rate, k_u, vs size, V for 18 fish by group, with estimated regression line for each group. (Symbols with group numbers: ▽, 1; ○, 2; x, 3.)

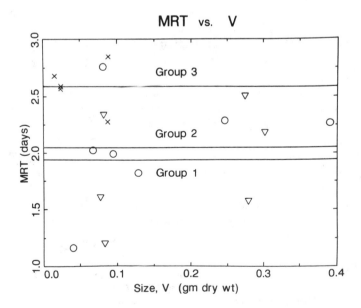

FIGURE 14. Scatterplot of estimated mean residence time, MRT, vs size, V, for 18 fish by group, with estimated regression line for each group. (Symbols with group numbers: ▽, 1; ○, 2; x, 3.)

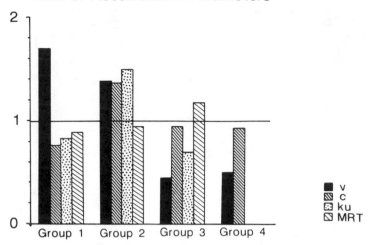

FIGURE 15.   Comparative ratios (group mean/overall mean) of estimated accumulation parameters for each group of fish.

Characterization of Fish in Various Accumulation Models

Some of the previous findings concerning the means in Table 6 may be combined to characterize the various accumulation models on the basis of fish size and accumulation parameters. The ratios of each group mean to the overall unweighted mean are graphed in Figure 15 for each of the four accumulation parameters. It is apparent, for example, that the fish in Group 4, with the initial convex accumulation, tended to be smaller in size with average equilibrium concentrations.

The fish in Group 3, with the initial lag in Hg depuration, were also smaller in size and had average equilibrium concentrations. Their intake rate was slower; however, this may be partially offset by a longer MRT.

The fish in Group 2, with the apparent storage compartment, were larger in size. These fish had a faster average intake rate with a moderate MRT. All of these characteristics combine conceptually to yield the highest equilibrium concentrations, as one might expect.

The fish in Group 1, with the simple uptake and elimination mechanisms, were larger in mean size. They had a slower intake and a shorter MRT which combine to give a relatively small equilibrium concentration.

DISCUSSION

The primary objective of this investigation is to explore the utility of compartmental modeling for describing Hg accumulation in fish. It seems evident in examining the fitted curves and the residuals for each of the 21 data sets that compartmental analysis does provide an adequate *empirical* curve for describing

Hg accumulation. In fact, the curves derived from the compartmental models tend to fit the data better than the Richard's curves. However, the question of greater subsequent interest is whether the compartmental modeling construct also provides a basis for constructing a more *mechanistic* model of Hg accumulation. If so, it could be used as a tool, as in other fields, to link physiological structure and to associate kinetic parameters with observed accumulation curves, perhaps produced under various designed experimental conditions.

This broad question of the value of compartmental modeling in a mechanistic context is explored in this study under the limiting constraints of only six or fewer data points for each of only 21 fish accumulation curves. Nevertheless, it does seem clear that compartmental modeling has indeed been useful in elucidating mechanistic structure. One can make a case that some fish have a "storage" component and that others have a "processing lag" in elimination, rather than merely simple uptake and elimination kinetics. Moreover, one can associate such additional structure to the size of the fish, which obviously is model independent.

Certainly, there are intriguing questions concerning Hg accumulation kinetics which remain unexplained. One is the mechanistic interpretation of the initial convex accumulation; this question requires further physiological input. However, another question which may be tackled by further modeling is an apparent subtle pattern in the residuals. It seems that, although each fitted curve is apparently unbiased at $t = 6$, there remains a pattern of negative residuals at $t = 3$ and positive residuals at $t = 4$. Perhaps this pattern is a manifestation of some further mechanistic structure, which might be slight but which is common to many fish.

There is a wealth of other available compartmental models to explore this particular question and, in general, to associate physiological parameters with observed accumulation curves. Most of the past models have been deterministic. More recent research has produced analogous stochastic models which yield such concepts as MRT and gamma compartments, both of which have been useful in the present study. The advantage of MRT as opposed to the commonly used half-lives is discussed elsewhere.[19] Techniques for the routine use of gamma compartments have also been developed.[17] For example, in one application related to the present problem, a three-compartment model was used to describe Ca clearance in humans.[17] The model has two storage compartments, one of which had gamma-type retention times, and six kinetic parameters which were estimated from only 19 data points from a single compartment. Methods have also been developed for the simultaneous fitting of data from many different compartments in order to validate the assumed model structure.[20,21] The main point is that many new models and tools for describing kinetic systems such as the present one may be obtained from the stochastic formulation of the compartmental model.

These new models could be relatively easily used in a simulation study to investigate various observed phenomena. The present authors prefer not to rely on the usual simulation studies alone but rather to validate proposed models through statistical analysis and hypothesis testing. One immediate, practical constraint to such a statistical study is the availability of computer software which would fit assumed regression models to data. Coding analytical solutions such as Equations (23) into nonlinear least-squares packages such as SAS has always presented a formidable burden, and fitting multicompartment models with $n > 2$ has been next to impossible. This study used a new software package called KINETICA, which requires only the symbolic representation of the system rather than the detailed analytical solutions, thereby greatly simplifying the effort.[7]

With the availability of new software such as KINETICA, the main constraint, at the present time, is the availability of data from suitably designed experiments. As previously mentioned, it would obviously be helpful to have more data points at crucial times during the accumulation process. It would also be extremely helpful to have observations over time from select organs of the fish anatomy. An appropriate experiment would be to serially sacrifice fish and determine metal content of individual organs. Such data could be described by a population compartmental model with heterogeneous kinetic rates, as recently given in Matis et al.[22] Another recent generalization is the addition of an "effect" compartment to the model which would then link the amount of metal in particular organs to observed toxicological end points.[23] Such models would, in principle, integrate bioaccumulation models with the toxicological studies reported in other chapters in this book. In summary, there are many possible generalizations to the compartmental models given in this study which are relatively easy to use in practice; we anticipate that such models will be very useful in the mechanistic modeling of metal accumulation in fish and small animals.

ACKNOWLEDGMENTS

The authors are greatly indebted to M. C. Newman (SREL), who brought the problem to our attention, made the data available, and encouraged the present investigation. We are also indebted to a referee who made a number of suggestions, including the reference to the example in Filov et al., and to Tim Matis for his help with the graphics. The research is supported in part by the 1990 Texas Advanced Technology Program Grant under Project #999902-123 to JHM.

REFERENCES

1. Box, G. E. P. Frequent quote.
2. Jacquez, J. A. *Compartmental Analysis in Biology and Medicine* (Ann Arbor, MI: University of Michigan Press, 1985), pp. 560.
3. Godfrey, K. *Compartmental Models and Their Applications* (London: Academic Press, 1983), pp. 293.

4. Newman, M. C., and D. K. Doubet. "Size-Dependence of Mercury (II) Accumulation Kinetics in the Mosquitofish, *Gambusia affinis* (Baird and Girard)," *Arch. Environ. Contam. Toxicol.* 18:819-825 (1989).

5. Filov, V. A., A. A. Golubev, E. I. Liublina and N. A. Tolokontsev. *Quantitative Toxicology* (New York: John Wiley & Sons, 1979), pp. 462.

6. SAS *User's Guide: Statistics, Version 5 Edition*. (Cary, NC: SAS Institute, Inc., 1985).

7. Allen, D. M. and J. H. Matis. *KINETICA, A Program for Kinetic Modeling in Biological Sciences,* Tech. Rep. 137, Dept. of Statistics, (Lexington, KY: University of Kentucky, 1990).

8. Chiang, C. L. *An Introduction to Stochastic Processes and Their Applications* (Huntington, NY: Krieger Publ. Co., 1980), pp. 517.

9. Parzen, E. *Stochastic Processes* (San Francisco: Holden-Day, 1962), pp. 324.

10. Johnson, N. L. and S. Kotz. *Continuous Univariate Distributions — 1* (New York: Interscience, 1970), pp. 300.

11. Matis, J. H. "An Introduction to Stochastic Compartmental Models in Pharmacokinetics," in *Pharmacokinetics,* A. Pecile and A. Rescigno, Eds. (New York: Plenum Press, 1988), pp. 113-128.

12. Brisbin, I. L., Jr., M. C. Newman, S. G. McDowell, and E. L. Peters. "Prediction of Contaminant Accumulation by Free-Living Organisms: Applications of a Sigmoidal Model," *Environ. Toxicol. Chem.* 9:141-149 (1990).

13. Spacie, A. and J. L. Hamelink. "Bioaccumulation," in *Fundamentals of Aquatic Toxicology,* G. M. Rand and S. R. Petrocelli, Eds. (Washington: Hemisphere Publ., 1985), pp. 495-525.

14. Gold, H. J. *Mathematical Modeling of Biological Systems* (New York: Interscience, 1977), pp. 357.

15. Matis, J. H., T. E. Wehrly, and C. M. Metzler. "On Some Stochastic Formulations and Related Statistical Moments of Pharmacokinetic Models," *J. Pharmacokinet. Biopharmacol.* 11:77-92 (1983).

16. Matis, J. H. and K. B. Gerald. "On Selecting Optimal Response Variables for Detecting Treatment Effects in a Two-Compartment Model," in *Modelling of Biomedical Systems,* J. Eisenfeld and M. Whitten, Eds. (North Holland, Amsterdam: Elsevier, 1986), pp. 153-159.

17. Matis, J. J. and T. E. Wehrly. "Generalized Stochastic Compartmental Models with Erlang Transit Times," *J. Pharmacokinet. Biopharmacol.* 18:589-607 (1990).

18. Dixon, P. M. and M. C. Newman. "Analyzing Toxicity Data Using Statistical Models for Time-to-Death: An Introduction," This volume, Chapter 8.

19. Matis, J. H., T. E. Wehrly, and K. B. Gerald. "Use of Residence Time Moments in Compartmental Analysis," *Am. J. Physiol.* 249 *(Endocrinol. Metab. 12)*:E409—E415(1985).

20. Grant, W. E., J. H. Matis, and T. H. Miller. "A Stochastic Compartmental Model for Migration of Marine Shrimp," *Ecol. Model.,* in press.

21. Matis, J. H., W. E. Grant, and T. H. Miller. *A Semi-Markov Process Model for Migration of Marine Shrimp.* Tech. Rep. 112, Dept. of Statistics, (College Station, TX: Texas A&M University, 1990), pp.23.

22. Matis, J. H., T. E. Wehrly, and W. C. Ellis. "Some Generalized Stochastic Compartment Models for Digesta Flow," *Biometrics* 45:703—720 (1989).
23. Sheiner, L. B. "Modeling Pharmacodynamics: Parametric and Nonparametric Approaches," in *Variability in Drug Therapy: Description, Estimation, and Control,* M. Rowland, L. B. Sheiner, and J. L. Steimer, Eds. (New York: Raven Press, 1985), pp. 139-152.

8

Analyzing Toxicity Data Using Statistical Models for Time-To-Death: An Introduction

Philip M. Dixon[1] and Michael C. Newman[2]

[1,2]Savannah River Ecology Laboratory, Drawer E,
Aiken, South Carolina 29802-0005
and
[1]Biomathematics Program, Dept. of Statistics,
North Carolina State University, Raleigh, North Carolina 27695
and
[2]Savannah River Ecology Laboratory, Drawer E,
Aiken, South Carolina 29802-0005

OVERVIEW

In their research toxicologists often use routine testing protocols mandated for regulatory purposes. However, other statistical techniques provide considerably more abundant and precise information than do the standard techniques. Such techniques (called survival analysis, failure time analysis, or life data analysis) are in common use in other disciplines such as epidemiology, clinical medicine, and engineering. They are readily implemented with several common software packages (SAS, BMDP, SYSTAT, S-plus). This review develops a conceptual understanding of these alternate techniques and provides examples of their use.

INTRODUCTION

Lord Rutherford, the eminent British physicist, is often quoted as saying, "If your experiment needs statistics, you ought to have done a better experiment." In some situations, the analysis of toxicity data does not need statistics; an answer is obvious when, for example, a large number of treated animals die while a large number of controls are still alive. In most other situations, however, some statistical analysis is essential. A good analysis both answers toxicological questions and quantifies the uncertainty in those answers. This chapter is concerned with the techniques available to answer some common toxicological questions.

These techniques can be divided into two groups. The first are the techniques mandated by regulation for use in routine toxicity testing. Standard bioassay procedure for a short-term, dose-response experiment is to expose animals for 96 h, count the number of death, and calculate LC_{50}'s and their 95% confidence intervals[1] (Am. Public Health Assoc., pp. 641-645). The focus in these studies is the routine toxicological evaluation of a new chemical or material of unknown constituents. Appropriate techniques are fast, easily performed, and not sensitive to violations of statistical assumptions. However, we find a common tendency for toxicological researchers to uncritically select these routine toxicity testing protocols in their research efforts. This chapter will serve to introduce researchers to some statistical techniques that provide considerably more abundant and precise information than do the standard techniques with only a small amount of additional effort. These techniques, called survival analysis, failure-time analysis, or life data analysis, are widely used in medical and engineering research.[24]

Standard toxicity testing analysis starts with data on the percentage of individuals that survive some period (e.g., 24 or 96 h). Survival analysis starts with the times at which individuals die. Collecting time-to-death data involves more work than recording survival to a fixed endpoint, but significant statistical benefits accrue from the small amount of additional work. Our goal in this chapter is to develop a conceptual understanding of the analysis of time-to-death data and provide examples and interpretation of data analysis using SAS[5,6] programs. Our examples were run using version 6.03 of SAS. Survival analysis programs are also available in the BMDP, SYSTAT, and S-plus systems (Table 1). More detailed and more theoretical treatments of survival analysis can be found elsewhere.[2,4,7] Additional discussion of the application of failure-time analysis in engineering and reliability studies can be found in Meeker and Hahn,[8] Nelson,[3] and Nelson.[9]

The toxicological problem is to describe the effects of factors that modify toxicity.[10] Typically, such questions involve the impact of differences in environmental conditions (water quality, pH), different test species, or individual differences (such as size, sex, or acclimation) on toxicity. Statistically, these questions can be answered by constructing a model that expresses how the time-to-death is influenced by various factors. Given a suitable model, the influences of factors can be estimated and hypotheses about them can be tested. Although we will present techniques for both, our focus will be on estimation, because

Table 1
Availability of Computer Software for Analysis of Censored Times-To-Death[a]

	BMDP	SAS	S-PLUS	SYSTAT
Estimate survival distribution	BMDP 1L	PROC LIFETEST	SURV.FIT	SURVIVAL
Test equality of survival	BMDP 1L	PROC LIFETEST	SURV.DIFF	SURVIVAL
Fit accelerated life model	—	PROC LIFEREG	—	SURVIVAL
Fit Cox model	BMDP 2L	PROC PHGLM	COXREG	SURVIVAL

[a] Absence of an analysis is indicated by a dash.

we find it more biologically informative than hypothesis testing. The biological interpretation of estimates is quite different from the interpretation of a hypothesis test, but estimation and hypothesis testing are closely linked in statistical theory.[11]

The effect of a factor can be expressed in various ways. Some typical examples include the average time to death, the median time-to-death (the time at which 50% of the group has died), the concentration at which 50% had died (the LC_{50}), or the hazard function (the instantaneous probability of dying). The effect of a factor can be expressed by a change in any one of these, not just the difference in LC_{50} typically used.

Data from three studies will be used as examples to introduce and clarify the concepts in survival analysis. We will reanalyze two classic data sets, one by Litchfield[12] on the survival of tuberculoid mice administered streptomycin or a placebo, and the second collected by Shepard[13] on the survival of speckled trout fry in water with various low concentrations of dissolved oxygen. We will also describe in some detail an analysis of the effects of individual covariates on survival of mosquitofish (*Gambusia holbrooki*) exposed to As.[14] These studies span a range of complexity from a comparison of 2 groups (mouse data), to a dose-response comparison of 10 groups (trout data), to an analysis of the effects of individual level covariates (mosquito fish data). Data for the mice and trout examples are published in the original papers and are repeated in Tables 2 and 3. Data for the mosquito fish example are available from the authors and discussed in this volume, Chapter 11.

DATA SETS

Consider the mouse data described by Litchfield,[12] a classic example of the estimation of mean time-to-death. Although it is a biomedical study, the principles can be easily applied to metal toxicology. Litchfield presents data on the survival of mice with tuberculosis. One group of 20 mice was treated with streptomycin, while a larger group of 80 was left untreated. The data consist of the number of days that each mouse survived (Table 2). It seems clear that treatment with streptomycin prolongs the life of these mice, but let us use these data to examine some different ways of describing and graphically presenting survival patterns.

The trout data set is one part of large study of the resistance and tolerance of speckled trout (*Salvelinus fontinalis*) to low oxygen levels.[13] Trout fry were acclimated to water containing 10.50 mg O_2/L; the water was then replaced with deoxygenated water containing from 0.77 to 1.77 mg O_2/L. The data are the number of minutes until the fish died (Table 3). The experiment was terminated at 5000 min; any fish still alive was recorded as censored at 5000 min.

The mosquitofish (*Gambusia holbrooki*) data set is part of a study of resistance to metal and metalloid intoxication. Field-collected mosquitofish (754) were acclimated to laboratory tanks and then exposed to 94 mg As/L in a continuous flow-through exposure system. Every 3 h, dead fish were collected, counted, sexed, and weighed. After 102 h, all remaining fish were collected, sexed, and weighed. Again, survivors at the end of the experiment are recorded as censored at 102 h. Genotypes of all fish at eight enzyme loci were determined with starch-gel electrophoresis.[14] Data are not presented here because of the large number of observations.

The routine toxicity testing approach to all of these data sets would be to calculate the percentage survivorship at some end point, often the end of the experiment (e.g., survival to 96 h). In the mouse data, none of the 80 untreated individuals were alive at 60 days, but 30% (6 out of 20) of the streptomycin-treated mice were. However, the choice of 60 days as the end of the experiment was arbitrary. If the experiment was ended at 20 days, the effect of streptomycin would have appeared larger: none of the 80 untreated mice survived, but all of

Table 2
Survival Time of Mice Infected with Tuberculosis[a]

	Number Dying in	
Day of Death	Untreated	Streptomycin Group
8	1	0
9	4	0
10	0	0
1	10	0
1	10	0
13	7	0
14	19	0
15	12	0
16	4	0
1	3	0
17	0	1
29	0	1
32	0	2
37	0	1
39	0	2
42	0	1
43	0	2
44	0	1
47	0	1
52	0	1
59	0	1
Still alive at 60 days	0	6

[a] One group of 20 was treated with streptomycin; the other group of 80 was left untreated. Experiment was terminated at 60 days, at which time six streptomycin-treated mice were still alive. Data from Litchfield.[12]

Table 3
Survival Times of Trout Exposed to Different Dissolved Oxygen Concentrations[a]

O$_2$ Concentration (mg/l)	Survival Times (min)
0.77	17, 18, 20, 20, 20, 21, 21, 22, 22, 23
0.94	20, 22, 24, 26, 26, 29, 29, 31, 34, 34, 41
1.10	25, 29, 33, 33, 37, 37, 37, 41, 48, 55
1.16	30, 30, 35, 35, 40, 45, 50, 58, 62, 70, 100
1.36	48, 52, 60, 85, 140, 160, 170, 190, 250
1.43	50, 50, 135, 175, 195, 215, 355, 405, 465, 600
1.55	165, 165, 195, 270, 270, 440, 440, 735, 865, 1400
1.69	195, 225, 270, 270, 440, 675, 995, 1150, 1150, 5000, 5000
1.77	240, 675, 995, 2080, 5000, 5000, 5000, 5000, 5000, 5000
1.86	400, 5000, 5000, 5000, 5000, 5000, 5000, 5000, 5000, 5000

[a] The experiment was terminated at 5000 min. From 9 to 11 fish were exposed to each concentration. Data from Shepard.[13]

the treated mice did. If the experiment had ended at 4 days, there would have appeared to be no effect of streptomycin: all of the mice in both groups were alive. What is needed is some way to describe patterns of survivorship and mortality at all times during the experiment. This can be done using either the survivor or hazard function.

SURVIVOR AND HAZARD FUNCTIONS

The survivor function [Equation (1)] describes the probability that an individuals survives longer than some time, t.

$$S(t) = P[\text{An individual dies after time } t] \tag{1}$$

When the times-to-death of all individuals in a group are known, $S(t)$ can be estimated by the proportion of individuals still alive at time t [Equation (2)].

$$\hat{S}(t) = \frac{\text{\# alive at time } t}{\text{total \# animals}} \tag{2}$$

At the start of an experiment, $S(0)$ is 1.00; over time, it decreases to 0.00 when the last individual dies. If the experiment is terminated before the last individual dies, the survivors are counted into the denominator, but not the numerator, so $S(t_e)$ remains above 0.00. Clearly, this function summarizes all information in a table of times-to-death, but it is not the only way to do so. The hazard function is a closely related alternative that describes different aspects of the pattern of mortality.

The hazard function, or force of mortality,[3] describes the probability of dying as a fraction of the number alive at the beginning of the period. It is mathematically related to the survival function [Equation (3)].

$$h(t) = -d \log S(t)/dt = \frac{-1}{S(t)} \frac{dS(t)}{dt} \tag{3}$$

It is often useful to know whether the hazard is constant over time (e.g., 10% of the current survivors will die during any time interval), increases over time (individuals are more likely to die at longer exposure durations), or decreases over time (individuals are less likely to die as exposure duration increases).[3] The shape of the hazard function is a useful tool to help choose a model for time-to-death (see verifying assumptions in Estimation and Hypothesis Testing section). The survivor and hazard functions for the untreated group of mice are given in Figure 1. For these mice, the hazard increases as the duration of exposure increases.

CENSORING

Computing the survival curve for the streptomycin-treated group of mice introduces one complication: not all the animals have died by the end of the experiment. We have partial information on these animals, because we know that they survived 60 days in this study, but we do not know exactly when they died. Censoring is characteristic of survival data, and sophisticated methods to handle very general censoring mechanisms have been developed in the biomedical literature.[2,4] The censoring in this study is very simple; all individuals are censored at the same fixed time, the length of the experiment. Under a general

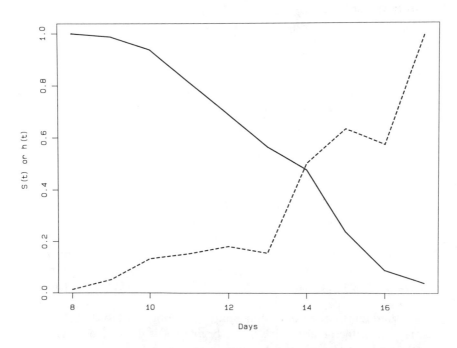

FIGURE 1.   Survivor (——) and hazard (- - -) functions for untreated tuberculoid mice. Data from Litchfield.[12]

censoring mechanism, the survival distribution can be estimated by the Kaplan-Meier (also called the product-limit) estimator[2] [Equation (4)].

$$\hat{S}_{KM}(t) = \prod_{j:t_j < t} \left(1 - \frac{d_j}{r_j} \right) \tag{4}$$

where d_j is the number of deaths occurring at time t_j among the r_j individuals alive just before time t_j.

If there is no censoring, this reduces to Equation (2) above. When censoring occurs only at the end of the experiment, the Kaplan-Meier equation reduces to Equation (2) at all times until the end of the experiment, and it has an undefined value at later times.

Observations can be censored for many reasons. Often, in biomedical studies, patients move away and are never heard from again, or they stop participating for some other reason, or, as in the Litchfield data, the study ends before the last animals die. The Kaplan-Meier estimate provides a consistent estimator of the survival distribution for quite general censoring patterns. In general, the data that we will analyze are pairs of numbers (T_i, C_i). T_i is the observed time of death or censoring, and C_i is a censoring flag. If C_i equals 0, T_i is a time-to-death, but if C_i equals 1, T_i is a censoring time.

Like every statistic calculated from data with variability, the estimated proportion surviving to a particular time is not known exactly. The variance around the survival curve can be approximated in several different ways, but Greenwood's formula (see Cox and Oakes[2] for details) is frequently used. For the special case of end point censoring, Greenwood's estimator reduces to Equation (5), the binomial variance, at all times before the end of the experiment.

$$\text{Var } \hat{S}(t) = \frac{\hat{S}(t)[1 - \hat{S}(t)]}{N} \tag{5}$$

where N is the number of individuals in the study
$\hat{S}(t)$ is the estimated survival at time t

The variance is smallest at either end of the survival curve, when nearly all individuals are still alive or when nearly all are dead. SAS computes approximate 95% confidence intervals around the survival curve as $1.96 \sqrt{\text{Var}}$. This confidence interval assumes that the survival estimate is normally distributed, an assumption which is reasonable for large samples. For small samples, the normal assumption may not be appropriate, especially for survival estimates close to 0 or close to 1. In particular, the computed upper of lower bound to the 95% confidence interval may be larger than 1 or smaller than 0, respectively. Other ways to calculate the confidence interval avoid these problems (see Kalbfleish and Prentice,[4] pp. 14-15 for details.)

PROC LIFETEST[6] will calculate and plot the Kaplan-Meier estimate of the survival curve. The following SAS code reads in the Litchfield data and computes the survival curve for each group. Part of the standard output includes a table of the survival curve, including the standard deviation at each time point, estimated by the square root of Greenwood's variance estimate.[6] If desired, the survival curves can be plotted by specifying PROC LIFETEST PLOTS = (S);. Estimates of the survival curve and its upper and lower confidence bounds can be stored in a new SAS data set for further manipulation. Inspection of the curves and the approximate 95% confidence intervals around each curve suggests that the survival distributions are quite different (Figure 2). SAS also computes two nonparametric tests of the equality of survival curves, which will be described in the next section.

```
data mice;
    input trt $ day dead @@;
    if day > = 60 then censor = 1;
        else censor = 0;
cards;
```

Untreat 8	1	Untreat 9	4	Untreat 10	10	Untreat 11	10	Untreat 12	10
Untreat 13	7	Untreat 14	19	Untreat 15	12	Untreat 16	4	Untreat 17	3
Strepto 26	1	Strepto 29	1	Strepto 32	2	Strepto 37	1	Strepto 39	2
Strepto 42	1	Strepto 43	2	Strepto 44	1	Strepto 47	1	Strepto 52	1
Strepto 59	1	Strepto 60	6							

```
/*Calculate Kaplan-Meier survival curve estimate */
proc lifetest plot = (s) method = km outs = curves;
    time day* censor(1);
    strata trt;
    freq dead;
    title 'Tuberculoid mice, data from Litchfield 1949';

/*superimpose survival curves with 95% confidence intervals */
proc plot data = curves;
    plot survival*day = trt sdf _ lcl*day = "l" sdf _ ucl* day = "u"/overlay;
```

NONPARAMETRIC TESTS FOR EQUALITY OF SURVIVAL CURVES

Survival curves computed from small numbers of observations can be quite variable. The size of the 95% confidence interval gives some idea whether the differences between two survival curves are large enough to be statistically significant, but, for technical reasons, comparing confidence intervals is not an appropriate test. Various nonparametric tests of the equality of two survival curves have been developed. SAS PROC LIFETEST calculates two tests, the log-rank[15] and the Gehan-Wilcoxon.[16] Both test the same null hypothesis: the observed times of death in two (or more) samples come from the same survival distribution. Both are frequently used in the biomedical literature, but they differ in some details.

The log-rank test can be derived as a comparison of the observed numbers of deaths to the expected number of deaths if every group had the same survival

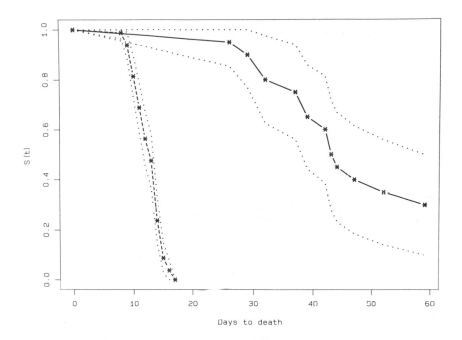

FIGURE 2. Kaplan-Meier estimates of the survival functions for treated (——) and untreated (- - -) tuberculoid mice. Dotted lines are approximate 95% confidence intervals calculated using Equation (5). See text for details. Data from Litchfield.[12]

curve.[17,18] If we consider only the number of animals still alive at the end of an experiment, a standard approach to answering the question, "Were the mortality rates in groups 1 and 2 different?" would be to construct 2 × 2 contingency table like Table 4. The log-rank test generalizes this approach to all times of death.

Consider Litchfield's tuberculoid mouse data (Table 2). On day 8, the first mouse died and the tabulation of animals is that in Table 4. On day 9, four more untreated animals died, and we can construct a second contingency table. In all, 21 tables are calculated, one for each of the 21 different times that mice died.

Table 4
Contingency Tables for
Log-Rank Test of Equality of
Survival Curves

Group	Died	Alive
Deaths on day 8		
Untreated	1	59
Streptomycin	0	20
Deaths on day 9		
Untreated	4	55
Streptomycin	0	20

Table 5
Results of Log-Rank and Wilcoxon Tests for the Equality of
Survival Distributions

	Log-Rank Test		Wilcoxon Test	
	X^2	$P > X^2$	X^2	$P > X^2$
Mouse data				
Streptomycin vs untreated mice (1 d.f.)	62.67	0.0001	41.26	0.0001
Trout data				
Different O_2 exposures (9 d.f.)	259.30	0.0001	217.4	0.0001
Mosquito fish data				
Male vs female (1 d.f.)	24.95	0.0001	45.83	0.0001
Among size groups, males only (5 d.f.)	37.02	0.0001	63.43	0.0001
Among size groups, females only (5 d.f.)	13.95	0.016	25.43	0.0001
Among GPI-2 genotypes (5 d.f.)	17.30	0.004	19.50	0.0015
Among genotypes, males only (5 d.f.)	8.74	0.12	5.41	0.37
Among genotypes, females only (5 d.f.)	9.32	0.097	13.33	0.02

From each table we can calculate the expected number of deaths in each group, just as in a Chi-square test of independence. The log-rank statistic combines information from all tables and all samples into an overall squared difference that measures the similarity between the two survival curves. If the true survival curves are identical, the observed log-rank statistic has an approximate Chi-square distribution with $k-1$ degrees of freedom, where k is the number of groups. The null hypothesis that the groups have the same survival curve can be tested by comparing the observed log-rank statistic to a critical value from the appropriate Chi-square distribution and rejecting the hypothesis if the observed value is too large. For the Litchfield data, the observed log-rank statistic is 62.7 with one degree-of-freedom (Table 5). This is extremely significant and confirms our initial impression that the two survival curves are different.

The Gehan-Wilcoxon test tests the same hypothesis but differs in some mathematical details. One practical difference is that the Gehan-Wilcoxon test is more sensitive to differences at earlier survival times. The log-rank test places more emphasis on differences at later survival times, so the numerical results of the two tests usually differ. Both tests can also be viewed as survival data analogs of familiar nonparametric tests. For example, the Gehan-Wilcoxon test is the censored data analog of the Kruskal-Wallis test.[4] Practically, the choice of log-rank or Gehan-Wilcoxon test makes little difference in the interpretation of any of the three data sets considered here (Table 5). In each study, the survival curves for different groupings of the data are significantly different from each other except for some sex-specific effects of PGI-2 genotype in the mosquitofish data (Table 5).

Both the log-rank and Gehan-Wilcoxon tests can be used to test whether some factor modifies the effect of a toxicant. However, they require that the data be grouped (e.g., treated with streptomycin/not treated), and they do not estimate the size of that effect. They test only whether the two (or more) curves are different. Many factors that might modify toxicant effects are naturally grouped (e.g., sex or species), but many more are continuous (e.g., size of animals, dose of toxicant, and aspects of water chemistry). Such continuous covariates may be artificially grouped, as was done in Table 5 for mosquito fish size. However, more detailed statistical models can be used to test whether these factors have any effect on the influence of a toxicant or to estimate the size of that effect.

MODELS FOR SURVIVAL TIMES

A second approach to testing the equality of survival distributions is to find a statistical model to describe the data, use that model to estimate the differences between the survivor functions, and then assess whether the differences are large enough to be statistically significant. Two models are in common use: the proportional hazards model [Equation (6)] and the accelerated failure time model [Equation (7)].

$$h(t,x_i) = e^{f(x_i)}h_0(t) \tag{6}$$

$$\log t_i = f(x_i) + \epsilon_i \tag{7}$$

The proportional hazards model describes the effect of a particular treatment by its influence on the hazard, the probability that a surviving individual will die during a small interval of time [Equation (6)]. If a reference group (e.g., control animals) has a baseline hazard function $h_0(t)$, then the hazard of another group is some multiple of the baseline hazard. The function $e^{f(x)}$ describes how the treatment (x) determines the multiplier. If $e^{f(x)}$ equals 2 for a particular group, then the hazard for an individual in that group is twice the baseline hazard. Some choices for $f(x)$ are described below.

The accelerated life model [Equation (7)] describes differences between individuals as effects on the distribution of times-to-death, rather than effects on the hazard. The differences between groups, $f(x)$, can take any of the same forms as in the proportional hazards model, but, in the accelerated time model, $f(x)$ acts on the log of the time-to-death. The accelerated life model [Equation (7)] can be converted into a model for the hazard [Equation (8)] that is slightly different from the proportional hazards model in that the effect of the covariate appears both in the multiplier of the hazard and inside the baseline hazard function.

$$h(t,x_i) = e^{f(x_i)}h_0(te^{f(x_i)}) \tag{8}$$

This parametric approach requires that we specify the two parts of the statistical model: the function $f(x)$, which describes how the groups differ from one another, and the error distribution, which describes the variability among individuals in a group. The function $f(x)$ is chosen to reflect our assumptions about how a particular covariate x influences the response of the animal. It may be any of the types of functions used in ordinary regression or analysis of variance.[19] Some possibilities are:

1. $f(X) = a + bX$ — linear response to a continuously measured covariate, e.g., water temperature

2. $f(X) = a + b \log X$ — linear response to the log of a covariate, e.g., size

3. $f(X) = a + bX + cX^2$ — polynomial response to a covariate

4. $f(X) = \begin{cases} m_1 & \text{if male} \\ m_2 & \text{if female} \end{cases}$ — different mean response in each group of individuals

The types of models and the methods for choosing independent variables used in linear regression[19] can also be applied to modeling of survival times.

In typical regression and analysis of variance applications, the error distribution is assumed to be a normal distribution. However, a normal distribution is inappropriate for most time-to-death data because they are not symmetrical around the mean. For example, the histogram of times-to-death of mosquito fish in an acute arsenic exposure experiment has a long right tail (Figure 3). Although many fish die in the first 40 h, some are still dying between 80 and 100 h, and many are still alive at 102 h when the experiment was terminated. Other statistical distributions that may be better descriptions of the distribution of times-to-death include the exponential, Weibull, log-normal, and log-logistic distributions. Each

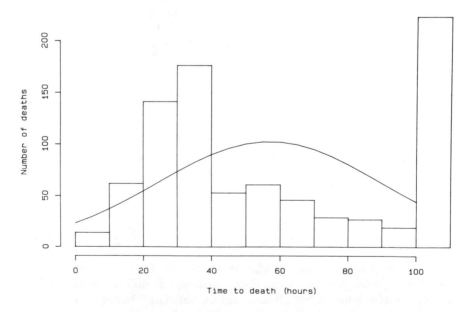

FIGURE 3. Times to death of mosquitofish in an acute arsenic exposure trial. The experiment was terminated at hour 102. The poor fit of the normal distribution (solid curve) is apparent. Data from Newman et al.[14]

Table 6
Survivor Function, Density Function, and Transformations to Linearity for Some Common Distributions of Time to Death[a]

Distribution	Survivor Function	Hazard Function	Transformation Y-axis	X-axis
Exponential	$e^{-\alpha t}$	α	$\log S(t)$	t
Weibull	$e^{-(\alpha t)^{\beta}}$	$\beta\alpha(\alpha t)^{\beta-1}$	$\log[-\log S(t)]$	$\log t$
Log-normal	$1 - \Phi\left(\dfrac{\log t - \mu}{\sigma}\right)$	$\dfrac{e^{-(\log t - \mu)^2/(2\sigma^2)}}{\sqrt{2\pi}\left[1 - \Phi\left(\dfrac{\log t - \mu}{\sigma}\right)\right]}$	$\text{Probit}[1 - S(t)]$	$\log t$
Log-logistic	$\dfrac{1}{1 + (t\alpha)^{\beta}}$	$\dfrac{\beta t^{\beta-1}\alpha^{\beta}}{1 + (t\alpha)^{\beta}}$	$\text{Log}\dfrac{S(t)}{1 - S(t)}$	$\log t$

[a] In the transformation, $S(t)$ is the survivor function, the observed proportion of individuals surviving to time t. Probit is the tabulated probit function. $\Phi(x)$ is the standard normal cumulative distribution function.

of these distributions has characteristic survival and hazard functions (Table 6).

The simplest of the distributions we will consider is the exponential distribution, which is characterized by a constant hazard function (Figure 4). A constant hazard means that the chance of a survivor dying is the same at all times. Radioactive decay is an example of a physical process with a constant hazard function. The density and survivor functions for the exponential distribution are negative exponential functions (Table 6). This distribution is described by one parameter, μ, the mean lifetime of an animal, which is equal to the reciprocal of the hazard. Larger values of μ correspond to lower hazards, more survivors, and fewer deaths in any time interval. The analysis of exponentially distributed data is relatively simple, and much early analysis of failure-time data was based on the exponential distribution,[3] but the assumption of a constant hazard function is appropriate for very few biological systems. For example, the hazard function from Litchfield's untreated mice (Figure 1) increases over time.

The Weibull distribution, a generalization of the exponential distribution, has a hazard function that can take a variety of shapes, not just a flat line like that of the exponential distribution. Weibull distributions are described by two positive parameters: α, the scale parameter that determines the spread and location of the values, and β, the shape parameter that determines the shape of the hazard or survivor functions. If $\beta = 1$, the Weibull reduces to the exponential distribution. If $\beta < 1$, the hazard is initially high and declines with time. If $\beta > 1$, the hazard increases with time, and the survivor function is S-shaped (Figure 5). An intuitive interpretation of the Weibull distribution and the role of β is that the Weibull describes the "weakest link" mode of failure.[3] Consider an individual to be composed of β parts, each of which has a constant hazard of failing. If the individual dies when any one of the β parts fails, then the time-to-death will fit a Weibull distribution. The Weibull distribution (and its special case, the exponential distribution) is the only distribution for which the accelerated time and the proportional hazards models are identical[4] because of the mathematical form of the hazard and survivor functions.

Although the Weibull distribution is quite flexible, its hazard function is monotonic: always increasing if $\beta > 1$ and always decreasing if $\beta < 1$. The log-normal and log-logistic distributions are similar distributions with hazard functions that may monotonically increase, monotonically decrease, or change directions over time (Figure 6). The hazard curves for both distributions are different for an accelerated time model and a proportional hazards model (Figure 7). For the proportional hazards model, the hazard of a treatment group is some constant proportion of the baseline hazard at all times. The hazard for the accelerated time model increases more quickly than the hazard under a proportional hazards mode, and then declines closer to the baseline level (Figure 7). Although the effect of the treatment is the same in the proportional hazard and accelerated failure models, the median time of death and the distributions of times-to-death will be different.

The choice of error distribution may or may not affect the estimation of treatment effects. For example, the shapes of the log-normal and log-logistic distributions are similar, so the substitution of one for the other usually makes little difference. However, changing the error distribution from log-normal to Weibull may substantially change the parameter estimates.

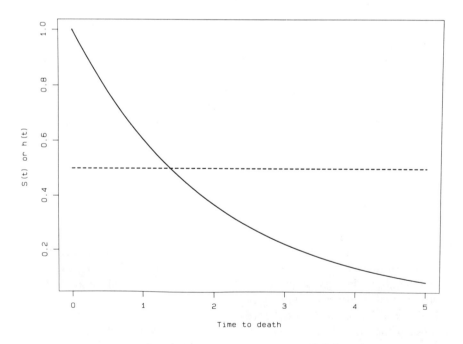

FIGURE 4.   Survival (——) and hazard (- - -) functions for a exponential distribution ($\alpha = 0.5$).

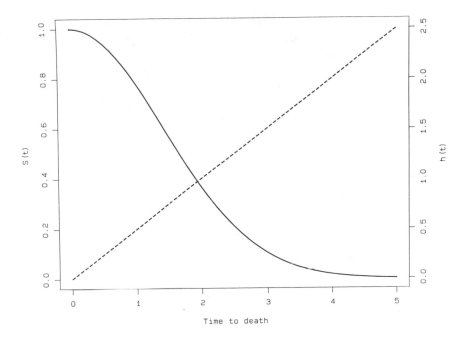

FIGURE 5. Survival (——) and hazard (- - -) functions for a Weibull distribution (α = 0.5, β = 2).

Two techniques can help in choosing an appropriate error distribution. The first is to compare maximum log-likelihoods for different models. These numbers are calculated by the common technique of fitting a model (see next section for details). For each model, the maximum value of the log-likelihood function is an index of relative fit. Larger values indicate better fit. For example, if we use a Weibull, log-normal, and log-logistic distribution to fit the mice data with MODEL DAY*CENSOR(1) = TRT, the maximized log-likelihoods are 9.74, 7.21, and 5.84, respectively. The Weibull fits better than the other two distributions because its log-likelihood is the largest. However, comparison of log-likelihoods does not show that the Weibull is a good fit. Plotting cumulative hazards, the second technique, can show that a particular distribution fits the data (see Miller[7], pp. 164-166 for details). In a cumulative hazard plot, a transformation of the observed survival is plotted against the time-to-death or the log of time-to-death (see Table 6 for details). The plot will be a straight line if the error distribution fits the data. An example and SAS code to graph a hazard plot is given in the section on Estimation and Hypothesis Testing.

Accelerated failure time models are models for time-to-death, but they can be related back to traditional toxicological models for dose-response curves. Consider an experiment in which time-to-death is recorded for individuals exposed to different doses. A model with a linear response to dose and log-logistic errors can be transformed into a logistic dose-response model. Similarly, a model with log-normal errors can become a probit dose-response model, and a model with Weibull errors can be transformed into a Weibull dose-response model.[20]

ESTIMATION AND HYPOTHESIS TESTING

A statistical model for times-to-death specifies how various covariates influence the median time-to-death, but it does not specify the magnitude of the influence. For example, a model for the Litchfield data might be

$$\log t_{ij} = \mu + \tau_i + \epsilon_{ij} \qquad (9)$$

where μ measures the average longevity in the control treatment the treatment effect,

τ_i, measures the change in longevity caused by streptomycin treatment

ϵ_{ij}, the errors, measure the differences among individuals in a group

Two models for the Shepard data might be

$$\log t_{ij} = \mu + \tau_i + \epsilon_{ij} \qquad (10)$$

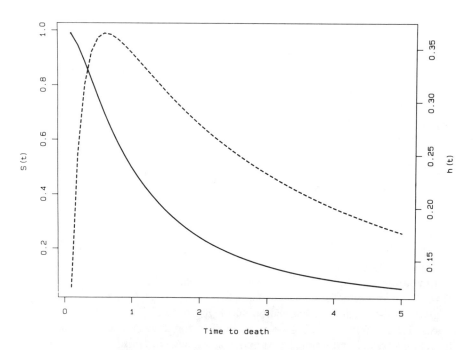

FIGURE 6. Survival (——) and hazard (- - -) functions for a log-normal distribution ($\mu =$ 40, $\sigma = 20$).

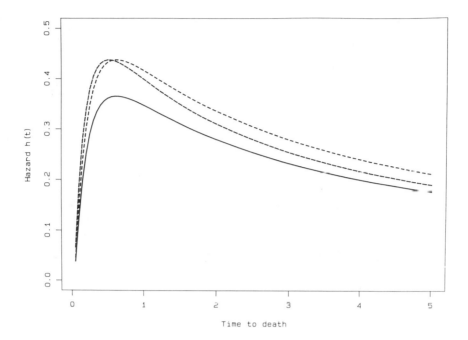

FIGURE 7. Hazard functions for the log-normal distribution: (A, ———) a baseline hazard, (B, ——————) an accelerated time model, and (C, - - - - -) a proportional hazard models. The hazard is increased by 0.2 units in (B); the median survival time is increased by 0.2 units in (C).

where the treatment effect, τ_i, measures the difference in longevity between a reference group and the group exposed to dose i
μ measures longevity in the reference group,
ϵ_{ij} measures the differences among individuals in a group

This model specifies that the groups exposed to various doses differ in their longevity but does not specify any dose-response relationship between dose and longevity. An alternate model that specifies a linear dose-response curve is

$$\log t_{ij} = \alpha + \beta x_{ij} + \epsilon_{ij} \tag{11}$$

Here, the dose effect, β, measures the slope of the relationship between the dose (x_{ij}) and the log-transformed time to death. A slope of 0.69 means that the median time-to-death doubles when the dose increases by 1 unit (natural log of 2 = 0.69). The intercept, α, measures the longevity for individuals with a dose of 0. Discrete and continuous parameters may be combined, as in the following model for the mosquito fish data

$$\log t_{ijk} = \mu + \alpha_i + \tau_k + \beta x_{ij} + \epsilon_{ijk} \tag{12}$$

where the α_i parameter measures the difference between sexes

τ_k parameters measure differences among genotypes

β parameter is a slope measuring the effect of size

Each model includes parameters (μ, α_i, β, τ_k) whose values must be estimated from the data. If the error distribution is specified, estimates can be obtained by maximum likelihood.[2] Briefly, maximum likelihood is a general procedure for statistical estimation.[21] The specified model, including the error distribution, is used to construct a likelihood function, a function of the unknown parameters. For any value of the unknown parameters, the likelihood function is proportional to the probability of observing the data. Intuitively, a good choice of estimate is the value which maximizes the likelihood. These are the maximum likelihood estimates (MLEs), which have many desirable statistical properties (see Edwards[21] or Mood et al.[11] for further details).

Three aspects of the likelihood function are useful. The MLEs are point estimates of the parameter, providing values for the mean of a group or the slope of a dose-response relationship, but they contain no information about variability. The precision of an estimate can be obtained from the curvature of the likelihood function around the maximum likelihood estimates. Usually, this curvature is expressed as the asympototic variance of an estimate. If the estimate is very precise, its asymptotic variance will be small, and the likelihood will decrease quickly as one considers estimates slightly different from the maximum likelihood estimate. Conversely, if the estimate is very poorly known, its asymptotic variance will be large, and estimates slightly different from the MLE will be almost as good. Finally, the log of the value of the likelihood function, calculated at the MLEs, provides a measure of fit that can be used to compare different models. This measure of fit is analogous to the model sum of squares used in regression analyses.

The likelihood function can be written down easily for any of the distributions considered here but finding its maximum is not as simple. Closed-form analytical expressions for the MLEs are available for an exponential error distribution but are not available for other distributions. Estimates for other error distributions have to be found by numerical iteration. SAS PROC LIFEREG[5] can be used to fit a wide variety of accelerated failure-time models using exponential, Weibull, log-normal, and other error distributions. If the exponential or Weibull distribution is used, then PROC LIFEREG is also fitting a proportional hazard model.

Details of the syntax of PROC LIFEREG can be found in the relevant SAS manual, but briefly: The MODEL statement specifies the desired model, including the response variable, a censoring indicator, the desired covariates, and the appropriate error distribution. Its syntax is

MODEL ttd*flag(number) = covariates/D = distribution;

where ttd is the name of the variable containing the times-to-death (or times-to-censoring) flag is a variable that identified whether that observation is a death or censored

The observation is treated as censored when the value of flag is the number in parentheses. The list of covariates to be included in the model is specified as a list of variable names to the right of the equals sign. Distribution is the name of the error distribution. For example, the model statement to fit Equation (12) with a log-normal distribution to the mosquitofish data is

MODEL TTD*CENSOR(1)
= SEX LOGSIZE GPI2/D = LNORMAL;

The variable CENSOR was created in a DATA step. CENSOR has the value of 0 if the fish died and 1 if the fish was still alive at 102 h. Multiple model statements can be included in one PROC. By default, SAS will treat all covariates as continuous linear variables. To declare a variable as a classification variable includes it in a CLASS statement. The following SAS code fits Equation (12) to a set of data. The CLASS statement is used to declare that SEX and GPI2 are classification variables. LOGSIZE, omitted from the CLASS statement, is a continuous variable (log size).

```
proc lifereg data = arsenic.all;
    class sex gpi2;
    model ttd * censor(1) = sex logsize gpi2/d = lnormal;
```

The following example of output from PROC LIFEREG includes a summary of the classification variables (if a CLASS statement was used), the dependent and censoring variables, counts of the number of noncensored and censored observations, and the error distribution. The maximized log-likelihood for the model is printed after the summary of variables. The parameter estimates, their standard errors, Chi-square statistics, and their tail probabilities are printed on a separate page. For each parameter, a 1 d.f. Chi-square statistic tests whether that parameter equals 0. If a CLASS statement is used, a $(k - 1)$ d.f. Chi-square statistic tests whether any of the k level of the classification variable differs from the other levels.

LIFEREG Procedure

Class	Levels	Values		
SEX	2	female	male	
GPI2	6	100/100	100/66	100/38
		66/66	66/38	38/38

Number of observations used = 751

Data Set	= WORK.ALL	
Dependent Variable	= Log (TTD)	
Censoring Variable	= FLAG	
Censoring Value(s)	= 1	
Noncensored Values	= 626 Right Censored Values	= 125
Left Censored Values	= 0 Interval Censored Values	= 0

Log Likelihood for LNORMAL − 808.9266927

Variable	d.f.	Estimate	SE	Chi-Square	Pr > Chi	Label/Value
INTERCEPT	1	3.67880488	0.195743	353.2178	0.0001	Intercept
SEX	1			12.93713	0.0003	
	1	0.22897392	0.06366	12.93713	0.0003	female
	0	0	0	.	.	male
LOGSIZE	1	0.19420818	0.039026	24.76389	0.0001	
GPI2	5		.	16.76149	0.0050	
	1	0.33847937	0.180507	3.51623	0.0608	100/100
	1	0.30274341	0.178672	2.871011	0.0902	100/66
	1	0.05067728	0.185455	0.07467	0.7847	100/38
	1	0.18769681	0.186692	1.010792	0.3147	66/66
	1	0.15954612	0.190794	0.699264	0.4030	66/38
	0	0	0	.	.	38/38
SCALE	1	0.70432209	0.020843			Normal scale parameter

INTERPRETING PARAMETER ESTIMATES

The estimates and their standard errors calculated by PROC LIFEREG can be interpreted in three different ways: as hypothesis tests, as shifts in median time-to-death, and as relative risks. As a hypothesis test, they are used to answer the question: does factor X have any influence on when individuals die? If it does, then changing the level of factor X will change the median times-to death. If X is a classification variable, then we can consider testing the hypothesis (H_0): Different levels of X have the same median time-to-death, against the alternate (H_1): At least one level of X has a different median time-to death. An approximate test of this hypothesis is calculated by SAS and presented as a $(k-1)$ degree-of-freedom Chi-square test associated with each factor X, where k is the number of levels of factor X. If X is a continuous variable, then the 1 d.f. Chi-square statistic tests the hypothesis that the slope of the relationship between X and the log time-to-death is zero. For example, using the previous SAS output, we find that the two sexes are significantly different in their time-to-death ($X^2 = 12.9$, $P = 0.0003$); size significantly affects time-to-death ($X^2 = 24.8, P = 0.0001$); and at least one GPI-2 genotype is significantly different from the rest ($X^2 = 16.8, P = 0.005$).

The tests calculated by SAS for each factor are approximate, because they assume that each estimate is normally distributed. Technically, they are Wald tests.[22] An alternate test for small samples is the likelihood ratio test,[22] which can be calculated using the output from two runs of PROC LIFEREG. To test whether factor X influences time-to-death, fit a model including factor X. Then, fit a model without factor X. If X has no influence on time-to-death, then the model without X will fit as or almost as well as the model with X. Conversely, if X has a large effect on time-to-death, then removing it from the model will increase the lack of fit. The lack of fit of a model is quantified by the log-likelihood statistic calculated by PROC LIFEREG for each model. A better fitting model has a larger (less negative) log-likelihood value. If X has no effect, then twice the difference between the two log-likelihood values is approximately distributed as a Chi-square random variable.[22] As in the Wald test, the degrees-of-freedom are $(k-1)$ if X is a classification variable with k levels and 1 if X is continuous.

The estimates themselves can be used to calculate the shift in median time-to-death. Remember the statistical model fit to the data is

$$\log t_{ij} = \mu + \beta X_{ij} + \epsilon_{ij} \tag{13}$$

If X is continuous, then a slope (β) of 0.5 means that the predicted log-transformed time-to-death increases 0.5 units for every increase of 1 unit in X. Transforming back from the log scale, a slope of 0.5 means that the predicted median time-to-death increases to $e^{0.5} = 165\%$ of the original baseline value. A slope of 0, indicating no effect, leads to a predicted median time-to-death of $e^0 = 1 = 100\%$ of the baseline value. A slope less than 0 means that the predicted time-to-death decreases as X increases. If X is a classification variable, then one of the groups is used as a reference group and the change in time-to-death is relative to that group. By default, SAS uses the largest value of the classification variable as the reference group. In the example given earlier, SEX and GPI2 are class variables. Males and the 38/38 genotype were used by SAS as the reference groups for SEX and GPI2, respectively. The estimate for females (0.229) is the difference between females and males, and the estimate for each GPI2 genotype is the different between that genotype and the 38/38 genotype.

If the model is a proportional hazards model, then estimates can be transformed into relative risks. Relative risk is the ratio of the probability of dying if an individual is in group X to the probability of dying if an individual is in the reference group. Alternatively, relative risk is the ratio of the hazard for individuals in group X to the baseline hazard. It can be calculated using the estimate from an accelerated failure time model by equation (14).

$$\text{Relative Risk} = \begin{cases} e^{-\tau_i/\sigma} & \text{for classification effects} \\ e^{-\beta_i \Delta x/\sigma} & \text{for continuous effects} \end{cases} \tag{14}$$

where τ_i and β_i are estimates of treatment differences and slopes, respectively
σ is the estimated scale parameter from the SAS output
ΔX is the desired difference between two values of a continuous variable

For the model and estimates given earlier, the relative risk of females (Relative to males of the same size and genotype) is $e^{-0.229/0.704} = 0.722$. The relative risk of an 0.15-g individual (logsize $= -1.90$), relative to an 0.10-g individual (logsize $= -2.30$) is $e^{-0.194[-1.90-(-2.30)]} = 0.92$. Relative risks greater than 1 mean that, at any point in time, individuals in group X are more likely to die than are individuals in the baseline group. Relative risks less than 1 mean that individuals in group X are less likely to die. Relative risk has a clear interpretation in proportional hazards models (e.g., accelerated failure-time models with exponential or Weibull distributions) where the ratio between the two hazards is the same at all times. It is less useful for other models.

CALCULATING MEDIAN TIMES-TO-DEATH

It is often useful to calculate the predicted median time-to-death for an individual having some combination of characteristics. This can be done using SAS PROC LIFEREG. The median time-to-death and its standard error can be calculated from characteristics of the individual (X, the parameter estimates ($\hat{\beta}$, $\hat{\tau}_k$, and $\hat{\alpha}_i$), the estimated scale parameter ($\hat{\sigma}$), and the choice of error distribution [Equation (15)].

$$\text{Median TTD} = \exp(\hat{\mu} + \hat{\alpha}_i + \hat{\tau}_k + \hat{\beta}x_{ijk} + \hat{\sigma}W_{0.5}) \qquad (15)$$

$W_{0.5}$, the median of the standardized error distribution, depends on the choice of distribution. For the log-normal distribution, $W_{0.5}$ is 0. For the exponential and Weibull distributions, $W_{0.5}$ is -0.3665. These calculations can be performed by SAS (see section on Case Study for an example).

VERIFYING ASSUMPTIONS

Any statistical model makes assumptions that should be checked as part of a careful analysis. If Weibull or exponential distributions are used in an accelerated failure time model, we assume that:

1.   Hazards are proportional.

Then an accelerated failure time model makes three major assumptions:

2.   Baseline hazard distribution is correctly specified.
3.   Response to the covariates is correctly specified.
4.   Observations are independent.

Most of these assumptions can be assessed using simple graphical tools. We will present methods for verifying that the first three assumptions are appropriate. Good experimental design helps justify the last assumption that observations are independent.

The proportional hazards assumption can be checked by dividing the data into groups of similar observations and plotting a cumulative hazard curve for each group (see test in section on Censoring and below). If the data contain continuous covariates, it will be necessary to divide the covariate into a small number of groups (2-4, depending on sample size). For each group, calculate the Kaplan-Meier estimate of the survival distribution, then plot $\log[-\log S(t)]$ against t or $\log t$. If the hazards are proportional, the curves will be parallel (Figure 8). If the sample sizes are small, the variability in each curve is high, and the eye can easily see nonparallel lines, even if the hazards are proportional. It is often helpful to compute the approximate 95% confidence interval around each survival estimate, then log-log transform the upper and lower bounds for the confidence interval to help judge whether the lines are parallel.

Hazard plots can help assess the choice of error distribution (Miller,[7] pp. 164-166). If the error distribution is Weibull (or exponential), the plot of $\log[-\log S(t)]$ vs $\log t$ will be a straight line (see Figure 8). Other transformations (Table 6) can be used to evaluate log-normal or log-logistic distributions. SAS will calculate and plot hazard curves in PROC LIFETEST, but one must do additional calculations to plot the confidence bounds. One can use PROC LIFETEST to estimate $S(t)$ and its standard error with PROC LIFETEST, use a DATA step to calculate and transform the confidence interval, and plot the curves (see the following code). For small samples (e.g., 50 individuals per group), the variation in the curves may mask a truly straight line. To continue the analysis of the mice data (started in the censoring section) the following SAS code plots hazard curves for treated and untreated mice.

```
/*Continuation of Analysis of Litchfield data to plot hazard curves */
data curves2;
  set curves;
  log _ ttd = log(day);
  lls = log( - log(survival));
  lower = log( - log(sdf _ lcl));
  upper = log( - log(sdf _ ucl));

plot plot;
  plot lower*log _ ttd = " - " upper*log _ ttd = " + " lls*log _ ttd = trt/overlay;
  title 'Log( - Log S(t)) vs Log TTD for each group';
  title2 'with confidence interval';
```

The form of response to covariates can be tested graphically or by fitting augmented models. Consider what might be a reasonable way to improve the functional form and see if that augmented model actually does fit the data better. If the original model includes a linear response to a continuous covariate, a possible augmented model is a quadratic, or perhaps a cubic, equation. If the model includes two continuous covariates, a possible augmented model includes the cross-product of the two variables. If the model includes two classification variables, a possible augmented model may include the interaction between the two variables. The augmented model will always fit the data better because we are using more variables. The relevant statistical question is whether the improvement in the fit is significant. The log-likelihoods reported by PROC LIFEREG for each model can be used to construct a likelihood ratio test[21] for the significance of the improvement.

We will test whether a linear dose-response model is adequate to describe the mortality patterns of trout exposed to low oxygen. In the subsequent SAS code (following the next paragraph), the first MODEL statement, labeled "Linear," fits a model with a linear response to oxygen level. The second, labeled "Quad," augments the linear model to include a quadratic response. The third, labeled "Loglin," fits a model with a linear response to log-transformed oxygen level. The last, labeled "Groups," fits a model with a different mean for each group.

If any of the first three models is appropriate, then that model will fit almost as well as the groups model. These calculations are repeated for just the intermediate group of oxygen levels in the second PROC LIFEREG step. Log-normal error distributions were chosen because plots of log $[-\log S(t)]$ vs log t were not straight lines.

Statistical tests of the improvements in fit can be constructed from the SAS output (Table 7). If the groups model fits just as well as a linear model, then twice the difference in likelihoods has a Chi-square distribution. The degrees-of-freedom for the Chi-square is the difference of the number of parameters in each model. For the complete data the Chi-square statistic to test the fit of the linear or loglin models has 8 d.f. because the "Groups" model has 11 parameters (10 means and 1 scale parameter), while the linear models each have 3 parameters (1 intercept, 1 slope, and 1 scale parameter). If twice the observed difference in log-likelihoods exceeds a critical Chi-square statistic, once concludes that the full model fits significantly better. For these data, the linear model is not sufficient to describe the response to all doses, but it is adequate for the intermediate doses (Table 7). This approach cannot be used to test between a linear response and a log-linear response because both models have 3 parameters. Technically, the

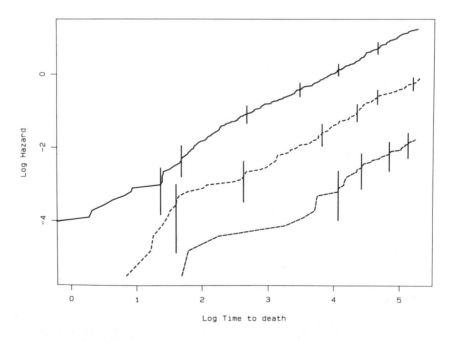

FIGURE 8.   Hazard plots for simulated data. Data were generated from a proportional hazards model with Weibull errors. Vertical bars indicate 95% confidence intervals for $S(t)$ at selected times. The relative risk of group 2 (- - -), relative to group 1 (————), was 4.48. The relative risk of group 3 (————), relative to group 1, was 20.

Table 7
Summary of Log-Likelihood and Tests of Fit of Linear Models for Trout Data

Model	Log-Likelihood (d.f. in model)	
	All Doses	Intermediate Doses
Groups	−112.48(11)	−67.73(7)
Linear response	−132.82(3)	−68.14(3)
Quadratic response	−115.45(4)	−68.06(4)
Log-linear response	−146.13(3)	−68.71(3)

Test of Fit	−2Δ Log-Likelihood (d.f.,p)	
	All Doses	Intermediate Doses
Linear response vs groups	40.68(8, 0.0001)	0.82(4, 0.94)
Quadratic response vs groups	5.94(7, 0.55)	0.66(3, 0.69)
Linear vs quadratic response	34.74(1, 0.0001)	0.16(1, 0.88)
Log-linear response vs groups	67.30(8, 0.0001)	1.96(4, 0.74)

likelihood ratio test is valid when one model is nested in the full model.[21] Both the linear and log-linear models are nested in the groups model, but the linear model is not nested in the log-linear model.

```
data shepard;
 infile cards missover;
 input o ttd @;
 log_o = log(o);
 o_group = o;
 o2 = o * o;
 do until (ttd =.);
   if ttd = 5000 then censor = 1;
     else censor = 0;
   output;
   input ttd @;
   end;
cards;
0.77  17  18  20  20  21  21  22  22  23
0.94  20  22  24  26  26  29  29  31  34    34  41
1.10  25  29  33  33  37  37  37  41  48  55
1.16  30  30  35  35  40  45  50  58  62  70  100
1.36  48  52  60  85  140  160  170  190  250
1.43  50  50  135  175  195  215  355  405  465  600
1.55  165  165  195  270  440  440  735  865  1400
1.69  195  225  270  270  440  675  995  1150  1150  5000  5000
1.77  240  675  995  2080  5000  5000  5000  5000  5000  5000
1.86  400  5000  5000  5000  5000  5000  5000  5000  5000  5000

proc lifereg;
 title 'all groups';
 class o_group;
 Linear: model ttd*censor(1) = o/d = lnormal;
 Quad: model ttd*censor(1) = o o2/d = lnormal;
 Loglin: model ttd*censor(1) = log_o/d = lnormal;
 Groups: model ttd*censor(1) = o_group/d = lnormal;
```

```
proc lifereg;
  title 'intermediate groups';
  where o between 0.95 and 1.76;
  class o_group;
  Linear: model ttd*censor(1) = o/d = lnormal;
  Quad: model ttd*censor(1) = o o2/d = lnormal;
  Loglin: model ttd*censor(1) = log_o/d = lnormal;
  Groups: model ttd*censor(1) = o_group/d = lnormal;
```

Unlike SAS PROC GLM, PROC LIFEREG will not automatically generate the quadratic, cross-product, or interaction terms. To fit such models, one should create a new variable for each new term in the model. For example, to test for possible interaction between two classification variables, we can create a new variable containing a unique value for each combination of the two original variables. An easy way to do this in SAS is to use the string concatenation operator, ||, if the two original variables are character variables. The importance of a quadratic or cross-product term can be tested by including new variables containing the square of the original variable or the product of two original variables, respectively.

GROUPED TIMES-TO-DEATH

All of the preceding analyses have assumed that the exact time-to-death was recorded; hence, time-to-death is a continuous variable. As in any measurement, the fineness of the measuring scale imposes some discreteness on what theoretically could be a continuous variable. In the mosquitofish data, dead fish were collected every 3 h, so that time-to-death is recorded as 9 h, or 12 h, but never as 10.3 h. Even so, time-to-death may be considered continuous in these data because 3 h is short relative to the 102-h duration of the experiment.

What if deaths were recorded weekly in a 12-week experiment (e.g., Quattro and Vrijenhoek[23])? Here the interval between measurements is a sizable fraction of the duration. An animal recorded as a death at week 3 actually died between the second and third weeks. Although we do not know the exact time-to-death, we know that it falls within a certain interval. PROC LIFEREG will fit parametric survival models to such interval censored data (see following code). Animals still alive at the end of the experiment are right censored, just as before. Parameter interpretations are the same as those for exact data.

```
/* SAS code to fit interval censored model to discrete times to death      */

/* experiment terminated at week 12,                                       */
/* survivors to week 12 are coded in the raw data as ttd = 13 and are      */
/* right censored,                                                         */
/* observed deaths occurred between observed ttd and previous week         */

data interval;
  input trt ttd;

/* If the animal was alive at the end of the experiment, code it as right censored at week
12                                                                        */
```

```
if ttd > 12 then do;
  lower = 12;
  upper = .;
  end;

/* But if it died during the experiment, code it as interval censored between the previous
week and this week                                                                    */

  else do;
  lower = ttd - 1;
  upper = ttd;
  end;

proc lifereg;
  class trt;
  model (lower,upper) = trt;
```

CASE STUDY: THE INFLUENCE OF SEX, SIZE, AND GENOTYPE ON ARSENIC INTOXICATION

As part of a larger study of metal tolerance in mosquitofish, we examined the roles of sex, size, and genotype as modifying factors affecting As intoxication. A summary of the data collection protocol is provided in Data Sets; further details are given in Newman et al.[14] The primary concern in this study was whether fish with particular genotypes were more sensitive than other genotypes to As intoxication. To increase the precision of comparisons between genotypes, it was necessary to control potential variation due to the effects of size and sex. We analyzed the effects of different genotypes at eight enzyme loci, but this analysis will be restricted to the effects at one locus, GPI-2. The questions we wish to answer are:

1. Are males and females equally sensitive to the effects of As?
2. Are larger individuals of either sex more resistant to As?
3. Does the genotype at the GPI-2 locus influence survival?

The first step in the analysis is to group individuals into size classes and calculate survival curves for each sex, size, and genotype class. Hazard plots for each sex (Figure 9a) are linear until approximately 42 h (log t = 3.75), when the slopes decrease, but are not parallel. Similar patterns are found among different sizes classes of male or female fish and among different genotypes (Figure 9b). A Weibull error distribution is not appropriate because the hazard plots are not linear and parallel across all the data. The Weibull would not be an appropriate distribution even if two modes of action were postulated for As, one responsible for deaths before 42 h and one after 42 h, because the lines are not parallel.

The differences between survival curves for different sexes, different size classes, and different genotypes are statistically significant by either the log rank test or the Wilcoxon test (Table 5). When each sex is considered separately, the

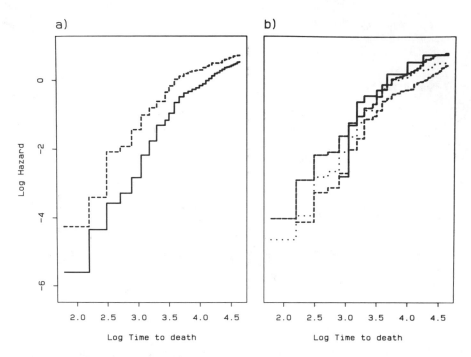

FIGURE 9.   Hazard plots for mosquitofish exposed to arsenic (a) males (- - -) and females (——) and (b) four of the six genotypes at the GPI-2 locus; 38/38 (——), 66/66 (....), 100/100 (- - -), 100/66 (—— ——).

survival curves for different GPI-2 genotypes are significantly different in females (using the Wilcoxon test) but are not significantly different in males (using either test). Fewer of the experimental animals were males, so the lack of significant differences between GPI-2 genotypes may just be a result of low statistical power. Accelerated time models can be used to verify that the effects of GPI-2 genotype are different in the two sexes.

Perhaps the simplest reasonable model that includes the effects of sex, size, and genotype is Equation (12), in which sex and genotype are classification variables and some function of size is a continuous variable. The form of the response to size can be estimated from plots of log time-to-death versus various transformations of size. There is considerable scatter in these plots, but the plots of log(size) appear to be more linear than those against untransformed size. A log transformation of size is commonly a better predictor of other measures of toxic potency, such as LD_{50} (ref. 24) (see also this volume, Chapter 4). The following SAS code fits this model using both Weibull and log-normal error distributions.

```
proc lifereg data = arsenic.all;
  class sex gpi2;
  model ttd*censor(1) = sex logsize gpi2;
  model ttd*censor(1) = sex logsize gpi2/d = lnormal;
  title 'Basic model — comparison of Weibull and lognormal errors';
```

The log-likelihood of the model using the log-normal error distribution (-808.9) is less negative than that from the Weibull distribution (-884.8), indicating that the log-normal is the more appropriate error distribution, as expected from the hazard plots (Figures 9a, b). This model assumes that the response to size and GPI-2 genotype is the same in male and females. These assumptions can be tested by fitting more general models. The following SAS code constructs new variables and fits two models to test whether the response to size differs between males and females.

```
data sexsize;
  set arsenic.all;
  mlsize = 0;
  flsize = 0;
  if sex = 'M' then mlsize = logsize;
    else flsize = logsize;
  sexsize = flsize

proc lifereg;
  class sex gpi2;
  2sex: model ttd*censor(1) = sex mlsize flsize gpi2/d = lnormal;
  sexdiff: model ttd*censor(1) = sex logsize sexsize gpi2/d = lnormal;
```

The log-likelihood for either model is -804.8, a statistically significant improvement from a model with the same slope for both sexes. The two models differ in how the differences between the sexes are parameterized. In the model labeled "2sex," separate slopes are fit for males and females. These estimates of these slopes can be directly interpreted, but the test of whether the two slopes are different must be hand calculated from the log-likelihood statistics. The model labeled "sexdiff" parameterizes the slope as the slope for males (LOGSIZE) and the difference between the male and female slopes (SEXSIZE). The SEXSIZE coefficient is an estimate of the difference between the slopes; the test of whether SEXSIZE equals 0 tests the hypothesis that the slopes are equal. The log-likelihoods of the two models are the same, because they are different forms of exactly the same model.

A similar procedure could be used to construct five new variables to test the SEX*GPI2 interaction (five variables are necessary because the GPI2 classification variable has five degrees-of-freedom), but an even more general model can be used to test whether there are any other interactions. The following SAS code fits separate models to the male and female data.

```
proc sort data = arsenic.all;
  by sex;

proc lifereg;
  by sex;
  class gpi2;
  model ttd*censor(1) = logsize gpi2 /d = lnormal;
```

The log-likelihood statistics (and degrees-of-freedom) for each sex can be added together to calculate the log-likelihood and degrees-of-freedom for a model in which each parameter (intercept, slope for log of size, effect of GPI-2 genotype, and scale) is allowed to differ between the sexes. The improvement in fit for this model is small (log-likelihood of full model is -803.6) and is not statistically significant (Table 8).

Estimates and their standard errors can be obtained directly from the SAS output (Table 9). Each estimate can be interpreted as the difference in log-transformed median time-to-death caused by a 1 unit change in that variable. Sizes of male and female fish are continuous variables, so the estimates can be interpreted like slope estimates from an analysis of covariance. For example, for a male fish, each 0.2-unit increase in log(size) increases the log-transformed median time-to-death by $(0.2)(0.53) = 0.106$ units. For females, the same change results in an increase of $(0.2)(0.157) = 0.031$ units. Transformed from the log scale, an 0.2-unit increase in size leads to $e^{0.106} = 1.11$ (or 11% larger) predicted median time-to-death in males and $e^{0.031} = 1.03$ (or 3% larger) in females. SEX and GPI-2 are classification variables that can be interpreted like groups in analysis of variance. The estimate of the SEX effect is the difference between the intercept for male fish and the intercept for female fish. The estimates for each GPI-2 effect are the differences between that genotype and the last genotype (in this case, the 38/38 genotype). For example, the estimated effect of the 100/100 genotype is 0.35. Hence, the 100/100 genotypes have a predicted median time-to-death that is $e^{0.35} = 1.42$ times as long as the median time-to-death for fish with the 38/38 genotype. Because the error distribution was log-normal rather than Weibull, the estimates cannot be interpreted as relative risks.

Rather than interpret estimates directly, it is possible to calculate the median times-to-death for individuals with specified characteristics and interpret them. For example, Table 10 presents characteristics of 12 individuals with different combinations of sexes, genotypes, and sizes that span realistic values. The

Table 8
Summary of Log-Likelihood Tests for Mosquito Fish Date

Model	Error Dist.	Log-Likelihood	d.f. in Model
1. Sex Log(size) GPI-2	Weibull	-884.8	9
2. Sex Log(size) GPI-2	Log-normal	-808.9	9
3. Sex Log(size) Sex*Log(size) GPI-2	Log-normal	-804.8	10
4. Both sexes: Log(size) GPI-2	Log-normal	-803.6	16
Females only: Log(size) GPI-2	Log-normal	-565.4	8
Males only: Log(size) GPI-2	Log-normal	-238.2	8

Tests of fit	-2Δ Log-likelihood	Δd.f.	$P > X^2$
Does response to size differ between sexes?			
Model 2 vs 3	8.2	1	0.0044
Do any other parameters differ between sexes?			
Model 3 vs 4	2.4	6	0.88

Table 9
Parameter Estimates and Hypothesis Tests for Model "2sex"
(see Case Study section) Fit to Mosquito Fish Data

Parameter	d.f.	Estimate(SE)	X^2	$P > X^2$
Intercept	1	4.30(0.29)	216.7	0.0001
Sex	1	−0.45(0.24)	3.3	0.0678
Log(size) for males	1	0.53(0.12)	18.4	0.0001
Log(size) for females	1	0.157(0.041)	14.8	0.0001
GPI-2 (overall)	5		17.1	0.0043
GPI-2, effect of 100/100	1	0.35(0.18)		
GPI-2, effect of 100/66	1	0.31(0.18)		
GPI-2, effect of 100/38	1	0.06(0.18)		
GPI-2, effect of 66/66	1	0.18(0.18)		
GPI-2, effect of 66/38	1	0.18(0.19)		
GPI-2, effect of 38/38	0	0.00(0.00)		

Table 10
Estimated Median Times to Death for Fish with Various
Combinations of Traits Using Model "2sex"
(see Case Study section) Fit to Mosquito Fish Data

Sex	Size(g)	GPI-2 genotype	Median TTD (h)	SE Median
Female	0.15	100/100	49.6	3.0
Female	0.25	100/100	53.8	2.9
Female	0.35	100/100	56.7	3.1
Male	0.15	100/100	38.2	2.6
Male	0.25	100/100	50.1	4.6
Male	0.35	100/100	59.6	7.4
Female	0.30	100/100	55.4	3.0
Female	0.30	100/66	53.2	2.6
Female	0.30	100/38	41.4	2.9
Female	0.30	66/66	47.1	3.4
Female	0.30	66/38	46.7	3.8
Female	0.30	38/38	39.1	6.7

following SAS code calculates predicted median times-to-death for individuals with those characteristics.

```
data controls;
  input sex $ size gpi2 $;
  logsize = log(size);
  mlsize = 0;
  flsize = 0;
  if sex = 'M' then mlsize = logsize;
    else flsize = logsize;
  predict = 1;
  ttd = .;
  censor = .;
cards;
F 0.15 100/100
F 0.25 100/100
F O.35 100/100
M 0.15 100/100
M 0.25 100/100
M 0.35 100/100
```

```
F 0.30 100/100
F 0.30 100/66
F 0.30 100/38
F 0.30 66/66
F 0.30 66/38
F 0.30 38/38
;
data all;
  set sexsize controls;

proc lifereg data = all;
  class sex gpi2;
  model ttd*censor(1) = sex mlsize flsize gpi2 /d = lnormal;
  output out = preds control = predict p = med _ ttd std = se _ med;

proc print;
  title 'Predicted median times to death';
  var sex size gpi2 med _ ttd se _ ttd;
```

The strategy in this SAS code is to create a data set containing a combination of characteristics for each new individual. The time-to-death for the new individuals is set to the SAS missing value code, so that the new individuals are not used to estimate the parameters of the model. The predicted median time-to-death and its standard error are written to a new SAS data set by the OUTPUT command. The CONTROL = option includes only the new individuals in the output data set. The relevant parts of this new data set are then printed onto the SAS listing file.

Median times-to-death (Table 10) are computed from the estimates, so they should show the same patterns as the parameter estimates; however, we feel that the median times-to-death are more easily interpreted. Small males die more quickly than either larger males or females (Table 10), and the response to size is different between the sexes. Increasing the size of female fish from 0.15 to 0.25 increases the median time-to-death by 4.2 h (Table 10). A similar increase in size of male fish leads to a much larger increase (11.9 h) in median TTD. Genotype has a large effect on either sex fish; there is a 16.3 h difference between the median time-to-death for the most resistant genotype (100/100) and the most susceptible genotype (38/38).

COX PROPORTIONAL HAZARDS MODEL

The accelerated life models fit by SAS PROC LIFEREG assume specific forms for the baseline hazard function and error distribution. An alternative class of models, called Cox proportional hazard models,[25] makes no assumptions about the form of the baseline hazard function. Instead, they make the proportional hazards assumption, i.e., that the effect of a covariate is to multiply the hazard function by a constant that does not change over time [see Equation (6)]. Such models are commonly used in the analysis of medical data, where the assumption of proportional hazards appears to be generally acceptable.[18] As discussed before (see Verifying Assumptions), plots of the hazard function can be used to verify the appropriate choice of error distribution. If none of the distributions adequately fits the data, a Cox model might be appropriate.

The Cox model has other advantages over accelerated life models. Often, one is more interested in describing the influence of covariates on survival than in describing the baseline hazard. By using a Cox model, one can estimate the effects of covariates while ignoring the unknown baseline hazard function. Also, the Cox model is less affected by outliers, unusually large or small failure-times, because the computations use only the rank ordering of failure and censoring times.[4] Cox proportional hazard models can be fit by the supplemental SAS procedure PHGLM,[26] which is available in some implementations of SAS, and by procedures in BMDP, SYSTAT, and S-Plus (Table 1). The covariates in a Cox proportional hazard model may be any combination of continuous or classification variables, like the covariates in an accelerated failure-time mode. The one difference is that a Cox model does not include an intercept.

If a Cox model is fit to the As data, the conclusions are similar to those obtained by fitting accelerated failure-time models. The parameter estimates (Table 11) are different from those obtained from an accelerated life model with a log-normal error distribution (Table 9), but the results of the hypotheses tests are the same, except for influence of size in females. In particular, there is still evidence that certain genotypes survive longer than do others. The difference in sign and magnitude of the estimates has two causes: (1) a mathematical difference in parameterization between proportional hazards and accelerated failure-time models[4] and (2) use of the log-normal error distribution (for which an accelerated time model is not a proportional hazards model) rather than the Weibull or exponential distributions.

Estimates from the Cox model can be interpreted as relative risks (see section Interpreting Parameter Estimates), because different hazard functions are as-

Table 11
Results from Fitting a Cox Proportional Hazards Model to the Arsenic Data[a]

Parameter	d.f.	Estimate (SE)	X^2	$P > X^2$
Sex	1	0.364(0.365)	0.99	0.32
Log(size) for males	1	−0.508(0.183)	7.67	0.0056
Log(size) for females	1	−0.129(0.067)	3.73	0.053
GPI-2 Overall	5		15.04	0.010
GPI-2, 100/100		−0.59(0.27)		
GPI-2, 100/66		−0.50(0.27)		
GPI-2, 100/38		−0.18(0.28)		
GPI-2, 66/66		−0.40(0.28)		
GPI-2, 66/38		−0.28(0.29)		
GPI-2, 38/38		0.00(0.00)		

[a] The model included the same covariates used in the model 2sex in Case Study section. Computations done with the COXREG and COXREG.PRINT functions in S-Plus. Overall test of GPI-2 genotypes computed from output of the COXREG function.

sumed to be proportional, so the relative risk is constant over time. For estimates from Cox models, the relative risk is computed using Equation (16).

$$\text{Relative Risk} = \begin{cases} e \ t_i & \text{for classification effects} \\ e^{\beta_i \Delta x} & \text{for continuous effects} \end{cases} \tag{16}$$

For example, the estimate of the effect of the 100/100 genotype at the GPI-2 locus is -0.586, so the relative risk for an individual with that genotype is $e^{-0.586} = 0.557$. In other words, such an individual is more likely to survive than is a reference individual.

Unlike the accelerated time models, the Cox model does not directly provide estimates of the median time-to-death. However, the entire survival curve (including the median) can be described if the baseline survival function is estimated. An estimate of the baseline survival curve is usually available in the output of the program that fits the Cox model.

The major assumption in a Cox model is that the baseline survival function is the same for all individuals. This assumption can be relaxed by stratifying the data into groups and estimating a separate baseline survival curve for each group. This is not quite the same as a separate analysis for each group; in a stratified analysis, a covariate is assumed to have the same effect in each stratum. Any potential covariate (e.g., GPI-2 genotype in the As data set), can be modelled it either as a covariate or as a variable that defines strata. When included as a covariate, its effect is estimated and tested, but the hazards are assumed to be proportional. When included as strata, the effects are not estimated, but the hazards do not have to be proportional.

Including stratification variables provides a way to check the proportional hazards assumption.[4] Calculate the baseline survival curve for each stratum, then plot $\log [- \log S(t)]$ against $\log t$ for each stratum. Just as in the hazard plots seen previously (in Verifying Assumptions), parallel lines on the plot suggest that the hazards are proportional. If a Weibull or exponential error distribution is appropriate, then the lines will also be straight. To check whether the proportional hazards assumption is appropriate for the effects of the GPI-2 genotype in the As data, a Cox model was fit using sex, male size, and female size as covariates and GPI-2 genotypes as strata. The hazard plots for each genotype are essentially parallel. Hence, we can assume that hazard functions for each genotype are proportional.

SUMMARY

There is a tendency for environmental toxicologists to uncritically select routine toxicity testing protocols in their research efforts. This can be unfortunate, as statistical techniques are available that provide considerably more abundant and precise information than do the standard techniques. One such class of techniques that uses time-to-death information to construct survival models is described.

A general review of survival analysis is provided using three data sets of increasing complexity. Modeling survival times with proportional hazards and accelerated failure-time models is introduced and compared briefly to standard techniques. The choice of distribution for survival analysis is discussed, with emphasis on the Weibull, exponential, and log-normal distributions. Cox proportional hazards models are described briefly. Techniques for parameter estimation and hypothesis testing are illustrated using an As toxicity data set.

These models provide a statistically powerful and conceptually easy way to assess the modifying effects of environmental conditions (e.g., water quality) or subject characteristics (e.g., fish size) in acute toxicity tests.

ACKNOWLEDGMENTS

The authors are grateful for comments and suggestions made to two anonymous reviewers. J. Quattro and R. Vrijenhock provided a fitness data set for analysis. Financial support was obtained from contract DE-AC09-76SROO-819 between the U.S. Department of Energy and the University of Georgia.

REFERENCES

1. American Public Health Association. *Standard Methods for the Examination of Water and Wastewater,* 15th ed. (Washington, DC: American Public Health Association, 1981), p. 1134.
2. Cox, D. R. and D. Oakes. *Analysis of Survival Data* (London: Chapman and Hall, 1984), 201 pp.
3. Nelson, W. *Applied Life Data Analysis* (New York: John Wiley & Sons, 1982), p. 634.
4. Kalbfleisch, J. D. and R. L. Prentice. *The Statistical Analysis of Failure Time Data* (New York: John Wiley & Sons, 1980), p.321.
5. SAS Institute Inc. *SAS/Stat User's Guide, Release 6.03 Edition* (Cary, NC:SAS Institute Inc, 1988), p. 1028.
6. SAS Institute Inc. *SAS Technical Report P-179, Additional SAS/Stat Procedures, Release 6.03 Edition* (Cary, NC:SAS Institute Inc, 1988), p. 255.
7. Miller, R. G. Jr. *Survival Analysis* (New York: John Wiley & Sons, 1981), p. 238.
8. Meeker, W. Q. and G. J. Hahn. *How to Plan an Accelerated Life Test — Some Practical Guidelines.* (Milwaukee: ASQC, 1985), p. 36.
9. Nelson, W. *How to Analyze Reliability Data* (Milwaukee: ASQC, 1983), p. 54.
10. Sprague, J. B. "Factors that Modify Toxicity," Fundamentals of Aquatic Toxicology, G. M. Rand, and S. R. Petrocelli, Eds., (Washington: Hemisphere Publ. Co., 1985), pp. 124-163.
11. Mood, A. M., F. A. Graybill, and D. C. Boes. *Introduction to the Theory of Statistics, 3rd ed.* (New York: McGraw Hill, 1974), p. 564.
12. Litchfield, J. T. Jr. "A Method for Rapid Graphic Solution of Time-Percent Effort Curves," *J. Pharmacol. Exp. Ther.* 97:399-408 (1949).
13. Shepard, M. P. "Resistance and Tolerance of Young Speckled Trout *(Salvelinus fontinalis)* to Oxygen Lack, with Special Reference to Low Oxygen Acclimation," *J. Fish Res. B. Can.* 12:387-446 (1955).

14. Newman, M. C., S. A. Diamond, M. Mulvey, and P. Dixon. "Allozyme Genotype and Time to Death of Mosquitofish, *Gambusia affinis* (Baird and Girard) during Acute Toxicant Exposure: A Comparison of Arsenate and Inorganic Mercury," *Aquat. Toxicol.,* 15:141-156 (1989).

15. Peto, R., M. C. Pike, P. Armitage, N. E. Breslow, D. R. Cox, S. V. Howard, N. Mantel, K. McPherson, J. Peto, and P. G. Smith. "Design and Analysis of Randomized Clinical Trials Requiring Prolonged Observation of Each Patient. II. Analysis and Examples," *Brit. J. Cancer* 35:1-39 (1977).

16. Gehan, E. A. "A Generalized Wilcoxon Test for Comparing Arbitrarily Singly-Censored Samples," *Biometrika* 52:203-223 (1965).

17. Mantel, N. "Evaluation of Survival Data and Two New Rank Statistics Arising in its Consideration," *Cancer Chemother. Rep.* 50:163-170 (1966).

18. Tibshirani, R. "A Plain Man's Guide to The Proportional Hazards Model," *Clin. Invest. Med.* 5:63-68 (1982).

19. Draper, N. R. and H. Smith. *Applied Regression Analysis, 2nd ed.* (New York: John Wiley & Sons, 1981), p. 709.

20. Christensen, E. R. "Dose-Response Functions in Aquatic Toxicity Testing and the Weibull Model," *Water Res.* 18:213-221 (1984).

21. Edwards, A. W. F. *Likelihood* (Cambridge: Cambridge Univ. Press, 1972), p. 235.

22. Rao, C. R. *Linear Statistical Inference and Its Applications, 2nd ed.* (New York: John Wiley & Sons, (1973), p. 625.

23. Quattro, J. M. and R. C. Vrijenhoek. "Fitness Differences among Remnant Populations of the Endangered Sonoran Topminnow," *Science* 245:976-978 (1989).

24. Anderson, P. D. and L. J. Weber. "Toxic Response as A Quantitative Function of Body Size," *Toxicol. Appl. Pharmacol.* 33:471-483 (1975).

25. Cox, D. R. "Regression Models and Life Tables," *J. R. Stat. Soc. B* 34:187-202, (1972).

26. SAS Institute Inc. *SUGI Supplemental Library User's Guide, Version 5 Edition* (Cary, N.C.:SAS Institute Inc., 1986), p. 437.

9

Trace Metals in Freshwater Sediments: A Review of the Literature and an Assessment of Research Needs

Alan McIntosh

Vermont Water Resources and Lake Studies Center, University of Vermont, Burlington, Vermont 05405-0088

OVERVIEW

Methods for estimating the effects of contaminated sediments on aquatic ecosystems are evolving rapidly. While approaches such as bulk sediment analyses and elutriate tests have long been used to evaluate dredge sites, tools designed to measure the response of organisms exposed to sediments have been a more recent development, with standardized EPA procedures being reported in the mid-1980s.

As is the case with standard water column toxicity tests, acute and chronic exposures of such species as the amphipod *Hyallela azteca* and the larval midge *Chironomus tentans* to sediments may indicate the potential for system-wide effects. However, the reliability of laboratory approaches in predicting field responses has not been proved. Recently, efforts to incorporate a greater degree of realism in test procedures have focused on multispecies testing; substantial uncertainty concerning the ultimate fate and system-wide effects of sediment-associated trace metals in rivers and lakes remains.

INTRODUCTION

Trace metals associated with sediments can pose a long-term threat to aquatic ecosystems. This generalization, only recently acknowledged by the scientific community, is now embraced by federal regulatory agencies as well. There are, for example, provisions in the Toxics Substances Control Act, the Clean Water Act, and the National Environmental Policy Act related to the assessment of sediment quality.[1] In addition, the EPA and several state regulatory agencies are evaluating various approaches in an attempt to develop sediment quality criteria. The EPA's objective is to develop criteria that can be applied nationwide.[2]

It is widely recognized by ecotoxicologists that an assessment of toxic trace metals in aquatic systems cannot be complete unless the role of sediments is considered. Despite this recent increase in interest, knowledge of the behavior of trace metals in freshwater sediments and subsequent effects on exposed biota is not extensive. As noted, steps are now being taken to establish national criteria for toxic substances, including metals, in sediments. Progress toward this objective, however, may be slowed by uncertainties regarding the applicability of criteria to all sediments.

In recent years, a number of review articles and several texts covering various aspects of sediment toxicology have been published.[3-6] This chapter will briefly review progress made to date in assessing the toxicology of trace metals associated with freshwater sediments and then consider several areas in need of further investigation.

The literature review portion of this chapter focuses on the following: (1) biological availability of sediment-associated trace metals; (2) laboratory approaches used to assess sediment toxicity; and (3) field investigations of trace-metal contaminated freshwater sediments.

PART 1. LITERATURE REVIEW OF TRACE METALS IN FRESHWA-TER SEDIMENTS
Biological Availability

Key to understanding the significance of sediment-associated trace metals in aquatic systems is a knowledge of the mechanisms involved in accumulation by exposed biota. To fully assess the uptake of trace metals by benthic fauna requires consideration of both the physicochemical behavior of the metals in sediments and the various routes by which biota may be exposed to these contaminants. The reader is referred to this volume, Chapter 10, for additional comments on this topic.

Physical/Chemical Aspects

The major sites for adsorption of trace metals in sediments have long been recognized. These include the oxides of Fe, Mn, Al, and Si, organic matter, and clays.[7] Experimental data have established a strong positive relationship between Fe/Mn and organic matter content of sediments and concentrations of various trace metals in benthic fauna exposed to these sediments.[8,9] Indeed,

Chapman[5] notes that an equilibrium partitioning approach based, in part, on the organic content of sediments is currently being considered in the establishment of national sediment quality criteria for trace metals.

Also of significance in some sediments are sulfides. In reduced sediments, precipitation of metal sulfides or association of metals with Fe sulfides may occur.[10] In fact, DiToro et al.[11] have theorized that solid-phase acid-volatile sulfide (AVS), because it is available to bind such metals as Cd and Ni in sediments, can be used to predict whether or not sediment-associated metals will be toxic.

Partitioning of trace metals among various substrates is influenced by a number of factors, including pH, Eh and the presence of various ligands in the interstitial, or pore, water. In addition, alterations in pH can affect the rate at which exposed organisms accumulate trace metals. Krantzberg and Stokes,[12] assessing the effects of pH on trace metal uptake by the midge *Chironomus* sp., noted that body burdens of several metals in both larvae and adults varied nonlinearly with hydrogen ion concentration. Results of the experiment suggested that accumulation of Cu, Cd, and Zn decreased at low pH's because of increased competition for absorption by hydrogen ions.

It is difficult to state with certainty how trace metals will be partitioned in any given sediment. Two methods commonly used to estimate metal distribution among sediment components are thermodynamic calculations under equilibrium conditions and selective chemical extractions.[13] Luoma[10] believes that the application of models to complex sediments is "...premature at best." Limitations also exist in the use of extractions. As Tessier and Campbell[13] note, because of inherent analytical problems, the partitioning of metals among various sediment fractions, as determined by sequential extractions, will necessarily be operationally defined. Luoma[10] adds that these approaches "...do not remove metals selectively from specific components of the sediment." Research to date, however, does suggest that biological availability can be related to relatively easily extracted fractions. For example, Förstner[14] found that a weak acid-reducing extractant yielded an estimate of the mobile and potentially available fraction of certain metals. An in-depth review of the partitioning of Cd in sediments and the various methods available for extracting sediment-associated Cd has been presented by Kersten and Förstner.[15]

There is substantial support among the research community for the concept that trace metal uptake by and toxicity to biota can be directly attributed to those metals found in pore waters.[6] Supportive evidence for the significance of pore water metals comes from the laboratory experiments of Schuytema et al.[16] and Cairns et al.,[17] who reported that the soluble forms of Cd and Cu, respectively, were primarily responsible for toxicity to exposed invertebrates and that metals associated with sediment particles apparently played little role in the responses noted. Nebeker et al.[18] also found that Cd adsorbed to organic materials in sediments was not available to the cladoceran *Daphnia magna* and the amphipod *Hyallela azteca*.

Other scientists, however, have underscored the role of sediment-associated metals in accumulation processes. Nalepa and Landrum[19] note that, since most metals require an active or facilitated process for assimilation by organisms, accumulation occurs primarily from the sediments through food ingestion. The reader is referred to Anderson et al.[20] for a review of methods available for predicting bioaccumulation of sediment-associated contaminants.

Biological Aspects

While the physicochemical aspects of trace metal distribution in sediments have received considerable attention, generally less effort has gone into assessing the role of biological variables in the uptake process, particularly among fresh-water organisms. Such factors as lipid content, reproductive condition, age, and weight have been cited as affecting contaminant accumulation by benthic fauna[19]; clearly, these parameters must be considered when assessing trace metal accrual by benthic organisms, as discussed in this volume, Chapter 4.

Also important to the accumulation process is the means by which benthic fauna obtain their food, as exposure may vary dramatically between groups of organisms such as filter feeders, including certain mollusks; grazers, including certain immature mayflies, beetles, and snails; and those organisms that ingest particles, such as oligochaetes and dipteran larvae. For example, analyses of clams exposed to trace metals indicate that high concentrations of metals occur in association with gill tissues.[13] This finding implies that absorption into the body of the clam via the gill may be a major avenue of uptake. In other organisms, however, such as oligochaetes and dipteran larva, ingestion of sediment particles represents a substantial route for uptake, and accumulation may be affected by such factors as gut pH and residence time of particles within the gut.[13] Finally, grazers relying on *aufwuchs* as a food source may be exposed to highly variable trace metal concentrations as the composition of the *aufwuchs* community undergoes seasonal changes.

An indication of the variability that can be expected in trace metal accumulation, even for closely related organisms, is seen in a study by Newman and McIntosh,[21] in which the accumulation of Pb by two snails, one a grazer (*Physa integra*) and the other an infaunal species (*Campeloma decisum*), was experimentally determined. Analyses indicated that the grazer rapidly accumulated Pb from *aufwuchs,* while the infaunal species took up little Pb, even when exposed to sediments containing high levels.

It is apparent that the biological and geochemical processes affecting the availability of sediment-associated trace metals are little understood. For a more detailed review of the subject, the reader is referred to Luoma.[10]

Not to be ignored in a discussion of biological availability is the transfer of these sediment-associated trace metals to higher levels in the food chain. The displacement of trace metals from sediments via migrating invertebrates, such as the opossum shrimp, *Mysis relicta,* and the amphipod *Diporeia* sp., has been shown to be significant in some waters. Several studies have demonstrated, for instance, that *Mysis* may facilitate the transport of metals such as Cd and Pb from sediments.[22-23] Evans and Lasenby,[22] in a study of Kootenay Lake (Br. Col.), speculated that, since mysids comprised 25 to 75% of the food supply of kokanee salmon, food chain-mediated transfer of Pb may be significant in that system.

Another factor to be considered when evaluating the bioavailability of trace metals is the effect of normal activities of benthic fauna on trace metal distribution. Through their feeding and burrowing activities, collectively known as bioturbation, organisms such as oligochaetes, larval chironomids, and mollusks may effectively redistribute trace metals in sediments and retard their burial in deeper layers.[19,24] There is also evidence that the activities of various benthic fauna may result in the movement of such elements as Hg from the sediments into the overlying water.[25]

Laboratory Approaches for Assessing the Toxicity of Trace Metals Associated with Freshwater Sediments

History

Much of the early work on contaminated sediment concerned the disposal of dredge spoils into surface waters. Lee and Jones[26] reviewed the development of guidelines by the Army Corps of Engineers in the 1970s for the disposal of such spoils into surface waters. The elutriate test, in which organisms such as the cladoceran *D. magna* are exposed to the elutriates of centrifuged sediments, was developed for use in dredge spoil assessment.

As recently as the 1970s, articles describing the effects of sediment exposure on benthic organisms were an unusual occurrence. Among the earliest assessments of freshwater sediment toxicity was a series of tests by Gannon and Beeton[27] on materials collected from Great Lakes harbors; significant mortality to the amphipod *Pontoporeia affinis* in many of the samples was noted. Related work was performed by Prater and Anderson,[28] who established the toxicity of Great Lakes sediments to a variety of organisms, including *D. magna* and the immature mayfly, *Hexagenia limbata*. Similar assessments were later performed on sediments[29] and elutriates[30] from Lake Michigan sites. In a series of toxicity tests utilizing metal-contaminated sediments from Palestine Lake, Indiana, Wentsel et al.[31-33] measured effects on the larval midge, *Chironomus tentans*. In general, however, progress during the 1970s and early 1980s in assessing contaminated sediments did not match that made in other areas of aquatic toxicology.

Recent Advances in Test Methodology

In 1984, an attempt was made to initiate standardization of freshwater sediment toxicity test methodology. Nebeker et al.[34] evaluated the use of *D. magna* and the amphipods *H. azteca* and *Gammarus lacustris, C. tentans,* and *H. limbata* to access acute and chronic effects of sediment exposure. They suggested that *Hyallela* and *Chironomus* were easy to rear and test, and were well suited for solid phase testing. They also suggested that a 48-h solid-phase beaker test with

Daphnia, in conjunction with an elutriate test, was a relatiely fast, simple, and inexpensive approach to initial screening for acute toxicity of sediments.

Since 1984, a number of acute and chronic toxicity tests on sediments have been conducted using the organisms suggested by Nebeker et al.[35-38] In addition, considerable effort has been expended to determine the suitability of other groups of organisms for use in sediment toxicity evaluations. Bacterial tests proposed include MICROTOX, a luminescence bioassay utilizing *Photobacterium phosphoreum.*[39-41] Others have proposed the use of measures of microbial activity to assess sediment toxicity. For example, Burton et al.[42] employed a suite of enzyme measurements to determine indigenous microbial activity in sediments from contaminated areas of Lake Michigan.

A variety of tests using phytoplankton, including the Algal Fractionation Bioassay, microcomputer-based toxicity testing, and *in situ* bioassays, have been described by Munawar and Munawar.[43] In particular, growth assays utilizing *Selenastrum capricornutum,* a green alga, have been gaining acceptance as a standard technique in sediment assessment tests.[6,42] Care must be taken in interpreting test results, since, as Burton et al.[42] noted, the algal assay has the potential to exhibit both inhibitory and stimulatory responses to contaminated sediments.

When evaluating various organisms for use in freshwater sediment toxicity tests, one must consider the substantial differences that exist in routes of exposure. For instance, cladocera inhabit the water column and may come into occasional contact with sediments, while organisms such as midge larva and amphipods have a much more constant association with sediments and their contaminants. Giesy and Hoke[6] provide a detailed evaluation of the organisms which have been used in assessing the toxicity of contaminated sediments. They recommend using the following battery of tests: 48-hr *D. magna* test; 7-day *Ceriodaphnia dubia* assay; MICROTOX; the 10-day *C. tentans* growth test; 7-day tests on the fathead minnow, *Pimephales promelas;* and a 96-h algal flask bioassay.

In fact, several recent efforts[40,42,44-46] have relied on a suite of methods, including *P. phosphoreum* assays; measurements of indigenous microbial activity; and acute or chronic responses of *S. capricornutum, D. magna,* the cladoceran *C. dubia, H. limbata, C. tentans,* and the aquatic oligocheate *Tubifex tubifex* to assess sediment toxicity. Results have shown considerable variability among organisms. For example, Wiederholm and Dove,[46] assessing the status of sediments from four Swedish lakes contaminated by effluents from steel mills, noted responses ranging from complete acute mortality to *D. magna* to 3-month survival for *T. tubifex.* Even among *T. tubifex,* however, growth and reproduction were retarded. Ankley et al.[45] found that *P. phosphoreum* were not as sensitive to exposure to sediment pore water from 13 sites within the lower Fox River/Green Bay watershed as other organisms tested. Ammonia was identified as a major contributor to toxicity noted among invertebrates tested. In contrast, Giesy et al.[40] found that the *P. phosphoreum* assay was the most sensitive of three approaches used in an assessment of the toxicity of Detroit River sediments. Burton et al.[42] found that microbial community assays demonstrated a greater range of response than did macrofauna upon exposure to contaminated sediments from Lake Michigan. Citing differences in degrees and/or pattern of response, they recommend the use of multitrophic level toxicity tests for sediment quality assessments.

Recent years have seen increasing attempts to achieve consistency in methodologies for sediment toxicity tests. Nebeker et al.[34] addressed such basic issues as rearing of test species and the effectiveness of existing test procedures. Various issues concerning the handling and preparation of test sediments remain, however. In particular, treatment of sediment samples prior to test initiation has been evaluated. Maleug et al.[47] argued against freezing of sediment samples prior to testing, citing a reduced mortality in *D. magna* exposed to frozen Cu-spiked sediments. Stemmer et al.[48] also recommended rapid processing of sediments. They assessed the effects of several methodological variables on the toxicity of Se to *D. magna*. Included in their study were comparisons of sediment spiking methods, variable time of shaking sediments, effects of sediment storage on toxicity, and variable sediment-to-water ratios. Their results led them to suggest that rapid processing, storage at 4°C for 48 h or less, and sediment-to-water volume ratios of approximately 1:4 would be ideal. They caution that, given the substantial effects resulting from alteration of test method variables, comparisons of experimental data should be done carefully.

Choice of the most appropriate method of exposure for test organisms, including exposure to spiked reference sediments, use of contaminated field material, and exposure to pore waters extracted from field sediments, is also of current interest. Giesy et al.,[45] in their study of sediments from the Detroit River, found substantial variability among the responses of *D. magna, C. tentans,* and *H. limbata* to pore waters versus whole sediment exposures.

Criteria Development

In addition to efforts to identify the most suitable methods and organisms for sediment toxicity assessments, there has been considerable discussion as to the most appropriate techniques for determining whether or not sediments are significantly contaminated and in need of further attention. Approaches proposed fall into two major categories: those that involve only chemical measurements of sediment-associated contaminants and those that incorporate a biological component. G. Chapman et al.[49] term these "numerical" criteria and "biological" criteria, respectively, while P. Chapman[5] considers them to be "chemical-chemical" and "chemical-mixture" approaches. The various methodologies relying on chemical analyses fall into two major classes: (1) bulk sediment analyses, in which concentrations of contaminants in sediments are measured and compared to levels in reference sediments and (2) the water quality criteria and sediment/water equilibrium partitioning approaches, in which pore water concentrations are compared to existing EPA water quality criteria.

The most commonly used technique among those methods which incorporate biological components are sediment toxicity tests, as outlined by Nebeker et al.[34] Several alternative approaches have been described by Chapman.[5] These include (1) the Sediment Quality Triad, in which data on sediment contaminant levels and results of laboratory toxicity tests and field assessments, such as enumeration of resident benthic infauna, are considered together; (2) the Apparent Effects Threshold (AET), which utilizes field data on contaminant levels in sediments and at least one indicator of biological effects to estimate the concentration of a particular contaminant likely to cause effects; and (3) the Screening Level Concentration approach (SLC), which estimates the level of contaminants in sediment that can be tolerated by 95% of the benthic community.

While each approach has distinct advantages and disadvantages, several generalizations can be stated. Although biological approaches are based on organism responses to contaminants, it is impossible, in most cases, to identify a specific causative agent. Hence, criteria established via such approaches for individual trace metals should be used with caution. Conversely, relying solely on chemical analyses of sediment or pore waters ignores the complex interactions between benthic fauna and sediment-associated contaminants.

Field Investigations

Two types of approaches have been used to investigate the effects of sediment-associated trace metals on biota under field conditions: (1) empirical studies of sites known to be contaminated by a trace metal or metals; and (2) sampling programs designed to validate laboratory experiments.

Identifying the effects of sediment-associated trace metals on benthic fauna is complicated by several factors. It is difficult to differentiate effects caused by metals in the water column from those related to sediment-associated metals.[25] In addition, co-occurrence of other contaminants and effects of variability in sediment parameters, such as organic content, make precise attribution impossible, in most cases. For example, Poulton et al.[50] noted a consistent positive or negative correlation between benthic composition and the organic fraction of several metals in sediments from Lake Ontario's Hamilton Harbour. They concluded, however, that they could not ascertain whether results indicated a response to metal toxicity or organic enrichment. After surveying the existing data base for toxic contaminants in Great Lakes sediments, Eadie et al.[24] stated that, although there is a common pattern in the distribution of benthic fauna in the most contaminated sediments of the lakes, establishing cause and effect relationships is very difficult.

There have been a number of studies in which trace metal levels in sediments have been related to the status of the benthic fauna. Changes in benthic community structure have, in some instances, been related to elevated concentrations of trace metals in sediments. Wentsel et al.[31] found diminished number of chironomid larvae in sediments from an Indiana lake receiving high levels of Cd, Cr, and Zn from an electroplating plant. Occhiogrosso et al.[51] attributed reduced benthic macroinvertebrate densities in the Foundry Cove region of the Hudson River to elevated concentrations of trace metals released by a Ni-Cd battery operation. Moore et al.[52] reported a decrease in the number of species and total densities of benthic fauna in Canadian lakes contaminated by mining waste containing As, Cu, Hg, and cyanide. Efforts to identify an indicator species were not successful, however.

In the above studies, it is likely that a number of factors, including synergistic/antagonistic effects of metal mixtures and the presence of unmeasured contaminants, may have contributed to the effect noted. In relatively few cases has it been possible to focus on a single metal. Winner et al.[53] noted that, following dosage of a stream with Cu, the number of species in the macroinvertebrate community was reduced. Condition of the benthic fauna in areas of Lake Superior enriched by Cu tailings has been assessed.[54-56] Kraft and Sypniewski[54] found a reduction in the number of benthic taxa in areas containing the highest levels of sediment-associated Cu. Maleug et al.[55] confirmed these field data and also conducted supporting laboratory toxicity tests using Cu-contaminated sediments. Finally, Maleug et al.[56] determined that high Cu concentrations in nearby Torch Lake reduced populations of benthic macroinvertebrates.

While outright mortality and subsequent loss of sensitive species of benthic fauna are possible in cases of gross metal contamination, of perhaps greater concern are subtle effects caused by long-term exposure of fauna to low levels of metal contaminants. Efforts to identify effects of such exposures have focused primarily on the identification of abnormalities and deformities in various benthic macroinvertebrates living in contaminated sediments. Milbrink[57] identified setal abnormalities in an oligochaete species exposed to high Hg levels in sediments of Lake Vanern in Sweden, while Wiederholm[58] documented the occurrence of deformed mouth parts in midge larvae collected from several Swedish lakes. He attributed the effects noted to the presence of metals. It would be useful to know whether or not the occurrence of such deformities might be utilized in field assessments to indicate the status of benthic communities.

In some sediment assessments, field sampling has been utilized to confirm results of laboratory tests. Wentsel et al.[31] found that adverse effects of sediment exposure on larval *C. tentans* noted in the laboratory were substantiated by field sampling in contaminated Palestine Lake, Indiana. Similar verification was obtained by Maleug et al.,[15] when they compared toxicity of Cu-contaminated sediments in laboratory tests to distribution of macroinvertebrates in heavily contaminated areas of the Keweenaw Waterway, Michigan. Unfortunately, such correspondence between laboratory and field findings is most easily established in those cases where metal contamination is extreme and effects are acute.

PART 2. ASSESSMENT OF RESEARCH NEEDS RELATED TO SEDIMENT-ASSOCIATED TRACE CONTAMINANTS

The following discussion includes comments that may pertain to both organic and inorganic trace contaminants in freshwater sediments. As noted in Part 1, substantial progress has been made in assessing the toxicity of contaminated sediments; much, however, remains to be learned. Most of the recent published literature in sediment toxicology has focused on the following: (a) attempts to develop and standardize laboratory test procedures (Nebeker et al.[34] and Giesy and Hoke[6] provide excellent summaries of such efforts); (b) evaluation of various general approaches (chemically based, biologically based, or a combination of the two) to be used in establishing national sediment criteria;[5] and (c) assessment of which suite of biological measurements, ranging from *P. phosphoreum* assays to fathead minnow tests, yields the most comprehensive screening of contaminated sediment.[40,41]

Despite these advances, a thorough understanding of the fate and effects of sediment-associated contaminants is absent. In particular, concentrated efforts are required in the areas discussed below.

Fate of Sediment-Associated Contaminants
Lotic Environments

There are substantial gaps in our knowledge of the behavior of sediment-associated trace contaminants in surface waters. While it has long been accepted that many persistent trace contaminants are associated with particulate matter during transport in surface waters,[59,60] patterns of contaminant accumulation over time in these waters are not well understood. For example, during low flow, trace contaminants introduced into a stream or river may rapidly associate with particles or form precipitates, and, depending on the flow regime, accumulate in depositional areas such as pools. If low-flow conditions persist, concentrations of contaminants may presumably reach the point where exposed benthic fauna are threatened. Subsequent storm events may flush the sediment-associated contaminants downriver into reservoirs, lakes, harbors or estuaries, thus, at least temporarily, reducing concentrations in scoured areas, while redepositing contaminants in new areas such as the depositional zones of estuaries.

An additional influence on the hydrodynamics of trace contaminants in flowing waters is the uptake and subsequent release of trace contaminants by living components of the *aufwuchs*. Seasonal changes in the composition of this complex community likely result in highly variable contaminant concentrations and, hence, exposure for organisms utilizing the *aufwuchs* for food. A review of contaminant-*aufwuchs* relationships has been prepared by Newman and McIntosh.[61] The effect of changing patterns of contaminant accumulation in both the biotic and abiotic components of substrates in flowing waters on stream biota should be more fully evaluated.

The nature of contaminant loads to rivers is important to consider as well. As efforts to control the release of contaminants from point sources proceed, the significance of nonpoint sources, such as urban runoff, will likely increase as a proportion of the total load to aquatic systems. The intermittent nature of nonpoint sources makes accurate assessments of the fate and effects of introduced contaminants difficult. For instance, sediment-associated trace metals, such as Pb and Cd, contained in urban stormwater move through surface water quickly during and following storm events. There is limited literature available to indicate that the ultimate fate of these substances and their ecological significance have been adequately investigated.

Lentic Environments

The fate of sediment-associated contaminants within lakes is complex as well. Robbins et al.[62] noted that physical disruption, resuspension, and horizontal redistribution of surface sediments can be significant in shallow lakes. In addition, the discharge of contaminants released from lake sediments to rivers may be substantial. Larsson et al.[63] estimated that 80% of the PCBs released from the sediments of a contaminated lake in Sweden resided at the mouth of an effluent river some 60 km downstream. In addition to physical processes, as previously noted, activities of benthic fauna may result in the redistribution of sediment-associated contaminants.[19] Knowledge of how these various processes alter exposure of benthic fauna to such contaminants is limited.

Another factor that must be considered is the heterogenous nature of lake sediments. As Marcus[64] noted, even within the same hydrologic environment, sediment composition may vary substantially. Since contaminants tend to be associated with small particles, sediment samples containing high proportions of silts or clays may exhibit substantially greater contaminant levels than surrounding materials. Obviously, such variability complicates the task of applying uniform sediment quality standards and will require regulators to give careful thought to sediment sampling strategies.

Finally, there has been relatively little research delineating the effect of localized sediment contamination within lakes on the distribution of trace contaminants in the system as a whole. For example, a harbor may be badly contaminated by trace metals; however, if the harbor is separated from the main body of water by a narrow channel, widespread dispersal of the contaminants may not occur. In other cases, such as shallow, well-mixed lakes, the possibility of substantial horizontal movement of contaminants and subsequent effects on biota exists. It would be useful, in particular, to have more data describing the extent of system-wide dispersal of contaminants entering near-shore waters from such sources as urban runoff.

Effects of Sediment-Associated Contaminants on Biota

Much remains to be learned about the impacts of sediment-associated contaminants on exposed organisms. For example, the chronic effects of sediment-associated contaminants on biota are poorly understood. While such responses as delayed emergence and reduced growth in larval *C. tentans* have been measured in chronic toxicity tests,[32,33,45] there is relatively little evidence from field or microcosm studies to corroborate findings of subacute effects.

Avoidance responses following exposure to contaminated sediments have been reported for larval *C. tentans,*[65] aquatic oligochaetes,[66] and *P. affinis.*[27] The significance of altered behavior of invertebrates exposed to sediment-associated contaminants in the field should be more fully explored. Also, when evaluating data on the status of field populations, it should be considered likely that many populations have become adapted to contaminants[67] (see this volume, Chapter 11), and thus may be found in sediments containing pollutants at levels higher than known tolerance limits. For example, a recent assessment of Lake Michigan sediments by Keilty and Landrum[68] suggests that populations of the aquatic oligochaete *Stylodrilus heringianus* may adapt to low-level, long-term chemical stresses.

While much attention in recent years have been focused on such benthic fauna as *C. tentans,* other organisms as well come into frequent contact with sediment-associated contaminants. Many fish are bottom feeders and may be exposed to such pollutants directly or through consumption of contaminated food items. Certain fish and amphibians deposit their eggs onto sediments in shallow areas of lakes where contaminants entering via runoff may have accumulated. Hence, critical formative stages of these organisms may be exposed to sediment-associated contaminants. As Birge et al.[69] noted, in epibenthic species, it is likely that early life stages are more vulnerable to effects of sediment-associated chemicals than mature stages. Birge et al.,[69] citing data from studies of the responses of trout embryo larval stages to metal-spiked sediments, proposed threshold concentrations for Cd, Hg, and Zn in sediments containing 4% total organic carbon.

An additional factor complicating the assessment of the potential threat that sediment-associated contaminants pose to early life stages of fish and other organisms is the coincidence of spring runoff with spawning activities. If trace metals and other substances in urban stormwater accumulate in nearshore sediments during periods of spring runoff, eggs recently deposited in these areas may be at risk.

Community and Ecosystem Concerns

There is critical need to develop *in situ* measurements to evaluate benthic community status. Schindler[70] discussed the inadequacies of existing approaches for monitoring stressed aquatic ecosystems and suggested that measures such as species changes in the phytoplankton community and life-table analysis of zooplankton may serve as early indicators of stress and, hence, be suitable biological monitors. To date, attempts to develop similar early warning systems for the benthos have been lacking.

Schindler also suggested that increased deformities among benthic fauna, such as those noted by Milbrink[57] and Wiederholm,[58] seem to occur before other symptoms of ecosystem stress are apparent; hence, such anomalies may be useful as early warning signs.

Another promising avenue may be the use of microbes; Burton et al.[42] found that indigenous microbial measurements were sensitive indicators of sediment contamination in Waukegan and Indiana Harbors in Lake Michigan. The authors noted that the utility of such measurements is due to the ability of microbes to respond quickly to changing environmental conditions and the fact that they play a major role in ecosystem biogeochemical cycling processes. They conclude that alterations of important enzyme systems such as hydrolases in microbial communities may be used to indicate that ecosystem degradation is occurring.

Many of the recent sediment toxicity assessments have occurred in heavily contaminated areas of such systems as Puget Sound, the Great Lakes, and the Hudson River. For example, in the Great Lakes, sediment-associated contaminants are a major concern in most of the 42 areas designated for clean-up as part of the RAP (Remedial Action Plan) program. Certainly, attention to sediment contamination in these areas is appropriate and necessary.

Not to be neglected, however, are other surface waters where extreme contamination is not an issue, but where gradual accumulation of contaminants in sediments has occurred. Examples include lakes where copper sulfate has been used repeatedly over a long period of time and systems exposed to substantial amounts of atmospherically deposited contaminants. Sediments in such bodies of water may eventually exhibit substantial accrual of potentially toxic contaminants. Given the gradual nature of the accumulation process, it is unlikely that dramatic effects would be detected. However, loss of sensitive species or alteration of critical ecosystem properties should not be discounted.

Also, as noted previously, many lakes may be exposed to intermittent introduction of sediment-associated contaminants from nonpoint sources. For example, frequent pulses of runoff-associated trace metals entering nearshore waters may lead to increases in sediment concentrations of these substances and, hence, the possibility of effects on various levels of the ecosystem must be considered. To date, the impact of such sporadic events on key ecosystem parameters has not been explored.

In summary, significant progress in defining appropriate laboratory tools for sediment toxicity assessment and in developing strategies for sediment quality criteria has occurred. There are, however, many gaps in our understanding of the behavior of sediment-associated contaminants and their impacts on aquatic ecosystems. For instance, the effects of these contaminants on entire ecosystems have not been substantively evaluated. While it is recognized that there are various physical, chemical, and biological means by which such contaminants may reenter the general circulation of a lake, a review of the literature reveals that there is little knowledge of the true significance of these processes to the ecosystem as a whole. Rapid progress in laboratory methodologies must be accompanied by basic field-oriented research designed to provide critically needed data on the role of sediment-associated contaminants in the ecosystem.

SUGGESTED AREAS FOR FURTHER RESEARCH

1.   Retention and transport of sediment-associated trace contaminants in lotic systems; in particular, effects of flow rate and biological interactions are poorly understood.

2.   Influence of sediment heterogeneity and localized "hot spots" on overall contaminant behavior in lakes and reservoirs.

3.   Significance of responses, such as avoidance and adaptation among benthic communities exposed to sediment-associated trace contaminants.

4.   Effects of trace contaminants accumulated in nearshore sediments on early life stages of amphibians and fish.

5.   Community-level measures designed to assess the health of contaminant-exposed benthic fauna.

6.   Responses of benthic communities in lakes exposed to long-term, low-level contamination.

ACKNOWLEDGMENTS

Thanks go to Marie MacLean for help with the manuscript and to Peter Chapman and an anonymous reviewer for critiques.

REFERENCES

1. Chapman, P. M., R. C. Barrick, J. M. Neff, and R. C. Schwartz. "Four Independent Approaches to Developing Sediment Quality Criteria Yield Similar Values for Model Contaminants," *Environ. Toxicol. Chem.* 6:723-725 (1987).

2. Shea, D. "Developing National Sediment Quality Criteria," *Environ. Sci. Technol.* 22(11):1256-1261 (1988).

3. Dickson, K., A. W. Maki, and W. A. Brungs, Eds. *Fate and Effects of Sediment-Bound Chemicals in Aquatic Systems* (New York:Pergamon Press, 1987), p. 449.

4. Thomas, R., R. Evans, A. Hamilton, M. Munawar, T. Reynoldson, and H. Sadar, Eds. "Ecological Effects of In-Situ Sediment Contaminants," *Hydrobiologia* 149:1-272 (1987).

5. Chapman, P. M. "Current Approaches to Developing Sediment Quality Criteria," *Environ. Toxicol. Chem.* 8(8):589-599 (1989).

6. Giesy, J. P. and R. A. Hoke. "Freshwater Sediment Toxicity Bioassessment: Rationale for Species Selection and Test Design," *J. Great Lakes Res.* 15(4):539-569 (1989).

7. Jenne, E. A. and J. M. Zachara. "Factors Influencing the Sorption of Metals," in *Fate and Effects of Sediment-Bound Chemicals in Aquatic Systems,* K. Dickson, A. Maki, W. Brungs, Eds. (New York: Permagon Press, 1987), pp. 83-98.

8. Jenne, E. A. "Controls of Mn, Fe, Co, Ni, Cu, and Zn Concentrations in Soils and Water: The Significant Role of Hydrous Mn and Fe Oxides," *Adv. Chem.* 73:337-387 (1968).

9. Newman, M. C., A. McIntosh, and V. Greenhut. "Geochemical Factors Complicating the Use of *Aufwuchs* as a Biomonitor for Lead Levels in Two New Jersey Reservoirs," *Water Res.* 17:(6):625-630 (1983).

10. Luoma, S. N. "Can We Determine the Biological Availability of Sediment-Bound Trace Elements?" *Hydrobiologia* 176/177:379-396 (1989).

11. DiToro, D. M., J. O. Mahony, K. J. Scott, M. S. Redmond, M. B. Hicks, S. M. Mayer, and D. J. Hansen, "Toxicity of Cadmium in Sediments: The Role of Acid Volatile Sulfide," Abstr. 253, 10th Annu. Meet. Soc. Environ. Toxicol. Chem. (1989).

12. Krantzberg, G. and P. M. Stokes. "The Importance of Surface Adsorption and pH in Metal Accumulation by Chironomids," *Environ. Toxicol. Chem.* 67(8):653-670 (1988).

13. Tessier, A. and P. G. C. Campbell. "Partitioning of Trace Metals in Sediments: Relationships with Bioavailability," *Hydrobiologia* 149:43-52 (1987).

14. Förstner, U. "Sediment-Associated Contaminants — An Overview of Scientific Bases for Developing Remedial Options," *Hydrobiologia* 149:221-246 (1987).

15. Kersten, M. and U. Förstner. "Cadmium Associations in Freshwater and Marine Sediment," in *Cadmium in The Aquatic Environment*, J. Nriagu and J. Sprague, Eds. (New York: John Wiley & Sons, 1987), pp. 51-88.

16. Schuytema, G. A., P. O. Nelson, K. W. Maleug, A. V. Nebeker, D. F. Krawczyk, A. K. Ratcliff, and J. H. Gakstatter. "Toxicity of Cadmium in Water and Sediment Slurries to *Daphnia magna*," *Environ. Toxicol. Chem.* 3(2):191-199 (1984).

17. Cairns, M. A., A. V. Nebeker, J. H. Gakstatter, and W. L. Griffis. "Toxicity of Copper-Spiked Sediments to Freshwater Invertebrates," *Environ. Toxicol. Chem.* 3(3):435-445 (1984).

18. Nebeker, A. V., S. T. Onjukka, M. A. Cairns, and D. F. Krawczyk. "Survival of *Daphnia magna* and *Hyallela azteca* in Cadmium-Spiked Water and Sediment," *Environ. Toxicol. Chem.* 5910:933-938 (1986).

19. Nalepa, T. F. and P. F. Landrum. "Benthic Invertebrates and Contaminant Levels in the Great Lakes: Effect, Fates, and Role in Cycling," in *Toxic Contaminants and Ecosystem Health: A Great Lakes Focus*, M. S. Evans, Ed. (New York: John Wiley & Sons, 1988), pp. 77-102.

20. Anderson, J., W. Birge, J. Gentile, J. Lake, J. Rodgers, Jr., and R. Swartz. "Biological Effects, Bioaccumulation, and Ecotoxicology of Sediment-Associated Chemicals," in *Fate and Effects of Sediment-Bound Chemicals in Aquatic Systems*, K. Dickson, A. Maki, and W. Brungs, Eds. (New York: Pergamon Press, 1987), pp. 267-296.

21. Newman, M. C. and A. McIntosh. "Slow Accumulation of Lead from Contaminated Food Sources by the Freshwater Gastropods, *Physa integra* and *Campeloma decisum*," *Arch. Environ. Contam. Toxicol.* 12:685-692 (1983).

22. Evans, R. D. and D. C. Lasenby. "Relationship Between Body-Lead Concentration of *Mysis relicta* and Sediment-Lead Concentration in Kootenay Lake. B.C.," *Can. J. Fish. Aquat. Sci.* 40:78-81 (1983).

23. Van Duyn-Henderson, J. A. and D. C. Lasenby. "Zinc and Cadmium Transport by the Vertically Migrating Opossum Shrimp, *Mysis relicta*," *Can. J. Fish. Aquat. Sci.* 43:1726-1732 (1986).

24. Eadie, B. J., T. F. Nalepa, and P. F. Landrum. "Toxic Contaminants and Benthic Organisms in the Great Lakes: Cycling, Fate and Effects," in *Toxic Contamination in Large Lakes, Proceedings of a Technical Session of the World Conference on Large Lakes*, May 1986, Mackinac Is. MI (Chelsea, MI: Lewis Publishers, 1988), pp. 161-178.

25. Reynoldson, T. B. "Interactions Between Sediment Contaminants and Benthic Organisms," *Hydrobiologia* 149:53-66 (1987).

26. Lee, G. F. and R. A. Jones. "Water Quality Significance of Contaminants Associated with Sediments: An Overview," in *Fate and Effects of Sediment-Bound Chemicals in Aquatic Systems*, K. Dickson, A. Maki, and W. Brungs, Eds. (New York: Permagon Press, 1987), pp. 3-34.

27. Gannon, J. E. and A. M. Beeton. "Procedures for Determining the Effects of Dredge Sediments on Biota-Benthos Viability and Sediment Selectivity Tests," *J. Water Pollut. Control Fed.* 43:392-398 (1971).

28. Prater, B. L. and M. A. Anderson. "A 96-hour Sediment Bioassay of Duluth and Superior Harbor Basins (Minnesota) Using *Hexagenia limbata, Asellus communis, Daphnia magna,* and *Pimephales promelas* as Test Organisms," *Bull. Environ. Contam. Toxicol.* 18:159-169 (1977).

29. Laskowski-Hoke, R. A. and B. L. Prater. "Multivariate Statistical Analyses of 96-Hour Sediment Bioassay and Chemistry Data," *Bull. Environ. Contam. Toxicol.* 33:400-409 (1984).

30. Laskowski-Hoke, R. A. and B. L. Prater. "Relationship of Mortality of Aquatic Biota from 96-hour Sediment Bioassays and the Change in Chemical Composition of the Test Water," *Bull. Environ. Contam. Toxicol.* 26:323-327 (1981).

31. Wentsel, R., A. McIntosh, and V. Anderson. "Sediment Contamination and Benthic Macroinvertebrate Distribution in a Metal-Impacted Lake," *Environ. Pollut.* 14(3):187-193 (1977).

32. Wentsel, R., A. McIntosh, and G. Atchison. "Sublethal Effects of Heavy Metal Contaminated Sediment on Midge Larvae *(Chironomus tentans),*" *Hydrobiologia* 56(2):153-156 (1977).

33. Wentsel, R., A. McIntosh, and W. McCafferty. "Emergence of the Midge *Chironomus tentans* when Exposed to Heavy Metal-Contaminated Sediment," *Hydrobiologia* 57(3):195-196 (1978).

34. Nebeker, A. V., M. A. Cairns, J. H. Gakstatter, K. W. Maleug, G. S. Schuytema, and D. F. Krawczyk. "Biological Methods for Determining Toxicity of Contaminated Freshwater Sediments to Invertebrates," *Environ. Toxicol. Chem.* 3(4):617-630 (1984).

35. Abel, P. D. and S. M. Garner. "Comparisons of Median Survival Times and Median Lethal Exposure Times for *Gammarus pulex* Exposed to Cadmium, Permethrin and Cyanide," *Water Res.* 20:579-582 (1986).

36. Henry, M. G., Chester, D. N., and Mauck, W. L. "Role of Artificial Burrows in *Hexagenia* Toxicity Tests: Recommendations for Protocol Development," *Environ. Toxicol. Chem.* 5(6):553-559 (1986).

37. Nebeker, A. V., S. T. Onjukka, and M. A. Cairns. "Chronic Effects of Contaminated Sediment on *Daphnia magna* and *Chironomus tentans,*" *Bull. Environ. Contam. Toxicol.* 41:574-581 (1988).

38. Nebeker, A. V. and C. E. Miller. "Use of the Amphipod Crustacean *Hyallela azteca* for Freshwater and Estuarine Sediment Toxicity Tests," *Environ. Toxicol. Chem.* 7(12):1027-1033 (1988).

39. Dutka, B. J., K. Jones, K. K. Kwan, H. Bailey, and R. McInnis. "Use of Microbial and Toxicant Screening Tests for Priority Site Selection of Degraded Areas in Water Bodies," *Water Res.* 22:503-510 (1988).

40. Giesy, J. P., R. L. Graney, J. L. Newsted, C. J. Rosiu, A. Benda, R. G. Kreis, Jr., and F. J. Horvath. "Comparison of Three Sediment Bioassay Methods Using Detroit River Sediments," *Environ. Toxicol. Chem.* 7(6):483-498 (1988).

41. Rosiu, C. J., J. P. Giesy, and R. G. Kreis, Jr. "Toxicity of Sediments in the Trenton Channel, Detroit River, Michigan to *Chironomus tentans* (Insecta: Chironomida)," *J. Great Lakes Res.* 15:570-580 (1989).

42. Burton, G. A., Jr., B. L. Stemmer, K. L. Winks, P. E. Ross, and L. C. Burnett. "A Multitrophic Level Evaluation of Sediment Toxicity in Waukegan and Indiana Harbors," *Environ. Toxicol. Chem.* 8(11):1057-1066 (1989).

43. Munawar, M. and I. F. Munawar. "Phytoplankton Bioassays for Evaluating Toxicity of *In-situ* Sediment Contaminants," *Hydrobiologia* 149:87-105 (1987).

44. Ankley, G. T., A. Katko, and J. W. Arthur. "Identification of Ammonia as an Important Sediment-Associated Toxicant in the Lower Fox River and Green Bay, Wisconsin," *Environ. Toxicol. Chem.* 9(3):313-322 (1990).

45. Giesy, J. P., C. J. Rosiu, R. L. Graney, and M. G. Henry. "Benthic Invertebrate Bioassays with Toxic Sediment and Pore Water," *Environ. Toxicol. Chem.* 9(2):233-248 (1990).

46. Wiederholm, T. and G. Dove. "Toxicity of Metal-Polluted Sediment to *Daphnia magna* and *Tubifex tubifex*," *Hydrobiologia* 176/177:411-417 (1989).

47. Maleug, K. W., G. S. Schuytema, and D. F. Krawczyk. "Effects of Sample Storage of a Copper-Spiked Freshwater Sediment," *Environ. Toxicol. Chem.* 5(3):245-253 (1986).

48. Stemmer, B. L., G. A. Burton, Jr., and S. Liebfritz-Frederick. "Effect of Sediment Test Variables on Selenium Toxicity to *Daphnia magna*," *Environ. Toxicol. Chem.* 9(3):381-389 (1990).

49. Chapman, G., W. Adams, H. Lee, W. Lyman, S. Pavlou, and R. Wilhelm. "Regulatory Implications of Contaminants Associated with Sediments," in *Fate and Effects of Sediment-Bound Chemicals in Aquatic Systems*, K. Dickson, A. Maki, and W. Brungs, Eds. (New York: Permagon Press, 1987), pp. 413-425.

50. Poulton, D. J., K. J. Simpson, D. R. Barton, and K. R. Lum. "Trace Metals and Benthic Invertebrates in Sediments of Nearshore Lake Ontario at Hamilton Harbour," *J. Great Lakes Res.* 14(1):52-65 (1988).

51. Occhiogrosso, T. J., W. T. Waller, and G. J. Lauer. "Effects of Heavy Metals on Benthic Macroinvertebrate Densities in Foundry Cove on the Hudson River," *Bull. Environ. Contam. Toxicol.* 122:230-237 (1979).

52. Moore, J. W., V. A. Beubien, and D. J. Sutherland. "Comparative Effects of Sediment and Water Contamination on Benthic Invertebrates in Four Lakes," *Bull. Environ. Contam. Toxicol.* 23:840-748 (1979).

53. Winner, R. W., T. Kelling, R. Yeager, and M. P. Farrell. "Response of a Macroinvertebrate Fauna to a Copper Gradient in an Experimentally-Polluted Stream," *Verh. Int. Ver. Limnol.* 19:2121-2127 (1975).

54. Kraft, K. J. and R. H. Sypniewski. "Effect of Sediment Copper on the Distribution of Benthic Macroinvertebrates in the Keweenaw Waterway," *J. Great Lakes Res.* 7:258-263 (1981).

55. Maleug, K. W., G. S. Schuytema, D. F. Krawczyk, and J. H. Gakstatter. "Laboratory Sediment Toxicity Tests, Sediment Chemistry and Distribution of Benthic Macroinvertebrates in Sediments from the Keweenaw Waterway, Michigan," *Environ. Toxicol. Chem.* 3(2):233-242 (1984).

56. Maleug, K. W., G. S. Schuytema, J. H. Gakstatter, and D. F. Krawczyk. "Toxicity of Sediments from Three Metal-Contaminated Areas," *Environ. Toxicol. Chem.* 3(2):279-291 (1984).

57. Milbrink, G. "Characteristic Deformities in Tubificid Oligochaetes Inhabiting Polluted Bays of Lake Vanern, Southern Sweden," *Hydrobiologia* 106:169-184 (1983).

58. Wiederholm, T. "Incidence of Deformed Chironomid Larvae (Diptera:Chironomidae) in Swedish Lakes," *Hydrobiologia* 109:243-249 (1984).
59. Lick, W. "The Transport of Sediments in Aquatic Systems," in *Fate and Effects of Sediment-Bound Chemicals in Aquatic Systems,* K. Dickson, A. Maki, and W. Brungs, Eds. (New York: Permagon Press, 1987), pp. 61-74.
60. Lau, Y. L., B. G. Oliver, and B. G. Krishnappan. "Transport of Some Chlorinated Contaminants by the Water, Suspended Sediments, and Bed Sediments in the St. Clair and Detroit Rivers," *Environ. Toxicol. Chem.* 8(4):293-301 (1989).
61. Newman, M. C. and A. McIntosh. "Appropriateness of *Aufwuchs* as a Monitor of Bioaccumulation," *Environ. Pollut.* 60(132):83-100 (1989).
62. Robbins, J. A., A. Mudroch, and B. G. Oliver. "Transport and Storage of 137Cs and 210Pb in Sediments of Lake St. Clair," *Can. J. Fish. Aquat. Sci.* 47:572-587 (1990).
63. Larrson, P., L. Okla, S. Ryding, and B. Westoo. "Contaminated Sediment as a Source of PCBs in a River System," *Can. J. Fish Aquat. Sci.* 47:746-754 (1990).
64. Marcus, W. A. "Regulating Contaminated Sediments in Aquatic Environments: A Hydrologic Perspective," *Environ. Manage.* 13(6):703-713 (1989).
65. Wentsel, R., A. McIntosh, W. McCafferty, G. Atchison, and V. Anderson. "Avoidance Response of Midge Larvae *(Chironomus tentans)* to Sediments Containing Heavy Metals," *Hydrobiologia* 55(2):171-175 (1977).
66. McMurtry, M. J. "Avoidance of Sublethal Doses of Copper and Zinc by Tubificid Oligochaetes," *J. Great Lakes Res.* 10:267-272 (1984).
67. Wentsel, R., A. McIntosh, and G. Atchison. "Evidence of Resistance to Metals in Larvae of Midge *(Chironomus tentans)* in a Metal Contaminated lake," *Bull. Environ. Contam. Toxicol.* 20:451-455 (1978).
68. Keilty, T. J. and P. F. Landrum. "Population-Specific Toxicity Responses by the Freshwater Oligochaete, *Stylodrilus heringianus,* in Natural Lake Michigan Sediments," *Environ. Toxicol. Chem.* 9:1147-1154 (1990).
69. Birge, W. J., J. A. Black, A. G. Westerman, and P. C. Francis. "Toxicity of Sediment-Associated Metals in Freshwater Organisms: Biomonitoring Procedures," in *Fate and Effects of Sediment-Bound Chemicals in Aquatic Systems,* K. Dickson, A. Maki, and W. Brungs, Eds. (New York: Permagon Press, 1987), pp. 199-218.
70. Schindler, D. W., "Detecting Ecosystem Responses to Anthropogenic Stress," *Can. J. Fish. Aquat. Sci.* 44(Suppl. 1):6-25 (1987).

10

Effects of Trace Metals on Aquatic Benthos

Samuel N. Luoma and James L. Carter

U.S. Geological Survey, Mail Stop 465, 345 Middlefield Road,
Menlo Park, California 94025

OVERVIEW

Examples of apparent adverse effects from metal exposure can be found at all levels of biological organization (biochemical, physiological, population, community) in moderately contaminated environments in nature. However, demonstrating the significance of metal responses observed at lower levels of organization, and conclusively proving that responses are attributable solely to metal exposures at higher levels of organization are important challenges in such studies. Most studies do an inadequate job of characterizing pollutant exposures, considering both the influences of environmental history and additional stressors other than metals. Biochemical responses to metal are relatively well understood, but rarely do studies consider their significance at higher levels of organization. Physiological responses to metals are usually nonspecific and separating metal effects from the influence of other factors requires an intensity of study rarely undertaken. Population-level responses to metals, other than the development of tolerance, have essentially not been studied. Community responses are difficult to detect because the patterns of change are complex and a multiple of factors can be involved in determining the response of a community. Studying biological responses across several levels of organization through long periods of time may be the most useful approach to ultimately understanding how metals affect aquatic benthos. The search for simplistic methods for detecting complex responses to metals may be fruitless and, in fact, may impede progress in understanding critical complexities.

INTRODUCTION

Criteria for protecting aquatic environments from trace metal contamination are necessary because of the potential toxicity of these pollutants. Development of effective criteria depends upon understanding how contaminant exposures influence biological processes in nature. At present, adverse effects of metal contaminants in nature are difficult to demonstrate except in the most extreme circumstances,[1-4] not because no effects are occurring but because of fundamental limiations in our ability to measure effects. The range of spatial, temporal, and organizational scales that must be considered, the lack of baseline data from natural systems, the complexity and variety of co-occuring anthropogenic and natural disturbances, and pervasive environmental stochasticity[5,6] all contribute to the challenge of detecting responses to metal pollutants.

A perception that simple methods and simplistic approaches should provide the ultimate support for protective criteria seems implicit in much of the study to date of metal effects. For example, single-species bioassays have historically provided the dominant scientific basis for protective criteria. At present, reviewers disagree about whether such criteria overprotect[7,8] or are inadequate to protect[9] aquatic life from pollutants. Despite their limited (or at least unproved) applicability to nature,[6,10,11] the use of simplistic approaches in interpreting metal effects is growing as the demand grows for tools to enforce regulations. The demand for simplicity may conflict with our crude, incomplete understanding of the complex processes by which contaminant effects are manifested in nature.[6,12] Moreover, if research efforts are dominated by a quest for simple answers, progress in understanding critical complexities could be impeded. The goal of this chapter is to expand appreciation for important complexities that may influence metal effects, especially under conditions of moderate contamination. We discuss the application to nature of some predominant approaches for studying the effects of metals and demonstrate the inadequacy of our present knowledge of metal influences in natural systems.

Attributing a biological change in a natural system to the specific influences of metals requires: (1) demonstrating which processes are sensitive to metals; (2) separating metal-induced changes in a process from background fluctuations; and (3) unambiguously relating the detected change to metal exposure rather than abiotic (e.g., temperature, salinity, oxygen, or physical processes) or biotic (e.g., species interactions, nutritional status) confounding variables.[13,14] We will discuss whether studies to date allow satisfaction of such criteria by approaching the questions: Have adverse effects of metals been adequately demonstrated in natural systems at different levels of biological organization? Is our knowledge adequate to describe the expected biochemical, physiological, life history, and community structural characteristics of a metal-contaminated locality? Implicit in our discussion is the question of whether understanding responses to metals might not be improved by less emphasis on developing methods to detect effects (a regulatory goal) and more on understanding metal-contaminated environments as a subset of ecosystems with at least some common, interrelated biochemical, physiological, population, and community characteristics (a scientific goal). To reduce the scope of this broad subject to manageable proportions, we restrict most of the discussion to benthic invertebrate communities.

THEORETICAL OVERVIEW

Some theoretical overview of the ecosystem-wide changes that result from metal contamination is necessary for establishing methods that measure metal effects. The chemical (or biogeochemical) reactions and physical processes that determine biological exposures to contaminant(s) must be appreciated, as well as the biological responses that occur across organizational levels from subcellular to the community.[15,16] Understanding the continuum of biochemical, physiological, population, and community responses to a contaminant exposure is critical to both proving cause/effect and to anticipating the most important functional impairments in an ecosystem. Yet the ecosystem view requires multidiscipline considerations that are difficult to achieve in any individual study of metal effects.

Bioaccumulation, resulting in exposure of subcellular processes to the potential metal toxicant, is the least ambiguous of the biological responses to measure and is a necessary prerequisite to any further response to a metal. Showing the significance of this unambiguous response is difficult, however. Detection of a metal response at a lower level of organization does not necessarily mean a response at a higher level is inevitable. Mechanisms of compensation for metal effects occur at each organizational level. Functional impairment within a level occurs after compensatory capabilities are overcome,[19] and it is likely that functional impairment at a lower level of organization may sometimes elicit only compensation at the next level.

Numeric responses are the "significant" effects (or disabilities[17]) against which aquatic communities should be protected.[20] These are defined for our purposes as toxicity to the individual organism, changes in population size (or genetic composition), changes in the species that comprise a community, or changes in one or more of the local communities within an ecosystem. These most significant effects are the most difficult to detect.[15] The variance associated with measures of biological effects increases as intrinsic biological interactions become more complex at higher organizational levels.[15] Alternative abiotic and biotic influences become more difficult to separate from specific metal influences as organizational level increases. In addition, responses at higher organization levels may be manifested over longer time scales than those at lower levels.[18,19]

Difficulties with the detection of significant effects emphasize the inherent limitations of studying biological response to metals at any individual level of organization. Understanding responses across several levels of organization is a feasible approach to reducing interpretive ambiguity and corroborating cause/effect conclusions with multiple lines of evidence.[16,20] Observation of a response at a lower level of organization may be the most clearly detectable signal that a numeric response at the next higher level is beginning or is imminent.[20] Simultaneous observation of change at a higher level of organization may verify the broader signifiance of the lower-level response. The most specific responses to metals may lie in patterns that exist in the continuum occurring across levels of organization, but the combination of breadth and intensity to test such a hypothesis is rare in any study of metal responses.

EXTERNAL INFLUENCES ON METAL EFFECTS

Defining dose-response is fundamental to understanding the toxicology of a chemical, but, in nature, both dose and response are affected by external processes. Substantial uncertainties surround several considerations of fundamental significance to defining how biological processes respond to metals in nature. Examples of such considerations include: geochemical influences on metal bioavailability; anthropogenic and physicochemical influences that make the spatial and temporal distributions of metals complex; the interactive effects of multiple pollutants; the effects of multiple disturbances; and the effects of environmental history.

Metal Bioavailability

Interpretations of biological responses to metals depend upon accurate determination of exposure to the contaminant.[16,21] The most important difficulty in defining exposures in nature is that total metal concentrations in solution and sediments are not the concentrations available to biota.[22-24] It is well established that geochemical and biological processes influence metal form which, in turn, influences metal bioavailability. Much remains to be learned about the relative importance of routes of metal exposure and the bioavailability of specific metal forms. Where the influence of form on bioavailability is understood (e.g., the association of bioavailability with the activity of free ion of some metals in solution), neither analytical or numeric approaches are presently adequate to define concentrations of those bioavailable forms in natural waters. These deficiencies in understanding fundamentally limit capabilities for predicting metal exposures and thus metal effects.[23-26]

Direct determination of the dose of a metal to which organisms are exposed in a natural system is not yet possible, but concentrations of metals in sediments, water, or biological tissues can be employed as indicators of exposure. All methods require special considerations during study design, analysis,[27] and interpretation.[28,29] Metal analyses of the tissues of indicator species are especially valuable as an indicator of bioavailable metal exposure.[25,30-32] Surprisingly, few studies of biological effects have taken advantage of tissue contaminant determinations, although the approach is widely used in studies of metal distributions in estuarine and coastal marine systems,[30,33-35] and, more recently, in fluvial environments.[36,37] Successful bioindicator studies require careful design, and consideration of factors such as analytical matrix effects,[38] sample care and preparation,[39] statistical design,[40-42] and biological factors such as choice of species,[30,35] animal size/age,[30,43-45] and reproductive cycle.[46-49] Hypotheses also must be carefully framed if biological exposure indicators are employed. For example, metals adsorbed to the exoskeleton of small arthropods (which dominate benthos of some streams) may comprise a large proportion of the body burden.[36,50] Thus, such animals may be adequate indicators of exposure to upper trophic levels, but their whole body concentrations will not necessarily be indicative of physiologically reactive exposure to the species itself. Used in combination with water and sediment analyses, tissue analysis has significant advantages for estimating biological metal exposures, despite the necessary caveats.

Spatial and Temporal Complexity of Exposures

Inadequate definition of metal exposure (at both treatment and control sites) is a common weakness of field studies of metal effects, partly because of the biological focus of such studies and partly because the complexity of metal distributions in nature is not adequately appreciated. Contamination gradients are not necessarily spatially simple nor is regional or widely scattered sampling necessarily adequate for the characterization of contamination. Available data suggest that patchy, irregular distributions are typical of contamination in most natural systems. Coastal marine and estuarine metal contamination is typically "a collection of local problems" at the subregional scale on the North Atlantic coast of North America[51] because of the many individual contaminant inputs. San Francisco Bay has more than one hundred contaminant "hot-spots" of varying size that occur near modern and historic sources of waste input and runoff.[3] These are superimposed upon a background of moderate contamination. Over reaches of less than 30 km, metal contamination in sediments and biota are highly variable in the Clark Fork River below a large Cu mine, especially near the mouths of tributaries.[52] Transport and depositional processes can also redistribute contamination away from input sources in estuaries,[53] and deposit contamination hundreds of kilometers downstream in rivers.[52]

Temporal variability is not well studied in many systems and is expensive to deal with in assessing metal exposures. Temporal variability in metal exposures of biota also may differ in amplitude and frequency among systems.[54] The few available temporally intensive studies suggest it may be a crucial consideration, however. For example, episodic, seasonal, and year-to-year changes in contaminant exposures are common in San Francisco Bay.[3,49,55,56] Seasonal redistribution of fine-grained, metal-enriched sediments may be characteristic of salt wedge estuaries in general.[57] Episodic or seasonal changes in exposure have affected interpretations of metal influences in a few studies where data are available.[48,58,59] Only more intensive consideration of the time variable will allow us to determine the magnitude of such effects.

Unambiguous observations of metal effects in nature are not likely if studies assume that metal concentrations are distributed in predictable gradients; that distance from the source of interest is adequate to locate "control" sites (without documentation of control exposures); that exposures will not change over periods of months or years; or that preexisting exposures were similar to those determined at the time of a study. Biological results from studies that employ overly simplified hydrogeochemical assumptions are likely to conflict with chemical data or necessitate speculation about the influences of unobserved factors. Intensive spatial and temporal characterizations of biologically available contaminant exposures are important to reduce such ambiguities.

Multiple Pollutant Exposures

Rarely are metals found individually in a system, and the multiple types of contamination that are found add to the difficulty of identifying the most active agents of toxicity.[16,51] Little consensus exists about how multiple contaminant exposures differ from single toxicant exposures, so interpretation of the significance of multiple contaminant influences in nature is especially difficult. Antagonism in bioaccumulation of some metal pairs (e.g., Cd and Zn) occurs at least during short-term exposures in benthic animals,[60-64] especially at high concentrations or when differences between the two are large.[65] On the other hand, bioaccumulation data from nature, reflecting lifetime exposures, show few such interactions.[66] The few available studies suggest adverse response to at least some combinations of metals are additive rather than antagonistic, in contrast to bioaccumulation[67-69] and some authors have speculated that effects they observe in nature appear to be the result of the joint action of several metals.[70,71] Interactions between metals and organic contaminants[72-74] undoubtedly add complexity to responses but also are poorly understood. In the absence of adequate understanding, documentation of the presence of multiple contaminants is a minimum step in assessing biological impacts of metals in a field study, but predicting the sum of effects from multiple contaminant exposures is not possible at this time.

Are Metal Effects Influenced by Simultaneous Abiotic Natural Disturbances?

Implicit in most strategies for studying metal effects is the assumption that metals are the only factor controlling biological response in an affected environment. The search for the "toxic" dose that generates each type of numeric change or functional impairment restricts consideration of any additional abiotic or biotic processes that contribute to biological phenomena. For example, if biological compensation mechanisms occur at a cost (some aspect of the biological process is weakened) and that cost is manifested when additional stress (e.g., natural disturbance) is imposed on the process, then functional impairment may occur in coincidence with a disturbance other than that which ultimately made the process vulnerable.

Controlled studies and models demonstrate that when chemicals affect filtration rate, digestion, or metabolism in starved mussels, the effects on population growth are greater than when individuals are adequately fed.[75] Modeling studies also suggest that fish populations subjected to intense harvesting are more vulnerable to the additional mortality imposed by a toxicant than populations harvested less intensively.[75] One effect of reduced growth rates is to reduce the average size of adults in zooplankton.[77] Smaller adults may then be more vulnerable to predation and to starvation during periods of natural catastrophe. Such interactions are probably common in nature, but are difficult to detect and rarely considered when the effects of metals are evaluated.

Environmental History

The characteristics of existing biological systems are at least partly determined by their history.[78] History affects the starting point from which dose-response within a contaminated ecosystem begins[79] by influencing the kinds of species present or the tolerance of populations. Thus the sensitivity of the biological systems may differ among or within estuaries, among rivers, or among coastal marine systems, depending upon historical environmental characteristics, disturbances, biological interactions, or previous metal exposures.[80-82] One aim of toxicological studies is to predict a concentration at which adverse biological effects begin in a system. Extrapolation of such studies to nature becomes more difficult if systems differ in influential historical characteristics and thus their inherent sensitivity. Protective criteria based on a single metal concentration assume systems with many different histories have the same sensitivity. If history is important, such criteria will inevitably overprotect some existing systems and underprotect others. Until we begin to understand system responses to metals we cannot approach the question of whether history influences metal responses or whether ecosystems and communities differ in their inherent susceptibilities to metals.

EFFECTS AT INDIVIDUAL LEVELS OF ORGANIZATION
Biochemical and Cytological Effects of Metals
Metal-Sensitive Biochemical Processes

The best studied biochemical responses of invertebrates to metals are associated with compensation mechanisms. Invertebrates, perhaps with a few exceptions,[83] respond to metal contamination with induction of specific metal-binding proteins (MBP) within the cell solution (cytosol[84]) and proliferation of subcellular inclusions that include compartmentalized structures such as granules, lysosomes, and vesicles.[74,85,86] (See this volume, Chapter 3.) Together, the inclusions and MBP conjugate, trap, and detoxify metals. They also provide the cellular basis for bioaccumulation.[85] The concentrations of metals associated with inclusions are lost from tissues slowly,[85] and their abundance may be an indicator of the history of an organism's metal exposure. Metal concentrations in the cytosol change more rapidly and may be indicative of more recent exposures.[74]

Toxicity occurs when compensatory mechanisms are overwhelmed, saturated, or damaged by metal influx.[86] In controlled studies, biochemical toxicity appears to occur when MBP cannot kinetically prevent metal from accumulating into other protein fractions.[87] Early studies with fish indicated that the cytosolic signal of stress was "spillover" of metal from MBP into operationally defined high-molecular weight fractions of protein that contain functional enzymes.[88] More recent studies with crab larvae and polychaetes indicate that the appearance of Cu in a very low-molecular weight fraction of the cytosol signals stress. Increase in that cytosolic fraction of Cu coincided with reduced growth, decreased fecundity, and poor reproductive success.[89,90]

Another signal of metal stress within cells is damage to lysosomes,[86] epithelial changes, or atrophy of cells rich in inclusions.[85,91] Damage to the lysosomes allows leakage of hydrolytic enzymes, which may be the ultimate cause of the cellular damage.[92] Lysosomal damage is detectable from the latency of some of the hydrolytic enzymes. This response is not metal-specific, however. Cu causes lysosomal damage in mussels (*M. edulis*) in controlled studies, but so do trace organics[93] and other environmental stressors such as adverse temperature changes.[94] Interactive effects among stressors add to the complexity of the response. For example, Cu exposure reduces the recovery of lysosomes damaged by toxic hydrocarbons.[73]

Observations of Biochemical Responses to Metals in Nature

Increases in metal inclusions and the presence of metal-specific binding proteins, the metal-sensitive biochemical compensatory responses defined in controlled studies, also have been described in natural systems.[58,85,95-100] Rarely is the significance of such responses evaluated, however. Biochemical stress signals, such as shifts in metal associations among cytosolic protein fractions, have been observed only in a few field studies. Johannson et al.[58] observed a shift of Cu and Ag into very low-molecular weight proteins, in association with an episode of increased Cu and Ag concentrations in tissues of the clam, *Macoma balthica,* at a site in San Francisco Bay. Frazier and George[95] demonstrated that the important difference between oysters collected from contaminated and uncontaminated environments was the high concentrations of Cu and Cd associated with the very low molecular weight proteins in the contaminated population. The significance of these responses was not determined in either study.

In a few studies, additional responses were considered along with the observation of biochemical compensation. Mason et al.[85] reported that renal epithelial cells of snails from a metal-contaminated estuary that were rich in inclusions also were abnormal in structure (Table 1), although that was not a major emphasis of their study. Roesijadi et al.[97] showed that metal uptake into both MBP and high-molecular weight protein in *Mytilus edulis* was accompanied by a decrease in body condition when the mussels were transplanted to a contaminated environment. A mutlidiscipline study of contaminant effects (metals and hydrocarbons) in a Norwegian fjord found that increased Cu concentrations in cytosol of

Table 1
Toxicity of Cu to Different Life-Stages and
Processes in the Marine Mussel, *Mytilus edulis*[a]

Process Affected	Toxic Dose	Source
Energetics		
Clearance rate	2	231
	80—230	232
Oxygen consumption	250—300	232
Growth		
Soft tissue	10[b]	233
Shell	3—10	234
Embryonic survival	5—6	118
Larval survival	400[c]	234
Whole organism LC_{50}		
4 d	200—300	232
10 d	90	233
14 d	15	114
30 d	2	233

[a] All values in $\mu g/L$ total Cu in solution.
[b] 30 d.
[c] 15 d.

mussels[99] was accompanied by damage to lysosomes,[91] damage to digestive cells,[101] and apparent impairment of some physiological processes.[92,102] Biochemical symptoms specific to effects from hydrocarbon exposure also were observed in this study, and the pattern of responses may have reflected hydrocarbon exposure more than (or as much as) metal exposure.

The above discussion indicates that a general understanding of how invertebrates respond to metals at the subcellular level is available. Metal exposures in at least some natural systems are sufficient to elicit biochemical compensation, and probably adversely impair processes at the biochemical level. However, the available field studies are too few to suggest consistent biochemical patterns of response to metal contamination, and insufficiently broad to suggest biochemical signals that accompany adverse effects at other levels of organization. Few studies of lysosomal and cellular damage in environments where metals are the predominant contaminant have been conducted. The exposure dependence of metal concentrations associated with MBP and other intracellular proteins has not been systematically characterized in nature for any species. MBP concentrations themselves appear to be highly variable in concentration,[103,104] but the causes of this variability are not understood. Metal distributions among protein fractions within the cytosol also sometimes differ between laboratory and field observations.[58,105] Differences between laboratory and field in the length and nature of the exposure, adaptation to historic exposures in nature,[95] effects from natural disturbance,[74] or the effects of multiple contaminants[74,106] are possible causes of such contradictions that demand further investigation.

Improved fundamental understanding of the specific mechanisms of compensation for metal exposure and the characteristic responses when compensation is overwhelmed would improve understanding of how invertebrates respond to metals. However, systematic, detailed study in nature of known subcellular responses to metals is also needed if biochemical responses are to be used to characterize environments influenced by metal contamination, or evaluate how widely such contaminant influences occur.

Physiological — Individual Organism
Metal-Sensitive Processes

A number of physiological and whole organism processes are sensitive to metal toxicity, although most responses are not metal-specific.[102,107] Processes that respond to metal exposure with some sensitivity in benthic invertebrates include feeding rate,[21,108] respiration,[109] and protein utilization (manifested as nitrogen excretion).[110,111] Reproductive processes are among the most sensitive to metals,[112] because of direct effects on sexual maturation, follicle[113] and gamete[114,115] development, embryogenesis, and larval development.[116-118] Solute concentrations at which the above effects are observed in toxicity tests with marine mussels are shown in Table 1, although it should be recognized that the sensitivity of processes varies widely among species. Metals also may cause morphological abnormalities,[119] and histopathological problems in adult invertebrates.[112,120] Effects of metals on organ function (heart rate, ventilation rate), behavior,[121] osmotic/ionic regulation, respiration in isolated tissues,[122] and hematology have been demonstrated, but these are too poorly known, too insensitive, or too variable in benthic invertebrates to be useful symptoms.[21]

Some of the best studied effects of contaminants are those expressed as changes in energetics. Effects on energetics may be determined as instantaneous changes in growth efficiency, a measure of the efficiency with which food is converted into body tissues,[123] or growth potential (determined as "scope for growth"). The latter is determined from estimates of feeding rate, oxygen consumption/nitrogen excretion or O:N, and assimilation.[124,125] Longer-term changes in energetics can affect reproduction, tissue growth, and body "condition" (e.g., measured as tissue mass relative to shell size). The potential significance of changes in scope for growth is shown by correlations with changes in processes at higher levels of organization, although not with metals as a causative factor. For example, in populations of *Mytilus edulis,* reduced scope for growth has been found coincident with population size class structure indicative of reduced growth rates[126] and reduced fecundity.[107]

Detection of Metal-Induced Physiological Impairment in Nature

Unambiguous relationships between physiological change and metal exposure in contaminated natural systems are difficult to obtain. One of the reasons is that physiological responses to metals differ among invertebrate species in sensitivity and variability. Thus the detection of a response is strongly affected by the choice of species and the choice of response. For example, the marine mussel *Mytilus edulis* maintains clearance and respiration rates characteristic of the field during scope for growth determinations and is especially well suited for determination of physiological impairment using this measure.[102] On the other hand, species with cyclic or erratic feeding and metabolic activities, or sensitivities to handling stress, are poorly suited for scope for growth determinations. Differences in the suitability of mussels and snails in scope for growth determinations were demonstrated in studies along a pollution gradient in Langesundfjord in Norway. Changes in scope for growth in *M. edulis* occurred along the contamination gradient, in agreement with bioaccumulation data and with measures of biochemical and cellular impairment.[102,127] However, no detectable changes in scope for growth were observed in the same area in *Littorina littorea*.[128] It was unclear whether the absence of a response was related to a lesser sensitivity to contamination in the snail or to the greater difficulty of determining scope for growth in this species. In any case, different conclusions about physiological effects of the contaminant would have resulted if either species had been employed alone.

A commonly employed simple measure of chronic energetic stress, condition index, may also differ in sensitivity to metals among species. Changes in condition index have been reported in a number of studies with the freshwater clam, *Corbicula fluminea*, from metal-contaminated environments[56,129,130] suggesting this is a consistent and sensitive response in this species. However, studies with benthos such as the deposit-feeding clam, *Macoma balthica*,[48,70] and the marine mussel, *M. edulis*,[59,97,99,131] in metal-contaminated environments where responses are occurring at other levels of organization, show changes in condition index in some circumstances and not in others.

Compensation is a second source of complexity in interpreting physiological responses to metals. For example, when lower concentrations of Cd inhibit follicle formation in the gonads of *M. edulis*, spawning frequency is increased. The result is little effect on gamete production.[113] Low-level preexposures of mussels (to Hg) or polychaetes (to Cu) may result in improved tolerance to more toxic levels,[132,133] although higher level preexposures have no effect or can weaken individuals.[81,132] General damping of physiological responses to metal exposure also may occur by induction of stress proteins.[134]

If physiological compensation occurs at a cost, it ultimately will reduce an individual's ability to survive additional environmental challenges.[19] Controlled studies have demonstrated that increased vulnerability to metals can result from simultaneous exposure to additional disturbance and conversely metal stress may increase vulnerability to such disturbances. For example, *M. edulis* is more sensitive to Cu after spawning than during spawning and least sensitive before spawning, when energy reserves are presumably high.[121] Many controlled studies have shown increased susceptibility to metals in suboptimal ranges of temperature, salinity, or food availability.[135-138] If the effects of metals and other disturbances have additive influences on physiological processes, then unambiguously relating a physiological response to a metal exposure could be difficult unless the environmental vulnerabilities of the organism, in addition to metal sensitivities, are understood.

The most important difficulty in unambiguously relating a physiological response to metal exposure lies in distinguishing among the factors that might cause the response. Most physiological and whole organism responses to metals are not metal-specific.[126] Thus, certain physiological symptoms may typify metal-contaminated populations but their occurrence is not restricted to metal contamination. For example, linkage of scope for growth with pollutant exposure alone is difficult to establish in food-limited populations of mussels,[139] and probably in any population living in marginal natural conditions. The longer-term effects of changes in growth potential, as evidenced by measures such as condition index, are also very sensitive to food availability and reproductive condition.

A number of field studies have suggested the possibility of metal effects on physiological processes, but, in almost all cases, inadequate attention was paid to assessing alternative explanations. Changes in condition index observed in association with contamination have not thoroughly considered natural factors that might vary coincidentally with contaminant exposures, such as nutritional status.[56,97] Histopathological lesions have been related to pollutant exposure in nature in species other than invertebrates[140,141] and should be a useful whole organism indicator of metal effects in invertebrates. Indications of the formation of granulocytomas in mussels in a metal-contaminated estuary[120] and a relationship between pollution history and pathological symptoms in the clam *Mya arenaria*[142] have been observed, but consideration of alternative explanations was insufficiently thorough to prove cause and effect.

Studies of energetic responses in mussels deployed in pollution gradients also illustrate the difficulties in unambiguously demonstrating metal effects in experiments of insufficient physiological and ecological breadth. Statistically significant reductions in scope for growth and growth efficiency were demonstrated in Narragansett Bay[123] and San Francisco Bay,[131] in mussels (*M. edulis*) deployed for 30 to 90 days. The physiological changes followed gradients in trace element and hydrocarbon contamination determined at the end of the experiment in both systems. The energetic abnormalities in San Francisco Bay were accompanied by reproductive impairment. Suspended solid concentrations were not considered in the San Francisco Bay study and could have explained at least part of the observed response. Temperature, salinity, dissolved oxygen, and total seston concentration were relatively constant at the one time they were sampled along the gradient studied in Narragansett Bay, however, leading the authors to conclude that anthropogenic stressors were the cause of the changes observed. A rigorous elimination of natural influences probably should include at least a more convincing evaluation of nutritional status, however. Thus, pollutants may be the most likely cause of the changes observed in both studies, but ambiguities cloud the certainty of the conclusions.

More powerful study designs better illustrate the complexities that can affect energetic responses in contaminated estuaries. For example, Widdows et al.[59] assessed physiological responses of *M. edulis* to reciprocal deployments between the Swansea and Tamar Estuaries, in the U.K. Animals were deployed for 6 months and their responses were compared to indigenous individuals. Chemical, bioaccumulation, and physiological data were collected monthly and illustrated the temporal variability of food availability, pollutant exposure, and physiological response that can occur over such an exposure. The study was also complicated by the fact that both estuaries were contaminated, but with different pollutants. In general, scope for growth and growth efficiency appeared to be higher in association with lower concentrations of bioavailable Cd, Cu, Zn, and hydrocarbons, but the effects of metals, hydrocarbons, and environmental conditions were difficult to convincingly separate. The principal conclusion of the authors was that environmental, rather than genetic, factors controlled physiological responses in these animals.[59] While such conclusions are valuable as an example for future deployment studies, the environmental and biological complexity encountered in the study meant that the specific physiological responses to contamination were left uncharacterized.

The most convincing characterizations of significant physiological impairment have occurred where responses are pollutant-specific. The best example was demonstration of the effects of alkylated tins on reproduction in snails[143,144] (see this volume, Chapter 12). Exposure to very low concentrations of alkylated tin (as low as 0.0024 µg/L tributyltin oxide[143]) were shown to cause specific morphological deformities (imposex, the induction of male sex characteristics in females) in the snail, *Nucella lapillus*,[119] and other gastropods.[145,146] These deformities were demonstrated to be Sn-specific, and to render females sterile.[119,143] Deployment of snails to Sn-contaminated sites caused induction of this functional impairment, in association with bioaccumulation of Sn. Specific reactions to tributyltins also have been linked to damage in other aquatic benthos[147] and the convincing nature of such observations has contributed to widespread regulation of the use of this compound. A broader coverage of the effects of organotin compounds on marine biota is given in this volume, Chapter 12.

In lieu of specific responses to most metals, one approach to reducing uncertainties in defining the causes of physiological responses is to simultaneously study responses at several levels of biological organization. Multidiscipline "workshop" experiments have recently been conducted in which a large group of investigators with different types of expertise have combined their capabilties in a one-time study at one location. The first such workshop was conducted in the contaminated Langesundfjord in Norway. Evaluation of physiological an' biochemical techniques was the goal of this study, but some conclusions also were possible about the suite of symptoms that typify responses to contamination.[2,99,127] The study was successful in demonstrating hydrocarbon-specific biochemical responses along a hydrocarbon and metal contamination gradient in all four species that were studied. Although hydrocarbons may have been the principal pollutant in the area, biochemical responses to metal exposures were indicated by metal uptake into cytosol in mussels. Metals also could have contributed to observations of lysosomal damage observed in mussels and snails at contaminated stations.[91,148-151] Physiological responses such as reduced scope for growth and growth efficiency,[127] gamete degeneration,[101] and a high incidence of granulocytomas[151] were observed within the contamination gradient, in co-incidence with the signals of biochemical impairment in *M. edulis*. However, as previously mentioned, species differed in their physiological responses (e.g., reduced scope for growth was not observed in snails) and some responses were contradictory within species (contaminant effects on condition index were not detectable in *M. edulis*). Thus, despite the unambiguous demonstration of responses to pollution at lower levels of organization and application of the highest quality biochemical and physiological methodology, the significance of the pollution to the continuum of processes occurring across levels of organization was not clear. The generally inadequate understanding of the differences in physiological response among species and sensitivity of responses within species contribute to such difficulties. In these specific studies, descriptions of the space/time dimensions of contaminant exposure also were limited by the one-time-only experimental design, affecting some interpretations. Another difficulty was that the goal of the study was an evaluation of techniques. The next step in such valuable efforts might include a more complete study of the physiological ecology of survivors in the contaminated environment, study of the adaptations that allowed them to survive (along with the lack of adaptation that prevented other species from surviving), and the development of general conclusions about the significance of the pollution within the context of other physiological and ecological influences. In any case, the magnitude of the challenge in understanding the biological effects of contaminants, even in one location, and the difficulties of meeting that challenge in a single study are clear.

The few available studies from nature suggest that physiological responses to metals within some benthic species are probably occurring in moderately contaminated environments in nature, but they are difficult to prove. The suite of typical symptoms of metal stress up to the level of the whole organism probably differs among species and can include changes in energetic balance, poor body condition, reduced reproductive capabilities, and, perhaps, histological anomalies. However, this suite of responses has been systematically studied in only a few species, and only a few studies have compared responses among major species in a typical community. Combining whole organism, physiological, and biochemical/subcellular studies may reduce uncertainties in evaluating the causes of physiological responses, but this approach may not be adequate to demonstrate cause and effect or the general significance of the pollution. A thorough understanding of the biological and environmental baseline conditions in the contaminated locality and, especially, the natural physiological ecology of the predominant species (including those that might be expected in the absence of contamination) seems a critical missing link in studies of metal effects at the physiological/whole organism level of organization.

POPULATION
Metal-Sensitive Population Processes

The persistence, abundance, and production of populations in metal-contaminated environments are of ultimate interest in biological conservation. In each population these are determined by the dynamics of mortality, fecundity, and migration rates. Numerous toxicity tests have assessed metal exposures at which survival (adult, larval, or juvenile), reproduction, or behavior are impaired in aquatic invertebrates. These tests have provided an important first step in understanding the vulnerability of species to metals. However, few such studies have successfully related bioassay predictions to the success of the populations of those species in nature.[6,8] Biological and geochemical simplifications (some inherent in the approach) limit the abilities of toxicity tests to predict population responses in nature,[6,10,11] but so do the complexity of the population-level responses. Given the paucity of studies of metal-sensitive processes in populations, we speculate below about some possible responses suggested from population models and observations.

Bioassays commonly find that metal sensitivity in a species can differ with age or size. For example, Giudici et al.[152] found that juvenile individuals of two species of benthic crustaceans were more vulnerable than older animals to short-term Cd exposure, suggesting that juveniles might be the first individuals affected in a short-term episode of Cd contamination in nature. Selective toxicity only to younger individuals could be fatal for a population if young age classes have the predominant influence on recruitment. However, population modeling studies show that if the species is capable of a density-dependent increase in the fecundity of older age classes when young age classes fail, survival is more likely.[77,153,154] Zajac and Whitlatch[155] demonstrated an example of this type of resilience in a study of polychaete (*Nephthys incisa*) survival during dredge spoil disposal. The disposals appeared to cause complete recruitment failure in the polychaetes, but the ability of older adults to survive a disposal episode and later reproduce allowed the local population to avoid extinction. Similarly, selective toxicity to older animals, which is a possible outcome of chronic metal contamination, could establish conditions that eventually lead to extinction of the population. Models of population dynamics suggest that the immediate effects on recruitment in a population may be less severe if older animals, rather than younger animals, are lost,[153] but the resilience that may allow survival during the inevitable periods of natural disturbance is lost. Thus age-class resilience is an example of a population process that could be as important as physiological sensitivity in determining a species' survival (or extinction) in a contaminated locality.

Toxicity tests also show that processes relevant to population success differ in their sensitivity to metals (although systematic, comparable compilations of such differences are surprisingly rare for benthic invertebrates). Adverse effects of metals on any population process can affect the success of a population. For example, adverse effects on recruitment can influence a population as readily as increased mortality.[144] Effects on one population process also might affect other processes. Metal-induced changes in age structure, longevity, growth, development period, or reproductive period will affect fecundity.[16] The response and the characteristics of the population in nature might be quite different depending upon the life stage affected and the life history characteristics of the species. For example, larval survival appears to be much less sensitive to Cu exposure than embryonic survival, physiological energetics, or even adult survival in *M. edulis* (Table 1). If so, one could speculate that physiologically stressed populations of mussels, incapable of reproduction, could persist in a Cu-contaminated environment as long as larvae from elsewhere could reach that locality. In fact, communities of stressed adults (limited longevity, tissue lesions, limited reproductive capacities) of benthic bivalves have been found in some highly metal-organic-contaminated environments where bioassays might predict toxicity to the most sensitive life processes.[112,156] Existence of such communities, of course, depends upon the health of their source populations.

The number of larvae that survive in the metal-contaminated environment also could be important. Recent studies[157] suggest that population densities and community processes in coastal rocky intertidal zones are influenced in important ways by the availability of larvae. When larval supply is unlimited in the rocky intertidal zone, population ecology is controlled by processes internal to the community (competition, predation). If larval supply is limited (in this case by physical factors), population dynamics are determined by the processes that control the rate of larval settlement at the site. In species in which larvae are relatively vulnerable to metal contamination, one might speculate that moderate metal contamination could affect the supply of larvae (through toxicity to the most vulnerable individuals), and thus important aspects of population and community dynamics. Thus, vulnerability of only a portion of the larval pool might seem unimportant in a toxicity test but have significant implications for some species in a natural setting. Inadequate appreciation of such complexities could result in substantial underestimation of the influences of metals in nature.

Observations of Population Effects in Nature

Studies that consider population dynamics of invertebrates in detail and assess metal contamination as one influence on those dynamics have not been conducted in natural systems. The most common approach to population studies in nature has been to assess the success (presence or absence) of species that show other indications of metal-specific stress. The sequence of studies by Bryan et al.[143,144] are an example of a successful implementation of this approach. They linked the progressive decline of *Nucella lapillus* throughout southwest England to increasingly frequent observation of Sn-specific inhibition of reproductive processes. As explained earlier, metal-specific responses of this intensity are rare, and broad-scale documentation of the disappearance of a species is rarely available. Furthermore, changes in population size may be indicative of an adverse effect, but such changes do not necessarily occur in the early stages of a metal effect (because of compensatory capabilities).[16] Thus, relating cause and effect from physiological responses and species abundance has rarely been this successful.

Some studies have sought a metal-specific response at the population level of organization. Luoma[80] suggested that elevated tolerance to a metal in one population, compared to other populations of the same species, might constitute such a response. Different sensitivity among individuals to environmental factors, including metal toxicity, is a common genetically driven characteristic of aquatic invertebrate populations.[158] Increased tolerance in a metal-contaminated locality develops as a result of selection for the most tolerant individuals in the population, which are the most fit in those circumstances.[80] However, genomes tolerant to metals have reduced fitness in uncontaminated environments. Thus if the contamination is eliminated, the population reverts to its dominance by less tolerant genomes (see review[80]), suggesting whenever tolerance is present it reflects an influence from the metal exposure. Recent studies rigorously demonstrated that a selective, genetic basis was the source of Cd tolerance in oligochaetes[100,159] and that individuals with enhanced tolerance can begin to dominate a population in a Cd-contaminated environment within a few generations.[100] Like all forms of compensation, tolerance probably comes at some cost (perhaps increased vulnerability to some natural stressors) even in adaptable species.[81,160] The specific "costs" to aquatic species have not been well studied, nor have differences in adaptability among species been studied. Co-tolerances, selection for generally improved hardiness, and physiological acclimation also must be considered when assessing tolerance as a metal-specific indicator of effects on populations.

Since first observed in polychaetes,[161,162] elevated metal tolerance in aquatic benthos has been described in a number of contaminated environments from nature,[80,81,156,159,163] and recently was used as an early warning indicator of metal-specific effects on populations.[160,164,165] The number of instances in which tolerant populations have been observed suggests that responses to metals at the population level may be widespread. Klerks and Weis[159] even suggest most populations occupying contaminated habitats may exhibit some tolerance but emphasize that many species that otherwise might be expected to occur could be absent from such environments (implying lesser abilities to adapt).

While metal tolerance appears to be widespread, evaluations of the significance of the response is difficult. Rarely is documentation of tolerance accompanied by thorough assessments of other responses in populations, individuals or communities, or thorough estimates of exposure. Therefore, symptoms of metal influence that might accompany tolerance, vulnerabilities of tolerant populations, and the exposure dependence of tolerance in different species remain relatively poorly understood. Tolerance may be a useful indicator, but, like all such tools, it is not very informative when employed in the absence of understanding of other contaminant responses or of the physiology/ecology of a variety of species in the area under study.

Rarely are thorough studies of population dynamics in invertebrates conducted in metal-contaminated environments. McGreer[70] assessed the distribution and population characteristics of *Macoma balthica* on a metal-contaminated mudflat in the Fraser River estuary (British Columbia) in one of the few studies where exposure, population processes, and responses at other levels of organization were considered. Closest to the contamination source he found an area devoid of *M. balthica*. Where the species first appeared, population densities were reduced, age structure differed from the most distant sites, and the author suggested effects on recruitment were indicated. Accumulative toxicity to older animals (1 + years) also was suggested, especially by the observation that nowhere in the study area were animals older than 2 years. Body condition, which was employed as an indicator of energetic effects, varied with contamination. Some alternative explanations for the observations also were considered. The effects observed were not correlated with oxygen consumption of the sediments, effluent toxicity to adults, chlorination, salinity, or sediment grain size on the mudflat. Many of the effects were correlated to metal concentrations in sediments, however. A rigorous analysis of this study would raise questions about the nature of the metal effects, despite the variety of processes considered, and the correlation with metal exposure. Organic carbon concentrations showed some distributional similarity to metal contamination, adding complexity to the age structure and condition interpretations. Also, sampling was conducted at only one point in time, and no information on food availability or growth rates was available; nor were the results compared to population parameters for this species on a mudflat not exposed to metal contamination. Clearly, if metals are affecting populations in complicated ways, intense and thoroughly designed studies will be necessary to provide convincing evidence that such effects are occurring.

A few studies in natural systems have demonstrated that metals probably affect fundamental population properties and the success of species in moderately contaminated environments. However, the complexity of population responses make unambiguous demonstration of effects difficult. If the earliest symptoms of stress in populations are compensatory responses, simplistic measures like changes in density are probably inadequate measures of such effects. Fundamental limitations of toxicity bioassays may preclude accurate conclusions from this approach about important effects of metals on populations in nature.[8-11,16] The temptation to draw such conclusions is great,[59,112,166,167] but disagreements between toxicity test results and observations of survival (or lack of survival) in nature[112] should not be surprising.

An important next step in improving understanding of metal-induced population effects is to better understand comparative life history sensitivities of key species. Each species approaches its habitat differently and solves the problems of that habitat in its own way.[154] Thus each species could have its specific metal-vulnerable processes, and the differential metal sensitivity among key species could be at least partly definable by these vulnerabilities. If such understanding can be combined with studies of population dynamics in metal-contaminated localities (along with controls), the symptoms of metal-induced stress at the population level may become more discernable.

COMMUNITY
Responses of Communities

Community responses to metal contamination reflect the impact of metals on the species that inhabit the contaminated ecosystem, changes in interactions within and among populations in the affected system, and possible changes in individual species' responses to the abiotic environment. In that a complex array of responses are possible, the patterns of change that result from progressively greater metal exposure have been difficult to identify. Some generalizations about community responses to metal contamination are beginning to develop, however.

Definitions of metal sensitivity must occur within an environmental context. Species distributions are controlled by abiotic and biotic factors. Throughout a species' range, locations exist where its abundance is relatively high and relatively low depending on the suitability of the interacting biotic and abiotic conditions in each habitat. When these factors are optimal, the species' population in a habitat is also optimized. Species-specific (and possibly population-specific) tolerances to metal contamination[168] are a function of the innate biochemistry and physiology of the species (population). However, the tolerance of a species to metal contamination in any locality is also a function of the intensity of natural stressors, i.e., how optimal the species environment is prior to metal contamination. Therefore, a range of interacting (and often poorly understood) factors from biochemical to ecological determine whether or not a species is "sensitive" in a metal-contaminated locality.

Metal-sensitive organisms, as defined above, are the most likely to respond to toxic metal contamination. Ford[4] explains conceptually what the consequences for community structure are of the loss of only the most sensitive species and how changes in community structure are greatly influenced by their role and abundance in the community. A numerically low-sensitive species may disappear or be replaced by numerically low-tolerant species with little detectable influence on the community structure. On the other hand, either the disappearance or replacement of a numerically abundant species would be more detectable and possibly more significant. Loss of a sensitive species whose presence influences the abundance of other species may result in changes that propagate throughout the system; loss of a species that plays a minor role or whose role is buffered by redundancy among more tolerant species will be of less significance. The ultimate trajectory of the community response depends upon individual physiological sensitivity; locality-specific habitat suitability, especially for the sensitive species; and the role and abundance of the sensitive species in that locality. If the contamination becomes more severe, more species are affected. When the loss of sensitive species begins to exceed the rate of replacement by tolerant species with similar community functions, then the most noticeable changes will begin to appear and larger changes are more likely to cascade through the community. At some point depauperate communities result, where only a few species survive.

Time scales, spatial scales beyond the local community, and nature's inherent variability are also important in determining the ultimate community that occupies a contaminated locality. Although they were not addressing metals, Sebastien et al.[169] showed drift to occur in all benthic taxa, regardless of feeding group immediately following the application of methoxychlor [2,2-bis(p-methoxy-phenyl)-1,1,1-trichloroethane] to the Souris River, Manitoba. Recolonization was a function of the ability of the taxa to drift from unexposed areas and the time in the life cycle when the impact occurred, with univoltine taxa most severely affected.

Recent reviews have emphasized the importance of environmental history and regional processes in understanding communities.[170,171] Obviously, these principles also apply to understanding community response to metals. The importance of temporal considerations have been illustrated in field studies. For example, years passed before the ultimate effects of experimental acidification were manifested throughout the food chain by the loss of upper trophic level fish in the Canadian experimental lakes program.[172] Alternatively, the total assemblage of opportunistic species at a metal-contaminated mudflat in San Francisco Bay stayed much the same over more than a decade, but the species that dominated the community differed greatly from year-to-year.[173,174]

Methods of Detecting Community Response

Ecosystems have both structure and function. Thus the methods used to detect responses of aquatic systems to contaminant inputs are dependent on whether the presumed change will be structural, functional, or both. Structural changes are represented by changes in the composition, number, and abundance of taxa that occur as a result of a change in contaminant input. Functional changes are represented by alteration of some process, such as primary productivity, decomposition rates, etc. Communities react both structurally and functionally to metal stress,[175] and the two reactions are interrelated, although these relationships are understood only in the most general sense.[175,176] No single structural, functional, or combined response has been identified that is specific to metal contamination. At present, functional measures often are more difficult to interpret, given functional redundancy and compensation that frequently occur in response to pollution impacts.[172,177,178] As a result, structural changes remain the most frequently measured response and will be emphasized here.

The search for simplified descriptive methods for determining the effects of discharges on aquatic community structure is well reviewed elsewhere,[4,16,179,180] and will only be briefly considered here. Kolkwitz and Marrson[181] (plant communities) and Kolkwitz and Marrson[182] (animal communities) first presented lists of taxa that were representative of conditions in streams and rivers receiving various waste inputs. This Saprobic System, is based on the presence of benthic species "uniquely dependent" upon the chemical composition of the environment and was derived from species responses to organic enrichment. With modification, the system is still used in Europe,[183] and similar systems of organism associations have been derived for areas in the United States.[184] Although describing associations of species which represent various environmental conditions is informative, it has proved geographically specific and dependent upon labor-intensive taxonomic judgments (many of which are subjective), which are difficult to explain to nonspecialists.[185]

The concept of indicator species was originally based upon observations that some species are particularly susceptible to water pollution. The focus on a few species also had the benefit of avoiding the problem of interpreting ambiguous results regarding community change derived from lists of unreliable species identifications. Recent studies have identified metal-sensitive taxa in different types of aquatic environments. Some types of copepods consistently disappear from metal-contaminated lakes,[4] and both mayfly species[186,187] and some chironomid species[188] disappear from metal-contaminated cobble-bed streams. Schindler[172] states that in the Experimental Lakes Area of northwestern Ontario, one of the earliest community responses appears to be subtle changes in species composition in rapidly reproducing organisms. In general, however, identification of which species are most sensitive has proved complicated.[189] In field studies, unidentified factors other than pollution influence the distributions of many indicator species,[190] and the absence of a species may not necessarily indicate that chemical conditions are unsuitable.[4] Data from toxicity tests are often not comparable to one another, and few investigators have systematically attempted to identify the mechanistic vulnerabilities of the species that are most affected by metals or the mechanistic strengths of the survivors. Thus, our understanding of why some species thrive and others die in moderately contaminated environments is limited.

Extensive use has been made of numerical techniques that attempt to define different aspects of community structure.[191-194] Simple, commonly employed measures are total number of individuals or total biomass. The trajectories of both are unpredictable as communities change, and neither has proved consistently useful in defining a community effect of metals.[4] The most often applied methods are diversity indices which (in various ways) assess the numbers of taxa present and the evenness of their numeric distributions in the community.[192] The most common application of diversity indices has been to evaluate organic enrichment, although some early studies suggested their use as an improved measure of contaminant effects in streams.[191] Recent studies demonstrate serious shortcomings in the diversity approach in assessing contaminant effects[180,195] or, for that matter, in the use of diversity as a concept in many aspects of ecological research.[4,180,195,196]

One component of the diversity index, species richness (i.e., the number of species) has proved more useful than the index itself. In two separate field experiments where Cu was added to streams, diversity was a less informative measure than species richness in depicting the metal gradient on community structure.[188,197] In an analysis of a number of disturbances to stream biota, Resh[185] also showed species richness to be one of the most consistent measures of community response. Changes in species richness, however, may be a community response that is representative of relatively short-term periods of intense contamination or disturbance. Hall and Ide[198] observed species replacements, but not significant changes in species richness, in aquatic insect communities from two low-alkalinity streams in Ontario subjected to decreasing pH over a 50-year period.

Multivariate methods also have been developed to summarize community structure. The available methods are numerous, but most are classification and ordination techniques[199-201] best applied in separating groups of organisms along gradients.[177,202-204] Interpretation of biological processes are not always easy to obtain from many numeric methods. The more complicated numeric approaches have special difficulties[195] unless questions and approaches are carefully developed. Questions also remain about which multivariate approach to employ, inconsistencies between different approaches, and the level of taxonomy necessitated by each approach to characterize community change.[204]

Recent research has emphasized development of biotic indices as methods to measure community structural change.[205] This development is an attempt to retain the single-number summary provided by diversity indices but include species (taxa)-specific information. The most common approach is to combine relative pollution sensitivities of various taxa with, in some cases, abiotic metrics that numerically define the magnitude of impact on a community. Thus, these new techniques combine certain aspects of indicator species lists (such as using many taxa as opposed to just a few representative taxa) with numerical methods. This advantage is balanced by the large uncertainties inherent in both components. Few of the biotic indices have been tested regarding their sensitivity to detecting metal-specific impacts, and important questions remain about their usefulness in identifying the processes that alter the structure of communities suffering the impacts of metal contamination. For example, most indices are unaffected if one species replaces another.[16] To be accurate in terms of real ecosystem changes, all require great understanding of the inherent sensitivity of species, populations, and ecological interactions. Finally, all consider only a limited number of interactions. Clements (this volume, Chapter 13) presents a similar technique using selected components of the community under study and sensitivities derived from community-based laboratory studies. This technique may be an improvement in establishing site-specific sensitivities of resident organisms.

Structural Responses in Nature

Even where available methods are appropriately employed, it can be difficult to unambiguously detect contaminant effects in a natural system.[206] Nevertheless, probable adverse effects of metals on benthic communities have been demonstrated in a few instances. In several studies, streams were purposefully dosed with metals and the reaction of the benthic community was followed over time. Winner et al.[188] continuously added Cu to Shayler Run, a second-order limestone stream in southwestern Ohio. Total aquatic insect densities decreased by 81%

just below the point of discharge, and the percentage of the insect fauna that were identified as members of the family Chironomidae increased. Mayflies were particularly sensitive to the Cu contamination. In the same study, Elam's Run was chronically affected by effluent containing Cu, Cr, and Zn. Although some patterns similar to those observed in Shayler Run emerged, it was noted that sampling uniformity and the level of taxonomic resolution may have affected the interpretation of results. Sheehan and Winner[175] also noted in a comparison of data from Shayler Run and Little Grizzly Creek that there was a better relationship between Cu level and species richness than between Cu level and diversity. They found a poor relationship between Cu level and total insect densities, which was attributed to the high variability of density measures.

In a study of low-level experimental Cu exposures in Convict Creek, Leland et al.[202] demonstrated community structural changes at nominal Cu concentrations as low as 5 µg/L by using ordination techniques. Measures of density or biomass were not good indicators of community change. In addition, attempts were made to relate structural changes to effects on different trophic levels.[197] They found that the response of herbivores and detritivores was greater than the response of predators. The difference in response of various orders of insects, which was also noted in this study, appeared to be a function of the trophic preferences of these taxa, with grazing mayflies severely impacted. Diversity was a poor measure of response to the Cu impact, whereas taxa richness and measures of similarity better represented a dose-response relationship.

Contamination gradients are more complex in marine and estuarine environments than in streams. Several studies have identified community changes coincident with metal contamination in such systems, but few have seriously studied alternative explanations of the changes. Lenihan et al.[147] demonstrated that the benthic assemblage of species in different harbors in San Diego Bay varied with the number of boats in the harbor and tributyltin concentrations in water. Rygg[207] demonstrated a correlation between Cu concentrations in Norwegian fjord sediments and benthic diversity. Species consistently absent at sediment concentrations greater than 200 µg/g (the concentration that correlated with a 50% reduction in diversity) were designated as nontolerant to the metal. Ward and Young[71,208] demonstrated a change in species composition of epifauna on both benthic seagrass and a bivalve near an Australian Pb smelter. This study was more informative because several alternative explanations for the differences in community structure were considered. Multivariate analyses showed that some community differences coincided with changes in sediment texture or water depth, but other changes correlated with metal contamination. For example, of the 55 taxa studied across the metal gradient, 13 disappeared coincident with metal exposure only and were classified as metal-sensitive. These authors concluded that one general effect of the pollution was to increase the similarity in community composition between habitats that otherwise might have contained more dissimilar communities. Sheehan and Winner[175] proposed a similar effect on stream benthos.

The above examples demonstrate that benthic community structure may be changed by metal exposure in natural systems, but there are many factors that limit our ability to draw conclusions from community-level studies of metal impacts. At no level of biological organization is it more difficult to adequately understand the dose of a metal to the system than at the level of community. Routes of uptake and their effects become progressively more complex as the interactions among species and between species and their environment become more complex. Not only can the metal be thought of as a physical control on the species — the species can also control the metal.[209] Alternative influences also add to the complexity of unambiguously relating community change to metal effects. For example, contamination gradients may be relatively simple in streams and rivers, but physical controls may play a major role in determining the distribution of organisms along the same gradient in such highly disturbed eco-systems.[210-212] Variations in substrate type[213] and presence or absence of impoundments[214] can have profound influences over the species which are present in a system. In order to reduce the variability in species distributions associated with physical controls and consequently increase our ability to identify and quantify metal affects, schemes have been developed to allow researchers to more uniformly choose habitats of similar quality.[215] Application of these schemes has, at least on a rather large geographical scale, been successful.[216] Taking into account the influence of gross physical differences is critical to our ability to detect community changes that are a function of metal stress.

The search for a consistent community response to contaminates has been driven largely by regulatory demands for simplistic methodology. This may conflict with the complicated circumstances in which structural changes occur. Existing methods for assessing pollutant impacts on communities are sensitive to different types of biological change (changes in rare species or in abundant species; changes at the phylum level or at the species level; changes in numbers of species or in the distribution of species among functional groups). A consensus has not yet been achieved as to which types of change best characterize a community that has been influenced by metals. The study of individual species autecology is an important consideration in the choice of methods and interpre-tation results,[217] as is taxonomy.[218] The validity of conclusions of many com-munity studies are often compromised by a lack of taxonomic and life history information available.[219] Attempts to design classification systems that incor-porate functional aspects of species as well as phylogenetic relationships have been generally well received.[220,221] Bahr[222] suggests that progress will be limited until classification systems are more functionally oriented. However, such higher level classifications may reduce the level of resolution that is available using classical taxonomic methods.[223]

CONCLUSIONS

Adverse effects occurring coincident with metal exposure in contaminated aquatic environments have been demonstrated at every level of biological organization. Understanding is growing of the general mechanisms of metal toxicity and broad responses to metal exposure at most levels of biological organization. However, conclusive determination of metal effects in any specific locality remains problematic. In most field studies conducted to date, the specific influences of metals could have been more convincingly proved. Existing methodologies for defining metal stress seem most successful in acutely or severely contaminated systems.[224] In this review, we suggest that approaches applied to date inadequately consider important complexities in the overall response of biological systems to metals. Conclusions about an absence of metal influences in moderately contaminated systems cannot be justified until those complexities are better understood. Inadequate consideration of exposures, multiple pollutant inputs, and alternative environmental influences have plagued many studies, as have poor understanding of linkages in response between levels of biological organization, nonmetal-induced abiotic and biotic response, the importance of historical occurrences, and spatial influences that extend beyond the local scale. Some influences (e.g., marginal environmental conditions) might enhance biological vulnerability, while other processes (e.g., compensation mechanisms; historical selection for inherently tolerant species) might reduce sensitivities of chronically contaminated systems. These poorly studied influences are major confounding factors in our ability to create the dose-response linkages between levels of biological organization that are the basis for realistic criteria to protect aquatic environments from metal toxicity.

Although our existing understanding of the effects of metals on ecosystems is limited, uncertainties can be reduced. Existing approaches often focus on the poorly defined regulatory goal of detecting *whether* metals have a biological effect. Changing focus to a more scientific goal of understanding *how* metals affect processes across levels of biological organization might be more profitable. Many existing approaches assume that pollutants are insignificant unless they are the single factor driving biological responses. Consideration of metals in moderately contaminated environments as one of several influences that might interact to determine the response of an organism, population, or community to its environment might be more realistic.

Bayne[225] called for a "coherent set of ideas" to guide the study of pollutant effects in nature. That set of ideas should include process-oriented questions as an alternative to simple "effect/no effect" questions. What processes determine the survivors in a metal-contaminated locality? Can some suite of consistent biochemical, physiological, and ecological attributes be identified that allow survival or confer vulnerability to species? Do responses of individuals, populations, and communities to the same metal exposure vary among habitats of different physical characteristics? Such questions must also consider that metal toxicity, at any biological level, might be meaningless outside of its abiotic and biotic context. Integration of questions across organizational levels also is important, recognizing that knowledge of responses at one level of organization can be critical to anticipating, predicting, or understanding responses at other levels.

Table 2
Concentrations of Trace Metals in Sediments in Studies that Observed Functional Impairment in a Natural System Compared to Sediment Criteria Suggested by the Apparent Effects Threshold Method[a]

| | Metals in Sediment | | | | | Trace | |
	Ag	Cd	Cu	Pb	Zn	Organics	Source
Biochemical/Cellular							
Protein shifts in cytosol							
San Fran. Bay — clam	2.0		80			Unk[b]	58
Fal Estuary — oyster			3000	400	3500	no	95
Cellular damage							
Norwegian Fjord — mussel	0.9		45	130	250	+ +	99
Fal Estuary — snail			3000	400	3500	no	85
Physiological/Individual							
SFG							
Narragansett Bay — mussel		2	385	70	300	+ +	23
San Fran. Bay — mussel	>1.0		90		193	Unk	131
Tamar	0.4	0.6	282	179	363	Unk	59
Condition Index							
San Fran. Bay — clam		0.6	60	50	120	Unk	56
Population							
Population absent							
Fraser River — clam	1.2	0.8	110	62	150	Unk	70
Age Structure/poor recruitment	1.0	0.4	30	18	73	Unk	70
Apparent Effects Threshold[c]	1.7	5.0	300	300	260		167

[a] Metal concentrations are $\mu g/g$ dry weight. Metal concentrations either determined in the study itself, or data are taken from earlier collections in that locality (3 or 66).
[b] Unk, unknown.
[c] Source: Ref. 167.

Uncertainties affect development of criteria to protect aquatic life from metal effects. Regulatory tools such as the "apparent effects threshold" or the "sediment quality triad,"[167,226,227] discussed in this volume, Chapter 9, illustrate the dilemma between the need to regulate and the inadequate knowledge available to do so accurately. These schemes could provide numeric criteria but are simplistic and insensitive in their application. Thus the criteria may fail to protect aquatic life. Many of the examples of probable effects in nature pointed out in our review occur at metal concentrations below some of the criteria values they suggest (Table 2). If criteria are to be based upon the actual responses of biological systems in nature, improvements in both biological and chemical understanding are critical, as are more accurate estimates of dose-response for metals in natural systems.

The above discussion suggests that we continue to face a difficult challenge in defining effects of pollutants in aquatic environments. Multilevel organizational studies have begun and are an important step toward making the challenge more tractable.[2,228,229] The next step may be to focus on why organisms survive where they do and to consider questions over longer time scales. One way to encourage these difficult studies might be to designate specific contaminated ecosystems as long-term study sites (similar to the LTER program[230]), where there should be long-term monitoring, accompanied by on-going process studies.

287

Just as long-term ecological research is a necessary part of piecing together the complexities of some fundamental ecological principles, long-term study at contaminated sites might be a necessary step in understanding how pollution works. Persistent study in representative environments might allow the breadth and depth of understanding necessary to proceed beyond the simplistic controversies that dominate current debate about metal effects. Better protective criteria will ultimately result, even if the immediate goals of such studies do not directly include development of criteria.

REFERENCES

1. Bryan, G. W. "Some Aspects of Heavy Metal Tolerance in Aquatic Organisms" in *Effects of Pollutants on Aquatic Organisms,* A. P. M. Lockwood, Ed. (London, U. K.: Cambridge Univ. Press, 1976), pp. 7-34.
2. Bayne, B. L., R. F. Addison, J. M. Capuzzo, K. R. Clarke, J. S. Gray, M. N. Moore, and R. M. Warwick. "An Overview of the GEEP Workshop," *Mar. Ecol. Prog. Ser.* 46:235-243 (1988).
3. Luoma, S. N. and D. J. H. Phillips. "Distribution, Variability and Impacts of Trace Elements in San Francisco Bay," *Mar. Pollut. Bull.* 19:413-425 (1988).
4. Ford, J. "The Effects of Chemical Stress on Aquatic Species Composition and Community Structure" in *Ecotoxicology: Problems and Approaches,* S. Levin, M. Harwell, J. Kelly, and K. Kimball, Eds. (New York:Springer-Verlag, 1989), pp. 99-129.
5. Kelly, J. R. and M. A. Harwell. "Indicators of Ecosystem Response and Recovery" in *Ecotoxicology: Problems and Approaches,* S. A. Levin, M. A. Harwell, J. R. Kelly, and K. D. Kimball, Eds. (New York:Springer-Verlag, 1989), pp. 9-32.
6. Cairns, J., Jr. "Gauging the Cumulative Effects of Developmental Activities on Complex Ecosystems" in *Ecological Processes and Cumulative Impacts: Illustrated by Bottomland Hardwood Wetland Ecosystems,* J. G. Gosselink, L. C. Lee, and T. A. Muir, Eds. (Chelsea, MI:Lewis Publishers, 1990), pp. 239-255.
7. Bascom, W. "The Effects of Waste Disposal on the Coastal Waters of Southern California," *Environ. Sci. Technol.* 16:226A-236A (1982).
8. Cairns, J., Jr. and D. I. Mount. "Aquatic Toxicology," *Environ. Sci. Technol.* 24:154-161 (1990).
9. Howarth, R. W. "Determining the Ecological Effects of Oil Pollution in Marine Ecosystems" in *Ecotoxicology: Problems and Approaches,* S. A. Levin, M. A. Harwell, J. R. Kelly, and K. D. Kimball, Eds. (New York:Springer-Verlag, 1989), pp. 69-87.
10. Spies, R. "Sediment Bioassays, Chemical Contamination and Benthic Communities: New Insights or Muddy Waters," *Mar. Environ. Res.* 27:73-75 (1990).
11. Luoma, S. N. "Sediment Bioassays," *Mar. Environ. Res.,* in press.
12. Harris, H. J., P. E. Sager, H. A. Regier, and G. R. Francis. "Ecotoxicology and Ecosystem Integrity: The Great Lakes Examined," *Environ. Sci. Techol.* 24:598-603 (1990).

13. Lee, R., J. M. Davies, H. C. Freeman, A. Ivanovici, M. N. Moore, J. Stegeman, and J. F. Uthe. "Biochemical Techniques for Monitoring Biology Effects of Pollution in the Sea" *Rapp. Process. Verb. Reunions Cons. Perma. Int. Explor. Mer.* 179:48-55 (1980).

14. Livingstone, D. R. "General Biochemical Indices of Sublethal Stress," *Mar. Pollut. Bull.* 13:261-263 (1982).

15. Stebbings, A. "A Possible Synthesis" in *The Effects of Stress and Pollution on Marine Animals,* (New York:Praeger, 1985), pp. 301-313.

16. Moriarty, F. *Ecotoxicology — The Study of Pollutants in Ecosystems* (London:Academic Press, 1988), p. 250.

17. Depledge, M. "The Rational Basis For Detection of the Early Effects of Marine Pollutants Using Physiological Indicators," *Ambio* 18:301-302 (1989).

18. Widdows, J. "Field Measurement of the Biological Impacts of Pollutants" in *Assimilative Capacity of the Oceans for Man's Wastes,* SCOPE/ICSU Academia Sinica, Taipei, Republic of China, pp. 111-122 (1982).

19. Gentile, J. H., G. G. Pesch, K. J. Scott, W. Nelson, W. R. Munns, and J. M. Capuzzo. "Bioassessment Methods for Determining the Hazards of Dredged Material Disposal in the Marine Environment" in *In Situ Evaluations of Biological Hazards of Environmental Pollutants,* S. S. Sandhu, Ed. (New York:Plenum Press, 1990), pp. 31-42.

20. Bayne, B. "Ecological Consequences of Stress" in *The Effects of Stress and Pollutants on Marine Animals,* B. L. Bayne, D. A. Brown, K. Burns, D. R. Dixon, and A. Invanovici, Eds. (New York:Praeger, 1985), pp. 141-157.

21. Bayne, B. L., J. Anderson, D. Engel, E. Gilfillan, D. Hoss, R. Lloyd, and F. P. Thurberg. "Physiological Techniques for Measuring the Biological Effects of Pollution in the Sea," *Rapp. Process. Verb. Reunions Cons. Perma. Int. Explor. Mer.* 179:88-99 (1980).

22. Sunda, W. G. and R. R. Guillard. "The Relationship Between Cupic Ion Activity and the Toxicity of Copper to Phytoplankton," *J. Mar. Res.* 34:511-529 (1976).

23. Luoma, S. N. "Bioavailability of Trace Metals to Aquatic Organisms — A Review," *Sci. Total Environ.* 28:1-22 (1983).

24. Luoma, S. N. "Can We Determine the Biological Availability of Sediment-Bound Trace Elements?" *Hydrobiologia* 176/177:379-396 (1989).

25. Campbell, P. G. C., A. Lewis, P. Chapman, A. Crowder, W. Fletcher, B. Imber, S. N. Luoma, P. Stokes, and M. Winfrey. "Biologically Available Metals in Seidments," NRCC 27694, NRCC/CNRC (Ottawa, Canada: National Research Council of Canada, 1988) 300 pp.

26. Campbell, P. G. C. and A. Tessier. "Geochemistry and Bioavailability of Trace Metals in Sediments" in *Aquatic Ecotoxicology: Fundamental Concepts and Methodologies, Vol. 1,* A. Boudou and F. Ribeyre, Eds. (Boca Raton, FL: CRC Press, 1989), pp. 125-150.

27. D'Elia, C. F., J. G. Sanders, and D. G. Capone. "Analytical Chemistry for Environmental Sciences," *Environ. Sci. Technol.* 23:768-773 (1989).

28. Salomons, W. and U. Forstner. *Metals in the Hydrocycle* (New York:Springer-Verlag, 1984), 349 pp.

29. Luoma, S. N. "Processes Affecting Metal Concentrations in Estuarine and Coastal Marine Sediments" in *Heavy Metals in the Marine Enviornment,* R. Furness and P. Rainbow, Eds. (Boca Raton, FL:CRC Press, 1990), pp. 51-66.

30. Phillips, D. J. H. *Qunatitative Aquatic Biological Indicators* (Barking:Applied Science Publishers, 1980), 488 pp.
31. Simkiss, K., M. Taylor, and A. Z. Mason. "Metal Detoxification and Bioaccumulation in Molluscs," *Mar. Biol. Lett.* 3:187-201 (1982).
32. Bryan, G. W., W. J. Langston, L. G. Hummerstone, and G. R. Burt. "A Guide to the Assessment of Heavy-Metal Contamination in Estuaries Using Biological Indicators," *Mar. Biol. Assoc. U.K. Occas. Publ. No. 4*, 92 pp. (1985).
33. Farrington, J. W., E. D. Goldberg, R. W. Risebrough, J. H. Martin, and V. T. Bowen. "U.S. Mussel Watch 1976-1978: An Overview of the Trace-Metal, DDE, PCB, Hydrocarbon, and Artificial Radionuclide Data," *Environ. Sci. Technol.* 17:490-496 (1983).
34. Borchardt, T., S. Burchert, H. Hablizel, L. Karbe, and R. Zeitner. "Trace Metal Concentrations in Mussels: Comparison Between Estuarine, Coastal and Offshore Regions in the Southeastern North Sea from 1983 to 1986," *Mar. Ecol. Prog. Ser.* 42:17-31 (1988).
35. Phillips, D. J. H. and P. S. Rainbow. "Barnacles and Mussels as Biomonitors of Trace Elements: A Comparative Study," *Mar. Ecol. Prog. Ser.* 49:83-93 (1988).
36. Cain, D. J., S. V. Fend, and J. L. Carter. "Temporal and Spatial Variability of Arsenic in Benthic Insects from Whitewood Creek, South Dakota" in *Toxic Substance Hydrology*, G. E. Mallard and S. E. Ragone, Eds., U. S. Geol. Surv. WRI Rept. 88-4220 (Menlo Park, CA: U.S. Geological Survey, 1989), pp. 257-268.
37. Moore, J. N., S. N. Luoma, and D. Peters. "Downstream Effects of Mine Effluent on an Intermontane Riparian System," *Can. J. Fish. Aquat. Sci.*, 48:222-232.
38. Galloway, W. B., J. L. Lake, D. K. Phelps, P. F. Rogerson, V. T. Bowen, J. W. Farrington, E. D. Goldberg, J. L. Laseter, G. C. Lawler, J. H. Martin, and R. W. Risebrough. "The Mussel Watch: Intercomparison of Trace Level Constituent Determinations," *Environ. Toxicol. Chem.* 2:395-410 (1983).
39. Uthe, J. F. and C. L. Chou. "Factors Affecting the Measurement of Trace Metals in Marine Biological Tissue," *Sci. Total Environ.* 71:67-84 (1988).
40. Gordon, M., G. A. Knauer, and J. H. Martin. "*Mytilus californianus* as a Bioindicator of Trace Metal Pollution: Variability and Statistical Considerations," *Mar. Pollut. Bull.* 11:195-198 (1980).
41. Wright, D. A., J. A. Mihursky, and H. L. Phelps. "Trace Metals in Chesapeake Bay Oysters: Intra-sample Variability and Its Implications for Biomonitoring," *Mar. Environ. Res.* 16:181-197 (1985).
42. Lobel, P. B. "Intersite, Intrasite and Inherent Variability of the Whole Soft Tissue Zinc Concentrations of Individual Mussels *Mytilus edulis:* Importance of the Kidney," *Mar. Environ. Res.* 21:59-71 (1987).
43. Boyden, C. R. "Trace Element Content and Body Size in Molluscs," *Nature* 251:311-314 (1974).
44. Bryan, G. W. and H. Uysal. "Heavy Metals in the Burrowing Bivalve *Scrobicularia plana* from the Tamar Estuary in Relation to Environmental Levels," *J. Mar. Biol. Assoc. U. K.* 58:89-108 (1978).
45. Strong, C. R. and S. N. Luoma. "Variations in Correlation of Body Size with Concentrations of Cu and Ag in the Bivalve Macoma balthica," *Can. J. Fish. Aquat. Sci.* 38:1059-1064 (1981).

46. Boyden, C. R. and D. J. H. Phillips. "Seasonal Variation and Inherent Variability of Trace Elements in Oysters and Their Implications for Indicator Studies," *Mar. Ecol. Prog. Ser.* 5:29-40 (1981).

47. Fischer, H. "Cadmium in Seawater Recorded by Mussels: Regional Decline Established," *Mar. Ecol. Prog. Ser.* 55:159-169 (1989).

48. Cain, D. J. and S. N. Luoma. "Effect of Seasonally Changing Tissue Weight on Trace Metal Concentrations in the Bivalve *Macoma balthica* in San Francisco Bay," *Mar. Ecol. Prog. Ser.* 28:209-217 (1986).

49. Cain, D. J. and S. N. Luoma. "Influence of Seasonal Growth, Age and Environmental Exposure on Cu and Ag in a Bivalve Indicator, *Macoma balthica* in San Francisco Bay," *Mar. Ecol. Prog. Ser.* 60:45-55 (1990).

50. Smock, L. A. "Relationships Between Metal Concentrations and Organism Size in Aquatic Insects," *Freshwater Ecol.* 13:313-321 (1983).

51. O'Connor, J. M. and R. J. Huggett. "Aquatic Pollution Problems, North Atlantic Coast, Including Chesapeake Bay," *Aquat. Toxicol.* 11:163-190 (1988).

52. Axtmann, E. V. and S. N. Luoma. "Large-Scale Distribution of Metal Contamination in the Fine-Grained Sediment of the Clark Fork River, Montana," *Appl. Geochem.* 6:75-88 (1990).

53. Uncles, R. J., J. A. Stephens, and T. Y. Woodrow. "Seasonal Cycling of Estuarine Sediment and Contaminant Transport," *Estuaries* 11:108-116 (1988).

54. Bryan, G. W., W. J. Langston, and L. G. Hummerstone. "The Use of Biological Indicators of Heavy-Metal Contamination in Estuaries," *Mar. Biol. Assoc. U. K. Occas. Publ. No. 1,* 73 pp. (1980).

55. Luoma, S. N., D. J. Cain, and C. Johansson. "Temporal Fluctuations of Silver, Copper and Zinc in the Bivalve *Macoma balthica* at Five Stations in South San Francisco Bay," *Hydrobiologia* 129:109-120 (1985).

56. Luoma, S. N., R. Dagovitz, and E. V. Axtmann. "Temporally Intensive Study of Trace Metals in Sediments and Bivalves From a Large River-Estuarine System: Suisun Bay/Delta in San Francisco Bay," *Sci. Total Environ.* 97/98:685-712 (1990).

57. Ackroyd, D. R., G. E. Millward, and A. W. Morris. "Periodicity in the Trace Metal Content of Estuarine Sediments," *Oceanol. Acta* 10:161-168 (1987).

58. Johansson, C., D. J. Cain, and S. N. Luoma. "Variability in the Fractionation of Cu, Ag, and Zn Among Cytosolic Proteins in the Bivalve *Macoma balthica,*" *Mar. Ecol. Prog. Ser.* 28:87-97 (1986).

59. Widdows, J., P. Donkin, P. N. Salkeld, J. J. Cleary, D. M. Lowe, S. V. Evans, and P. W. Thomson. "Relative Importance of Environmental Factors in Determining Physiological Differences Between Two Populations of Mussels *(Mytilus edulis),*" *Mar. Ecol. Prog. Ser.* 17:33-47 (1984).

60. Oakden, J. M., J. S. Oliver, and A. R. Flegel. "EDTA Chelation and Zinc Antagonism with Cadmium in Sediment: Effects on the Behaviour and Mortality of Two Infaunal Amphipods," *Mar. Biol.* 84:125-130 (1984).

61. Simkiss, K. and A. Z. Mason. "Cellular Responses of Molluscan Tissues to Environmental Metals," *Mar. Environ. Res.* 14:103-118 (1984).

62. Devineau, J. and C. A. Triquet. "Patterns of Bioaccumulation of an Essential Trace Element (Zinc) and a Pollutant Metal (Cadmium) in Larvae of the Prawn *Palaemon serratus,*" *Mar. Biol.* 86:139-143 (1985).

63. Elliott, N. G., D. A. Ritz, and R. Swain. "Interaction Between Copper and Zinc Accumulation in the Barnacle *Elminius modestus* Darwin," *Mar. Environ. Res.* 17:13-17 (1985).

64. Elliott, N. G., R. Swain, and D. A. Ritz. "The Influence of Cyclic Exposure on the Accumulation of Heavy Metals by *Mytilus edulis planulatus* (Lamarck)," *Mar. Environ. Res.* 15:17-30 (1985).

65. Fischer, H. "*Mytilus edulis* as a Quantitative Indicator of Dissolved Cadmium. Final Study and Synthesis," *Mar. Ecol. Prog. Ser.* 48:163-174 (1988).

66. Luoma, S. N. and G. W. Bryan. "A Statistical Study of Environmental Factors Controlling Concentrations of Heavy Metals in the Burrowing Bivalve *Scrobicularia plana* and the Polychaete *Nereis diversicolor*," *Estuarine Coastal Shelf Sci.* 15:95-108 (1982).

67. Voyer, R. A., J. A. Cardin, J. F. Heltshe, and G. L. Hoffman. "Variability of Embryos of the Winter Flounder *Pseudopleuronectes americanus* Exposed to Mixtures of Cadmium and Silver in Combination with Selected Fixed Salinities," *Aquat. Toxicol.* 2:223-233 (1982).

68. Sprague, J. B. and B. A. Ramsay. "Lethal Levels of Mixed Copper-Zinc Solutionsfor Juvenile Salmon," *J. Fish. Res. Bd. Can.* 22:425-432 (1965).

69. de March, B. G. E. "Acute Toxicity of Binary Mixtures of Five Cations (Cu, Cd, Zn, Mg, and K) to the Freshwater Amphipod *Gammarus lacustris* (Sars): Alternative Descriptive Models," *Can. J. Fish. Aquat. Sci.* 45:625-633 (1988).

70. McGreer, E. R. "Factors Affecting the Distribution of the Bivalve, *Macoma balthica* (L.) on a Mudflat Receiving Sewage Effluent, Fraser River Estuary, British Columbia," *Mar. Environ. Res.* 7:131-149 (1982).

71. Ward, T. J. and P. C. Young. "Effects of Metals and Sediment Particle Size on the Species Composition of the Epifauna of *Pinna bicolor* Near a Lead Smelter, Spencer Gulf, South Australia," *Estuarine Coastal Shelf Sci.* 18:79-95 (1984).

72. Fair, P. A. and L. V. Sick. "Accumulation of Naphthalene and Cadmium After Simultaneous Ingestion by the Black Sea Bass, *Centropristis striata*," *Arch. Environ. Contam. Toxicol.* 12:551-557 (1983).

73. Moore, M. N., J. Widdows, J. J. Cleary, R. K. Pipe, P. N. Salkeld, P. Donkin, S. V. Farrar, S. V. Evans, and P. E. Thomson. "Responses of the Mussell *Mytilus edulis* to Copper and Phenanthrene: Interactive Effects," *Mar. Environ. Res.* 14:167-183 (1984).

74. George, S. G. "Biochemical and Cytological Assessments of Metal Toxicity in Marine Animals" in *Heavy Metals in the Marine Environment*, R. W. Furness and P. S. Rainbow, Eds. (Boca Raton, FL: CRC Press, 1990), pp. 123-132.

75. Kooijman, S. A. L. M. and J. A. J. Metz. "On the Dynamics of Chemically Stressed Populations: The Deduction of Population Consequences from Effects on Individuals," *Ectoxicol. Environ. Safety* 8:254-274 (1984).

76. Barnthouse, L. W., G. W. Suter, II, and A. E. Rosen. "Risks of Toxic Contaminants to Exploited Fish Populations: Influence of Life History, Data Uncertainty and Exploitation Intensity," *Environ. Toxicol. Chem.* 9:297-311 (1990).

77. Meyer, J. S., C. G. Ingersoll, and L. L. McDonald. "Sensitivity of Population Growth Rates Estimated from Cladoceran Chronic Toxicity Tests," *Environ. Toxicol. Chem.* 6:115-126 (1987).

78. Diamond, J. D. "Bob Dylan and Moas' Ghost," *Nat. Hist.* 10/90:26-31 (199).

79. Odum, E. P. "The Effects of Stress on the Trajectory of Ecological Succession" in *Stress Effects on Natural Ecosystems,* G. W. Barrett and R. Rosenberg, Eds. (New York:John Wiley & Sons, 1981), pp. 43-46.

80. Luoma, S. N. "Detection of Trace Contaminant Effects in Aquatic Ecosystems," *J. Fish. Res. Bd. Can.* 34:436-439 (1977).

81. Weis, J. S. and P. Weis. "Tolerance and Stress in a Polluted Environment: the Case of the Mummichog," *Bioscience* 39(2):89-95 (1989).

82. Tedengren, M., M. Arner, and N. Kautsky. "Ecophysiology and Stress Response of Marine and Brackish Water *Gammarus* species (Crustacea, Amphipoda) to Changes in Salinity and Exposure to Cadmium and Diesel-Oil," *Mar. Ecol. Prog. Ser.* 47:107-116 (1988).

83. Langston, W. J. and M. Zhou. "Cadmium Accumulation, Distribution and Elimination in the Bivalve *Macoma balthica:* Neither Metallothionein nor Metallothionein-Like Proteins are Involved," *Mar. Environ. Res.* 21:225-237 (1987).

84. Roesijadi, G. "The Significance of Low Molecular Weight, Metallothionein-Like Proteins in Marine Invertebrates: Current Status," *Mar. Environ. Res.* 4:167-179 (1980).

85. Mason, A. Z., K. Simkiss, and K. P. Ryan. "The Ultrastructural Localization of Metals in Specimens of *Littorina littorea* Collected from Clean and Polluted Sites," *J. Mar. Biol. Assoc. U. K.* 64:699-720 (1984).

86. Moore, M. N. "Cellular Responses to Pollutants," *Mar. Pollut. Bull.* 16:134-139 (1985).

87. Roesijadi, G. and P. L. Klerks. "A Kinetic Analysis of Cd-Binding to Metallothionein and Other Intracellular Ligands in Oyster Gills," *J. Exp. Zool.* 251:1-12 (1989).

88. Brown, D. A. and T. R. Parsons. "Relationship Between Cytoplasmic Distribution of Mercury and Toxic Effects to Zooplankton and Chum Salmon (*Oncorhynchus keta*) Exposed to Mercury in a Controlled Ecosystem," *J. Fish. Res. Bd. Can.* 35:880-884 (1978).

89. Sanders, B. M. and K. D. Jenkins. "Relationships Between Free Cupric Ion Concentrations in Seawater and Copper Metabolism and Growth in Crab Larvae," *Biol. Bull.* 167:704-712 (1984).

90. Jenkins, K. D. and A. Z. Mason. "Relationships Between Subcellular Distributions of Cadmium and Perturbations in Reproduction in the Polychaete *Neanthes arenaceodentata,*" *Aquat. Toxicol.* 12:229-244 (1988).

91. Moore, M. N. "Cellular and Histopathological Effects of a Pollutant Gradient — Summary," *Mar. Ecol. Prog. Ser.* 46:109-110, (1988a).

92. Moore, M. N. "Cytochemical Responses of the Lysosomal System and NADPH-Ferrihemoprotein Reductase in Molluscan Digestive Cells to Environmental and Experimental Exposure to Xenobiotics," *Mar. Ecol. Prog. Ser.* 46:81-89 (1988b).

93. Moore, M. N., R. K. Pipe, and S. V. Farrer. "Lysosomal and Microsomal Responses to Environmental Factors in *Littorina littorea* from Sullom Voe," *Mar. Pollut. Bull.* 13:340-345 (1982).

94. Moore, M. N. "Lysosomes and Environmental Stress," *Mar. Pollut. Bull.* 13:42-43 (1982).

95. Frazier, J. M. and S. G. George. "Cadmium Kinetics in Oysters — A Comparative Study of *Crassostrea gigas* and *Ostrea edulis,*" *Mar. Biol.* 76:55-61 (1983).

96. Engel, D. W. and M. Brouwer. "Cadmium-Binding Proteins in the Blue Crab, *Callinectes sapidus:* Laboratory-Field Comparison," *Mar. Environ. Res.* 14:139-151 (1984).

97. Roesijadi, G., J. S. Young, A. S. Drum, and J. M. Gurtisen. "Behavior of Trace Metals in *Mytilus edulis* During a Reciprocal Transplant Field Experiment," *Mar. Ecol. Prog. Ser.* 18:155-170 (1984).

98. Langston, W. J. and M. Zhou. "Evaluation of the Significance of Metal-Binding Proteins in the Gastropod *Littorina littorea,*" *Mar. Biol.* 92:505-515 (1986).

99. Viarengo, A., G. Mancinelli, G. Martino, M. Pertica, L. Canesi, and A. Mazzucotelli. "Integrated Cellular Stress Indices in Trace Metal Contamination: Critical Evaluation in a Field Study," *Mar. Ecol. Prog. Ser.* 46:65-70 (1988).

100. Klerks, P. L. and J. S. Levinton. "Effects of Heavy Metals in a Polluted Aquatic Ecosystem" in *Ecotoxicology: Problems and Approaches,* S. A. Levin, M. A. Harwell, J. R. Kelly, and K. D. Kimball, Eds. (New York:Springer-Verlag, 1989), pp. 41-60.

101. Lowe, D. M. "Alterations in Cellular Structure of *Mytilus edulis* Resulting from Exposure to Environmental Contaminants Under Field and Experimental Conditions," *Mar. Ecol. Prog. Ser.* 46:91-100 (1988).

102. Capuzzo, J. M. "Physiological Effects of a Pollutant Gradient — Summary," *Mar. Ecol. Prog. Ser.* 46:147-148 (1988).

103. Engel, D. W. "The Effect of Biological Variability on Monitoring Strategies: Metallothioneins as an Example," *Water. Res. Bull.* 24:981-987 (1988).

104. Viarengo, A. "Heavy Metals in Marine Invertebrates: Mechanisms of Regulation and Toxicity at the Cellular Level," *Rev. Aquat. Sci.* 1:295-317 (1989).

105. Engel, D. W. and M. Brouwer. "Detoxification of Accumulated Trace Metals by the American Oyster," *Crassostrea virginica:* Laboratory vs. Environment, in *Physiological Mechanisms of Marine Pollutant Toxicity,* W. B. Vernber, A. Calabrese, F. P. Thurberg, and F. J. Vernberg, Eds. (San Diego:Academic Press, 1982), pp. 89-107.

106. Brown, D. A., S. M. Bay, and R. W. Gossett. "Using the Natural Detoxification Capacities of Marine Organisms to Assess Assimilative Capacity," in *Aquatic Toxicology: 7th Symposium ASTM,* S. T. P. 854, (Philadelphia, PA: ASTM, 1984), pp. 364-382.

107. Bayne, B. L., M. N. Moore, J. Widdows, D. R. Livingstone, and P. Salkeld. "Measurement of the Responses of Individuals to Environmental Stress and Pollution: Studies with Bivalve Molluscs," *Trans. R. Soc. London B* 286:563-561 (1979).

108. Capuzzo, J. M. and J. J. Sasner, Jr. "The Effect of Chromium on Filtration Rates and Metabolic Activity of *Mytilus edulis* L. and *Mya arenaria* L." in *Physiological Responses of Marine Biota to Pollutants,* F. J. Vernberg, A. Calabrese, F. P. Thurberg, and W. Vernberg, Eds. (San Diego:Academic Press, 1977), pp. 225-237.

109. Calabrese, A., F. P. Thurberg, and E. Gould. "Effects of Cadmium, Mercury, and Silver on Marine Animals," *Mar. Fish. Rev.* 39:5-11 (1977).

110. Widdows, J. "Physiological Responses to Pollution," *Mar. Poll. Bull.* 16:129-133 (1985).

111. Carr, R. S., J. W. Williams, F. I. Saksa, R. L. Buhl, and J. M. Neff. "Bioenergetic Alterations Correlated with Growth, Fecundity and Body Burden of Cadmium for Mysids *(Mysidopsis bahia),*" *Environ. Toxicol. Chem.* 4:181-188 (1985).

112. Langston, W. J. "Toxic Effects of Metals and the Incidence of Metal Pollution in Marine Ecosystems" in *Heavy Metals in the Environment,* R. W. Furness and P. S. Rainbow, Eds. (Boca Raton, FL:CRC Press, 1990), pp. 101-122.

113. Kluytmans, J. H., F. Brands, and D. I. Zandee. "Interactions of Cadmium with the Reproductive Cycle of *Mytilus edulis* L.," *Mar. Environ. Res.* 24:189-192 (1988).

114. Myint, U. M. and P. A. Tyler. "Effects of Temperature, Nutritive and Metal Stressors on the Reproductive Biology of *Mytilus edulis,*" *Mar. Biol.* 67:209-223 (1982).

115. Earnshaw, M. J., S. Wilson, H. B. Akberali, R. D. Butler, and K. R. M. Marriott. "The Action of Heavy Metals on the Gametes of the Marine Mussel, *Mytilus edulis* (L.) — III. The Effect of Applied Copper and Zinc on Sperm Motility in Relation to Ultrastructural Damage and Intracellular Metal Localisation," *Mar. Environ. Res.* 20:261-278 (1986).

116. Eyster, L. S. and M. P. Morse. "Development of the Surf Clam *(Spisula solidissima)* Following Exposure of Gametes, Embryos, and Larvae to Silver," *Arch. Environ. Contam. Toxicol.* 13:641-649 (1984).

117. MacInnes, J. R. and A. Calabrese. "Response of Embryos of the American Oyster, *Crassostrea virginica* to Heavy Metals at Different Temperatures," in *Physiology and Behaviour of Marine Organisms,* D. S. McLusky and A. J. Berry, Eds. (Oxford:Pergamon Press, 1978), pp. 175-190.

118. Martin, M., K. Osborn, P. Billing, and N. Glickstein. "Toxicity of Ten Metals to *Crassostrea gigas* and *Mytilus edulis* Embryos and *Cancer magister* Larvae," *Mar. Pollut. Bull.* 12:305-308 (1981).

119. Gibbs, P. E. and G. W. Bryan. "Reproductive Failure in Populations of the Dog-Whelk, *Nucella lapillus,* Caused by Imposex Induced by Tributyltin from Antifouling Paints," *J. Mar. Biol. Assoc. U. K.* 66:767-777 (1986).

120. Lowe, D. M. and N. M. Moore. "The Cytology and Occurrence of Granulocytomas in Mussels," *Mar. Pollut. Bull.* 10:137-141 (1978).

121. Akberali, H. B. and E. R. Trueman. "Effects of Environmental Stress on Marine Bivalve Molluscs," *Adv. Mar. Biol.* 22:101-198 (1985).

122. Bayne, B. L. and F. P. Thurberg. "Physiological Measurements on *Nucula tenuis* and on Isolated Gills of *Mytilus edulis* and *Carcinus meanas,*" *Mar. Ecol. Prog. Ser.* 46:129-134 (1988).

123. Widdows, J., D. K. Phelps, and W. Galloway. "Measurement of Physiological Condition of Mussels Transplanted Along a Pollution Gradient in Narragansett Bay," *Mar. Environ. Res.* 4:181-194 (1980-81).

124. Bayne, B. L. and J. Widdows. "Physiological Ecology of Two Populations of *Mytilus edulis* L.," *Oceologia (Berl.)* 37:137-162 (1978).

125. Widdows, J. "Physiological Indices of Stress in *Mytilus edulis,*" *J. Mar. Biol. Assoc. U. K.* 58:125-142 (1978).

126. Bayne, B. L. and C. M. Worral. "Growth and Production of Mussels *Mytilus edulis* from Two Populations," *Mar. Ecol. Prog. Ser.* 3:317-328 (1980).

127. Widdows, J. and D. Johnson. "Physiological Energetics of *Mytilus edulis:* Scope for Growth," *Mar. Ecol. Prog. Ser.* 46:113-121 (1988).

128. Bakke, T. "Physiological Energetics of *Littorina littorea* Under Combined Pollutant Stress in Field and Mesocosm Studies," *Mar. Ecol. Prog. Ser.* 46:123-128 (1988).

129. Belanger, S. E., J. L. Farris, D. S. Cherry, and J. Cairns, Jr. "Growth of Asiatic Clams *(Corbicula sp.)*, During and After Long-Term Zinc Exposure in Field-Located and Laboratory Artificial Streams," *Arch. Environ. Contam. Toxicol.* 15:427-434 (1986).

130. Foe, C. and A. Knight. "A Thermal Energy Budget for Juvenile *Corbicula fluminea*," *Am. Malacol. Bull.* 2:143-150 (1986).

131. Martin, M., G. Ichikawa, J. Goetzl, M. de los Reyes, and M. D. Stephenson. "Relationships Between Physiological Stress and Trace Toxic Substances in the Bay Mussel, *Mytilus edulis* from San Francisco Bay, California," *Mar. Environ. Res.* 11:91-110 (1984).

132. Roesijadi, G., A. S. Drom, J. M. Thomas, and G. W. Fellingham. "Enhanced Mercury Tolerance in Marine Mussels and Relationship to Low Molecular Weight, Mercury-Binding Proteins," *Mar. Pollut. Bull.* 13:250-253 (1982).

133. Pesch, C. E. and G. L. Hoffman. "Adaptation of the Polychaete *Neanthes arenaceodentata* to Copper," *Mar. Environ. Res.* 6:307-317 (1982).

134. Sanders, B. M. "The Role of the Stress Protein Response in Physiological Adaptation of Marine Molluscs," *Mar. Environ. Res.* 24:207-211 (1988).

135. Theede, H. "Physiological Responses of Estuarine Animals to Cadmium Pollution," *Helgoleander Wiss. Meeresunters.* 33:25-35 (1980).

136. Cotter, A. J. R., D. H. J. Phillips, and M. Ahsanullah. "The Significance of Temperature, Salinity and Zinc as Lethal Factors for the Mussel *Mytilus edulis* in a Polluted Estuary, *Mar. Biol.* 68:135-144 (1982).

137. Chapman, P. M., M. A. Farrell, and R. O. Brinkhurst. "Relative Tolerances of Selected Aquatic Oligochaetes to Combinations of Pollutants and Environmental Factors," *Aquat. Toxicol.* 2:69-78 (1982).

138. Chandini, T. "Effects of Different Food *(Chlorella)* Concentrations of the Chronic Toxicity of Cadmium to Survivorship, Growth and Reproduction of *Echinisca triserialis* (Crustacea:Cladocera)," *Environ. Pollut.* 54:139-154 (1988).

139. Riisgard, H. U. and A. Randlov. "Energy Budgets, Growth and Filtration Rate in *Mytilus edulis* at Different Algal Concentrations," *Mar. Biol.* 61:227-234 (1981).

140. Malins, D. C., B. B. McCain, D. W. Brown, M. S. Myers, M. M. Krahan, and S-. L. Chan. "Toxic Chemicals, Including Aromatic and Chlorinated Hydrocarbons and Their Derivatives, and Liver Lesions in White Croaker *(Genyonemus lineatus)* from the Vicinity of Los Angeles," *Environ. Sci. Technol.* 21:765-770 (1987).

141. Bengtsson, B. E., A. Bengtsson, and M. Himberg. "Fish Deformities and Pollution in Some Swedish Waters," *Ambio* 14:32-41 (1985).

142. Walker, H. A., E. Lorda, and S. B. Saila. "A Comparison of the Incidence of Five Pathological Conditions in Soft-Shell Clams, *Mya arenaria,* from Environments with Various Pollution Histories," *Mar. Environ. Res.* 5:109-123 (1981).

143. Bryan, G. W., P. E. Gibbs, L. G. Hummerstone, and G. R. Burt. "The Decline of the Gastropod *Nucella lapillus* around South-West England: Evidence for the Effect of Tributyltin from Antifouling Paints," *J. Mar. Biol. Assoc. U. K.* 66:611-640 (1986).

144. Bryan, G. W. this volume, Chapter 12.

145. Smith, B. S. "The Estuarine Mud Snail, *Nassarius obsoletus:* Abnormalities in the Reproductive System," *J. Molluscan Stud.* 46:247-256 (1980).

146. Short, J. W., S. D. Rice, C. C. Brodersen, and W. B. Stickle. "Occurrence of Tri-*n*-butyltin-Caused Imposex in the North Pacific Marine Snail *Nucella lima* in Auke Bay, Alaska," *Mar. Biol.* 102:291-297 (1989).

147. Lenihan, H. S., J. S. Oliver, and M. A. Stephenson. "Changes in Hard Bottom Communities Related to Boat Mooring and Tributyltin in San Diego Bay: A Natural Experiment," *Mar. Ecol. Prog. Ser.,* in press.

148. Livingstone, D. R. "Responses of Microsomal NADPH-cytochrome c Reductase Activity and Cytochrome P-450 in Digestive Glands of *Mytilus littorea* to Environmental and Experimental Explosure to Pollutants," *Mar. Ecol. Prog. Ser.* 46:37-43 (1988).

149. Stegeman, J. J., B. R. Woodin, and A. Goksoyr. "Apparent Cytochrome P-450 Induction as an Indication of Exposure to Environmental Chemicals in the Flounder *Platichthys flesus*," *Mar. Ecol. Prog. Ser.* 46:55-60 (1988).

150. Capuzzo, J. M. and D. F. Leavitt. "Lipid Composition of the Digestive Glands of *Mytilus edulis* and *Carcinus maenas* in Response to Pollutant Gradients," *Mar. Ecol. Prog. Ser.* 46:139-145 (1988).

151. Auffret, M. Histopathological Changes Related to Chemical Contamination in *Mytilus edulis* from Field and Experimental Conditions," *Mar. Ecol. Prog. Ser.* 46:101-107 (1988).

152. Giudici, M. de N., L. Migliore, and S. M. Guarino. "Effects of Cadmium on the Life Cycle of *Asellus aquaticus* and *Proasellus coxalis* (Crustacea, Isopoda)," *Environ. Technol. Lett.* 7:45-54 (1986).

153. Schaaf, W. E., D. S. Peters, D. S. Vaughan, L. Coston-Clements, and C. W. Krouse. "Fish Population Responses to Chronic and Acute Pollution: The Influence of Life History Strategies," *Estuaries* 10:267-275 (1987).

154. Partridge, L. and P. H. Harvey. "The Ecological Context of Life History Evolution," *Science* 241:1449-1455 (1988).

155. Zajac, R. N. and R. B. Whitlatch. "Natural and Disturbance-Induced Demographic Variation in an Infaunal Polychaete, *Nephtys incisa*," *Mar. Ecol. Prog. Ser.* 57:89-102 (1989).

156. Bryan, G. W. and P. E. Gibbs. "Heavy Metals in the Fal Estuary, Cornwall: A Study of Long-term Contamination by Mining Waste and Its Effects on Estuarine Organisms," *Mar. Biol. Ass. U. K. Occas. Publ. No. 2,* 111 pp. (1983).

157. Roughgarden, J., S. Gaines, and H. Possingham. "Recruitment Dynamics in Complex Life Cycles," *Science* 241:1460-1466 (1988).

158. Levinton, J. S. "Genetic Divergence in Estuaries" in *Estuarine Perspectives,* V. S. Kennedy, Ed. (San Diego:Academic Press, 1980), pp. 509-520.

159. Klerks, P. L. and J. S. Weis. "Genetic Adaptation to Heavy Metals in Aquatic Organisms: A Review," *Environ. Pollut.* 45:173-205 (1987).

160. Grant, A., J. G. Hateley, and N. V. Jones. "Mapping the Ecological Impact of Heavy Metals on the Estuarine Polychaete *Nereis diversicolor* Using Inherited Metal Tolerance," *Mar. Pollut. Bull.* 20:235-238 (1989).

161. Bryan, G. W. and L. G. Hummerstone. "Adaptation of the Polychaete *Nereis diversicolor* to Estuarine Sediments Containing High Concentrations of Heavy Metals. I: General Observations and Adaptation to Copper," *J. Mar. Biol. Assoc. U. K.* 51:845-863 (1971).

162. Bryan, G. W. and L. G. Hummerstone. "Adaptation of the Polychaete *Nereis diversicolor* to Estuarine Sediments Containing High Concentrations of Zinc and Cadmium," *J. Mar. Biol. Assoc. U. K.* 53:839-857 (1973).

163. Krantzberg, G. and P. M. Stokes. "Metal Regulation, Tolerance and Body Burdens in the Larvae of the Genus *Chironomus,*" *Can. J. Fish. Aquat. Sci.* 46:389-398 (1989).

164. Blanck, H. and W. A. Wangberg. "Induced Community Tolerance in Marine Periphyton Established Under Arsenate Stress," *Can. J. Fish. Aquat. Sci.* 45:1816-1819 (1988).

165. Ben-Shlomo, R. and E. Nevo. "Isozyme Polymorphism as Monitoring of Marine Environments: The Interactive Effect of Cadmium and Mercury Pollution on the Shrimp, *Plaemon elegans,*" *Mar. Pollut. Bull.* 19:314-317 (1988).

166. Klapow, L. A. and R. H. Lewis. "Analysis of Toxicity Data for California Marine Water Quality Standards," *J. Water Pollut. Control Fed.* 51:2054-2070 (1979).

167. Long, E. R. and L. G. Morgan. "The Potential for Biological Effects of Sediment-Sorbed Contaminants Tested in the National Status and Trends Program," NOAA Tech. Mem. NOS OMA No. 52 Seattle, WA, (1990), 175 pp.

168. Pilli, A., D. O. Carle, E. Kline, Q. Pickering, and J. Lazorchak. "Effects of Pollution on Freshwater Organisms," *J. Water Pollut. Control Fed.* 60:994-1065 (1988).

169. Sebastien, R. J., R. A. Burst, and D. M. Rosenberg. "Impact of Methoxychlor on Selected Nontarget Organisms in an Riffle of the Souris River, Manitoba," *Can. J. Fish. Aquat. Sci.* 46:1047-1061 (1989).

170. Rickleffs, R. E. "Community Diversity: Relative Roles of Local and Regional Processes," *Science* 235:167-171 (1987).

171. Swanson, F. J. and R. E. Sparks. "Long-Term Ecological Research and the Invisible Place," *Bioscience* 40:502-523 (1990).

172. Schindler, D. W. "Detecting Ecosystem Responses to Anthropogenic Stress," *Can. J. Fish. Aquat. Sci.* 44:6-25 (1987).

173. Nichols, F. H. "Abundance Fluctuations Among Benthic Invertebrates in Two Pacific Estuaries," *Estuaries* 8:136-144 (1985).

174. Nichols, F. H. and J. K. Thompson. "Time Scales of Change in the San Francisco Bay Benthos," *Hydrobiologia* 129:121-138 (1985).

175. Sheehan, P. J. and R. W. Winner. "Comparison of Gradient Studies in Heavy Metal-Polluted Streams" in *Effects of Pollutants at the Ecosystem Level,* P. J. Sheehan, D. R. Miller, G. C. Butler, and Ph. Bourdeau, Eds. (New York:John Wiley & Sons, 1984), pp. 255-271.

176. Cairns, J., Jr. and J. R. Pratt. "On the Relation between Structural and Functional Analyses of Ecosystems," *Environ. Toxicol. Chem.* 5:785-786 (1986).

177. Leland, H. V. and J. L. Carter. "Use of Detrended Correspondence Analysis in Evaluating Factors Controlling Species Composition of Periphyton," ASTM S.T.P. No. 894, (Philadelpha, PA:American Society for Testing and Materials, 1986), pp. 101-117.

178. Levine, S. N. "Theoretical and Methodological Reasons for Variability in the Responses of Aquatic Ecosystem Processes to Chemical Stresses," in *Ecotoxicology: Problems and Approaches,* S. A. Levin, M. A. Harwell, J. R. Kelly, and K. D. Kimball, Eds. (New York:Springer-Verlag, 1989), pp. 145-174.

179. Hellawell, J. M. *Biological Indicators of Freshwater Pollution and Environmental Management (Pollution Monitoring Series)* (London:Elsevier/Applied Science Publishers, 1986), p. 546.

180. Ramade, F. "The Pollution of the Hydrosphere by Global Contaminants and Its Effects on Aquatic Ecosystems," in *Aquatic Ecotoxicology: Fundamental Concepts and Methodologies, Vol. I,* A. Boudou and F. Ribeyre, Eds. (Boca Raton, FL:CRC Press, 1989), pp. 151-184.

181. Kolkwitz, R. and M. Marsson. "Ecology of Animal Saprobia," *Int. Rev. Hydrobiol.* 2:126-152 (1909).

182. Kolkwitz, R. and M. Marsson. "Ecology of Plant Saprobia," *Rep. German Botan. Soc.* 26a:505-519 (1908).

183. Metcalfe, J. L. "Biologial Water Quality Assessment of Running Waters Based on Macroinvertebrate Communities: History and Present Status in Europe," *Environ. Pollut.* 60:101-139 (1989).

184. Patrick, R. "A Proposed Biological Measure of Stream Conditions, Based on a Survey of the Conestoga Basin, Lancaster County, Pennsylvania," *Proc. Acad. Nat. Sci. Philadelphia* 101:277-341 (1949).

185. Resh, V. H. "Variability, Accuracy, and Taxonomic Costs of Assessment Approaches in Benthic Monitoring," paper Presented at the 1988 North American Benthological Society Technical Information Workshop, Tuscaloosa, AL, May, 1988.

186. Clements, W. H., D. S. Cherry, and J. Cairns, Jr. "Structural Alterations in Aquatic Insect Communities Exposed to Copper in Laboratory Streams," *Environ. Toxicol. Chem.* 7:715-722 (1988).

187. Specht, W. L., D. S. Cherry, R. A. Lechleitner, and J. Cairns, Jr. "Structural, Functional, and Recovery Responses of Stream Invertebrates to Fly Ash Effluent," *Can. J. Fish. Aquat Sci.* 41:884-896 (1984).

188. Winner, R. W., M. W. Boesel, and M. P. Farrell. "Insect Community Structure as an Index of Heavy-Metal Pollution in Lotic Ecosystems," *Can. J. Fish. Aquat. Sci.* 37:647-655 (1980).

189. Cairns, J., Jr. "The Myth of the Most Sensitive Species," *Bioscience* 36:670-672 (1986).

190. Gaufin, A. F. and C. M. Tarzwell. "Aquatic Invertebrates as Indicators of Stream Pollution," *Public Heal. Rep.* 67:57-64 (1952).

191. Wilhm, J. L. and T. C. Dorris. "Species Diversity of Benthic Macroinvertebrates in a Stream Receiving Domestic and Oil Refinery Effluents," *Am. Midl. Natur.* 76(2):427-449 (1966).

192. Washington, H. G. "Diversity, Biotic and Similarity Indices: A review with Special Relevance to Aquatic Ecosystems," *Water Res.* 18(6):653-694 (1984).

193. Kaesler, R. L., E. E. Herricks, and J. S. Crossman. "Use of Indices of Diversity and Hierarchical Diversity in Stream Surveys," Biological Data in Water Pollution Assessment ASTM S.T.P. 652 (Philadelphia, PA:American Society for Testing and Materials, 1978), pp. 92-112.

194. Wilhm, J. L. "Range of Diversity Index in Benthic Macroinvertebrate Populations," *J. Water Pollut. Control Fed.* 42(5):R221-R224 (1970).

195. Gray, J. S. and T. H. Pearson. "Objective Selection of Sensitive Species Indicative of Pollution-Induced Change in Benthic Communities. I. Comparative Methodology," *Mar. Ecol. Prog. Ser.* 9:111-119 (1982).

196. Hurlburt, S. H. "The Nonconcept of Species Diversity: A Critique and Alternative Parameters," *Ecology* 52(4):577-586 (1971).

197. Leland, H. V., S. V. Fend, T. L. Dudley, and J. L. Carter. "Effects of Copper on Species Composition of Benthic Insects in a Sierra Nevada, California, Stream," *Freshwater Biol.* 21:163-179 (1989).

198. Hall, R. J. and F. P. Ide. "Evidence of Acidification Effects on Stream Insect Communities in Central Ontario Between 1937 and 1985," *Can. J. Fish. Aquat. Sci.* 44:1652-1657 (1987).

199. Boesch, D. F. "Application of Numerical Classification in Ecological Investigations of Water Pollution," Ecological Research Series, EPA-600/3-77-033 (U.S. Environmental Protection Agency, Washington, D.C., 1977), pp. 114.

200. Gauch, H. G., Jr. *Multivariate Analysis in Community Ecology* (Cambridge, U. K.: Cambridge Univ. Press, 1982), p. 298.

201. Pielou, E. C. *The Interpretation of Ecological Data* (New York:John Wiley & Sons, 1984), p. 263.

202. Leland, H. V., J. L. Carter, and S. V. Fend. "Use of Detrended Correspondence Analysis to Evaluate Factors Controlling Spatial Distribution of Benthic Insects," *Hydrobiologia* 132:113-123 (1986).

203. Swartz, F. C., F. A. Cole, D. W. Schults, and W. A. DeBen. "Ecological Changes in the Southern California Bight Near a Large Sewage Outfall: Benthic Conditions in 1980 and 1983," *Mar. Ecol. Prog. Ser.* 31:1-13 (1986).

204. Warwick, R. M. "A New Method for Detecting Pollution Effects on Marine Macrobenthic Communities," *Mar. Biol.* 92:557-562 (1986).

205. Plafkin, J. L., M. T. Barbour, K. D. Porter, S. K. Gross, and R. M. Hughes. "Rapid Bioassessment Protocols for Use in Streams and Rivers: Benthic Macroinvertebrates and Fish," Office of Water, U. S. EPA Rep. 444/4-89-001 (1989).

206. Roper, D. S., S. F. Thrush and D. G. Smith. "The Influence of Runoff on Intertical Mudflat Benthic Communities," *Mar. Environ. Res.* 26:1-18 (1988).

207. Rygg, B. "Effect of Sediment Copper on Benthic Fauna," *Mar. Ecol. Prog. Ser.* 25:83-89 (1985).

208. Ward, T. J. and P. C. Young. "Effects of Sediment Trace Metals and Particle Size on the Community Structure of Epibenthic Seagrass Fauna Near a Lead Smelter, South Australia," *Mar. Ecol. Prog. Ser.* 9:137-146 (1982).

209. Krantzberg, G. and P. M. Stokes. "Benthic Macroinvertebrates Modify Copper and Zinc Partitioning in Freshwater-Sediment Microcosms," *Can. J. Fish Aquat. Sci.* 42:1465-1473 (1985).

210. Hynes, H. B. N. *The Ecology of Running Waters* (Canada:University of Toronto Press, 1970), p. 555.

211. Vannote, R. L., G. W. Minshall, K. W. Cummins, J. R. Sedell, and C. E. Cushing. "The River Continuum Concept," *Can. J. Fish Aquat. Sci.* 37:130-137 (1980).

212. Statzner, B. and B. Higler. "Stream Hydraulics as a Major Determinant of Benthic Invertebrate Zonation Patterns," *Freshwater Biol.* 16:127-139 (1986).

213. Simpson, K. W., J. P. Fagnani, R. W. Bode, D. M. DeNicola, and L. E. Abele. "Organism-Substrate Relationships in the Main Channel of the Lower Hudson River," *J. N. Am. Benthol. Soc.* 5(1):41-57 (1986).

214. Ziser, S. W. "The Effects of a Small Reservoir on the Seasonality and Stability of Physicochemical Parameters and Macrobenthic Community Structure in a Rocky Mountain Stream," *Freshwater Invertebr. Biol.* 4(4):160-177 (1985).

215. Hughes, R. M. and D. P. Larsen. "Ecoregions: an Approach to Surface Water Protection," *J. Water Pollut. Control Fed.* 60(4):486-493 (1988).

216. Whittier, T. R., R. M. Hughes, and D. P. Larsen. "Correspondence Between Ecoregions and Spatial Patterns in Stream Ecosystems in Oregon," *Can. J. Fish. Aquat. Sci.* 45:1264-1278 (1988).

217. Resh, V. H. "Sampling Variability and Life History Features: Basic Considerations in the Design of Aquatic Insect Studies," *J. Fish. Res. Bd. Can.* 36:290-311 (1979).

218. Resh, V. H. and J. D. Unzicker. "Water Quality Monitoring and Aquatic Organisms: The Importance of Species Identification," *J. Water Pollut. Control Fed.* 47(1):9-19 (1975).

219. Wiggins, G. B. "Systematics of North American Trichoptera: Present Status and Future Prospect," in *Systematics of the North American Insects and Arachnids: Status and Needs,* M. Kosztarab, and C. W. Schaefer, Eds., Va. Agr. Exp. Sta. Inform. Ser. 90-1. (Blacksburg, VA:Virginia Polytechnic Institute and State University, 1990), pp. 203-210.

220. Cummins, K. W. "Trophic Relations of Aquatic Insects," *Annu. Rev. Entomol.* 18:183-206. (1973).

221. Cummins, K. W. and M. J. Klug. "Feeding Ecology of Stream Invertebrates," *Annu. Rev. Ecol. Syst.* 10:147-172 (1979).

222. Bahr, L. M., Jr. "Functional Taxonomy: an Immodest Proposal," *Ecol. Model.* 15:211-233 (1982).

223. Grassle, J. P. and J. F. Grassle. "Sibling Species in the Marine Pollution Indicator *Capitella* (Polychaeta)," *Science* 192:567-569 (1976).

224. Nriagu, J. O. "Preface" in *Aquatic Ecotoxicology: Fundamental Concepts and Methodologies,* A. Boudou and F. Ribeyre, Eds. (Boca Raton, FL:CRC Press, 1989), pp. 1-3.

225. Bayne, B. L. "The Biological Effects of Contaminants: Are There Any General Rules?," *Mar. Environ. Res.* 24:251 (1988).

226. Chapman, P. M., R. N. Dexter, and E. R. Long. "Synoptic Measures of Sediment Contamination, Toxicity and Infaunal Community Composition (The Sediment Quality Triad) in San Francisco Bay," *Mar. Ecol. Prog. Ser.* 37:75-96 (1987).

227. Chapman, P. M. "Current Approaches to Developing Sediment Quality Criteria." *Environ. Toxicol. Chem.* 8:589-599 (1989).

228. Underwood, A. J. and C. H. Peterson. "Towards an Ecological Framework for Investigating Pollution," *Mar. Ecol. Prog. Ser.* 46:227-234 (1988).

229. Adams, S. M., K. L. Shepard, M. S. Greely, J., B. D. Jiminez, M. G. Ryan, L. R. Shugart, and J. F. McCarthy. "Bioindicators for Assessing the Effects of Pollutant Stress on Fish," *Mar. Environ. Res.* 28:459-464 (1989).

230. Callahan, J. T. "Long-Term Ecological Reserarch," *BioScience* 34(6):363-367 (1984).

231. Grace, A. L. And L. F. Gainey, Jr. "The Effects of Copper on the Heart Rate and Filtration Rate of *Mytilus edulis,*" *Mar. Pollut . Bull.* 18:87-91 (1987).

232. Abel, P. D. "Effect of Some Pollutants on the Filtration Rate of *Mytilus,*" *Mar. Pollut. Bull.* 7:228-231 (1976).

233. Beaumont, A. R., G. Tserpes, and M. D. Budd. "Some Effects of Copper on the Veliger Larvae of the Mussel *Mytilus edulis* and the Scallop *Pecten maximus* (Mollusca, Bivalvia), *Mar. Environ. Res.* 21:299-309 (1987).

234. Stromgrem, T. "Effect of Heavy Metals (Zn, Hg, Cu, Cd, Pb, Ni) on the Length Growth of *Mytilus edulis,*" *Mar. Biol.* 72:69-72 (1982).

11

Genetic Factors and Tolerance Acquisition in Populations Exposed to Metals and Metalloids

Margaret Mulvey[1] and Stephen A. Diamond[2]

[1]Savannah River Ecology Laboratory, University of Georgia,
Drawer E, Aiken, South Carolina 29802
and
[2]Department of Zoology, Miami University, Oxford, Ohio 45056

OVERVIEW

Genetic factors can significantly affect the response of individuals or populations exposed to metal pollution. We discuss the differential response of individuals and populations exposed to heavy metals or metalloids and evidence for the evolution of tolerance in exposed populations. Also discussed are methods used to detect selection and evolution of tolerance in populations of aquatic organisms.

INTRODUCTION

Evolution of populations in changing environments can be both dramatic and rapid. Such evolutionary potential was clearly evident where human perturbations of the environment have been followed by evolution and adaptation of organisms to novel environmental conditions. Industrial melanism in moths and other insects in response to atmospheric pollution is perhaps the best known example.[1,2]

Further evidence is provided by the rapid evolution of pesticide resistance in insect[3] and rodent[4] populations subjected to intensive control programs. Indeed, Mallet[5] recently argued that high initial mortalities associated with the application of new pesticides may actually accelerate the evolution of resistance by efficiently eliminating all but the most resistant individuals.

Metal pollution represents a significant environmental challenge to organisms. Although some heavy metals (e.g., Cu, Fe, Mn, Zn) are essential micronutrients, they can be toxic at elevated concentrations; others (e.g., As, Au, Cd, Hg, Pb) have no known metabolic function and can be toxic at very low levels. High concentrations of metals occur naturally in some environments (i.e., the Zambian copper belt of Central Africa).[6] Organisms in these environments have had long periods of time to evolve tolerance mechanisms and such sites are often characterized by endemic metalliferous forms.[7] Populations occurring in mining or industrial sites may have had scores of years to evolve tolerance.[8,9] Elevated metal concentrations are also associated with recent pollution events; exposed populations must evolve tolerance or be eliminated from those sites. Thus, populations have had varying lengths of time during which mechanisms to cope with heavy metal stress might have developed. As with other pollution stressors, the expectation is that populations inhabiting metal-contaminated environments would evolve toward higher tolerance levels.

Although metal contamination is widespread and affects all habitat types, the evolution of tolerance has been most intensively investigated in plants. This literature has been summarized[10,11] and will not be included here. Considerable emphasis also has been placed on aquatic systems where discharge of metals associated with human activities has been common.[12,13] This chapter will summarize information concerning the evolution of metal tolerance in aquatic organisms (exclusive of plants and insects). We will also discuss approaches used to detect natural selection and the evolution of metal tolerance. As the literature regarding the evolution of metal tolerance is large, we have chosen representative material rather than attempt an exhaustive survey. Considerable emphasis will be placed on electrophoretic approaches, as our own work and that of several other laboratories have recently used these techniques for studying the evolution of tolerance in populations.

TOLERANCE EVOLUTION IN CONTAMINATED ENVIRONMENTS

Metals released into aquatic systems can stress the organisms that live there. Metal stress might be associated with individual mortality and local extinction or persistence of populations. Any stress, whether it is caused by natural or anthropogenic factors, shifts the energy budget of an organism.[14] Energy used to cope with environmentally stressful conditions is not available for growth and/or reproduction. The degree of stress experienced and the response to stress can vary among individuals in a population. This variation in response is often associated with differences in sex, size, or life stage of individuals[15] and interactions between these factors. Response differences can result from microhabitat differences in toxicant concentrations or other ecological factors. Because individuals in populations are genetically different from each other (except in special circumstances), additional and significant variation in individual response can result from genetically based differences in physiological and/or biochemical characteristics. Individual variation in response to metal exposure has often been treated as troublesome "noise" in data analysis. For example, the average response is usually reported in a toxicity test or bioassay. However, it is the genetic variation in response that provides the basis for the evolution of tolerance.

Tolerance can be broadly defined as the ability of individuals to cope with the stress associated with exposure to metal concentrations that are inhibitory or lethal to nontolerant individuals. Mechanisms associated with tolerance include (but are not limited to) changes in metal uptake or elimination rates, the ability to bind or sequester metals, and differences in enzyme sensitivity to inhibition by metals. Tolerance can be achieved by physiological acclimation during low-level exposures. Such phenotypically plastic responses are exemplified by metallothionein (MTN) induction[16-18] or intracellular granule formation.[19,20] This type of tolerance is not inherited by offspring and is lost when individuals are transferred to unpolluted environments. Alternatively, tolerance can be achieved by genetically based mechanisms. Genetically based tolerance is not lost when individuals are placed into clean environments and is inherited by offspring regardless of whether they are reared in polluted or nonpolluted environments. The distinction between physiologically and genetically based tolerance is somewhat arbitrary, as the capacity for a physiologically plastic response must ultimately have an underlying genetic basis. However, it is the genetic response that will be critical when considering the evolution of tolerance in exposed populations.

Factors Influencing the Evolution of Tolerance

The aquatic organisms considered here often have wide geographical distributions with distinct local populations occurring in lakes, streams, or rivers. Local populations may be more or less isolated from one another and may or may not be impacted by metal pollution. Thus local populations will vary in the factors influencing the evolution of tolerance. Nevertheless, the frequent occurrence of tolerance in populations inhabiting metal-polluted environments and the expectation that such tolerance should evolve suggest a need to understand the factors that contribute to the ability of organisms to evolve tolerance. Many factors can influence the acquisition of tolerance and will interact to determine the outcome of metal exposure (Table 1).

The acquisition of tolerance is not a predictable consequence of exposure to metal stress. A comparison of species composition, numbers, and/or diversity reveals that polluted sites are often biologically depauperate.[21,22] Many species present in noncontaminated areas are absent from nearby contaminated areas. This is true even for species with obvious means of movement or dispersal into the polluted sites. Presumably these species are not present because they have not or cannot achieve the appropriate tolerance. From this perspective, the evolution of tolerance would be the exception and most species are excluded from contaminated sites.

Metals, because they often induce high mortality, can have an intense selective effect on exposed populations. Although selection (natural or human-induced) acts on individual phenotypes and not directly on genotypes, it is the genetic changes that determine the characteristics of succeeding generations. If tolerance to metal exposure is genetically based, the survivors will constitute a population with an altered genetic composition relative to the preexposure population (Figure 1). The frequencies of genetically determined characteristics that reduce the effects of metals are expected to increase in populations inhabiting polluted environments. In addition, the genetic variance is expected to decrease following selection.

Acquisition of tolerance may be largely due to directional selection, which occurs when one extreme of the phenotypic distribution (e.g., individuals with the greatest tolerance) has the highest fitness in each generation (Figure 1). Thus the mean tolerance of the population is shifted following selection. For heritable traits, changes in phenotypic distributions temporally (before vs after pollution) or spatially (clean vs polluted environments) can reflect the operation of selection even if changes in gene frequencies have not been directly demonstrated. The

Table 1
Factors that Can Influence the Acquisition of Tolerance in Populations Exposed to Metals

Major Factor	Influence
Genetic factors	
Mutation	Establishes variation on which selection can act; rate $\simeq 10^{-4}$ to 10^{-6} per gene per generation
Dominance	Tolerance controlled by dominant alleles will be selected more rapidly in early generations following metal exposure
Mono- vs Polygenic	Selection will act more rapidly on traits under monogenic control
Fitness	Selection is rapid when fitness differences between tolerant and sensitive individuals are large
Reproductive factors	
Generation time and rate of increase	Shorter generation times and rapid population growth rate may be associated with rapid response to selection
Population size	Small populations may have less genetic variation than large populations
Mating system	Affects recombination of tolerant genotypes
Ecological factors	
Emigration/immigration	Matings with nontolerant gentoypes will slow the evolution of tolerance
Microhabitat	Selection can be slowed if sensitive genotypes persist in noncontaminated refugia
Life stages	Most sensitive life stage will influence the effectiveness of selection

Source: Modified from Wood and Bishop.[81]

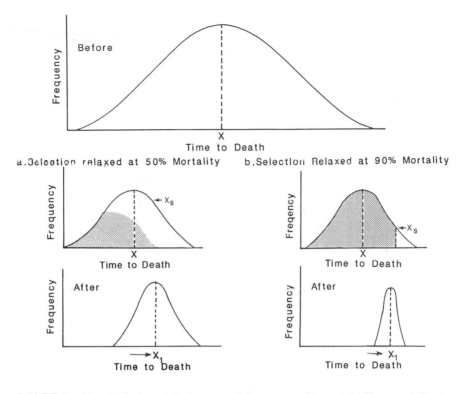

FIGURE 1. Hypothetical mortality in a population exposed to metals. The population is assumed to have a typical range of genetic variation and rare mutations at a few loci that confer tolerance. Not only the amount of mortality but also which individuals are killed will determine the rate of evolution of tolerance. (a) With 50% mortality, differential mortality may reflect normal polygenic variation; response to selection would be slow. (b) With 90% mortality, only tolerant genotypes survive; these occupy the extreme of the phenotypic distribution; response to selection would be rapid. Response to selection (R) is equal to the difference between the mean tolerance before selection (x) and the mean tolerance after selection (x₁). xₛ is the mean tolerance of the selected (surviving and reproducing) individuals. x₁ is often less than xₛ because tolerant genotypes may be reorganized by genetic recombination. (Modified from Endler.[82])

response to selection (R) is the difference between the mean phenotype before selection (x) and the mean phenotype following selection (x₁) (see Hartl and Clark,[23] p. 443). Because selected individuals have tolerant genotypes (e.g., longer times to death) and pass them to their offspring, x₁ is greater than x. Generally, x₁ is less than xₛ (mean of the selected individuals) because, in sexually reproducing species, the most tolerant genotypes are disrupted by Mendelian segregation and assortment.

The underlying genetic control of tolerance may be quantitative (influenced by many genes with small effects) or polymorphic (influenced by one or a few genes with major effects). Traits under polygenic control usually have a continuous distribution of phenotypes (Figure 2a) (i.e., from very sensitive to very tolerant of metals). Traits controlled by single gene loci with major effects usually have discrete phenotypic classes which reflect the underlying genotype (Figure 2b).

The evolution of tolerance is ultimately dependent on a genetically determined differential response. Individuals with tolerant genotypes are better adapted to the contaminated environment; they have a greater relative fitness. Darwinian fitness has two general components: viability and fecundity. Selection imposed by metals will act to increase the frequency of any genetically determined trait that increases the probability of surviving and reproducing in the contaminated environment.

Fisher's[24] fundamental theorem of natural selection states that, in order for evolutionary change to occur, there must be additive genetic variation in fitness-related characters. Such variation arises through mutation-heritable changes in

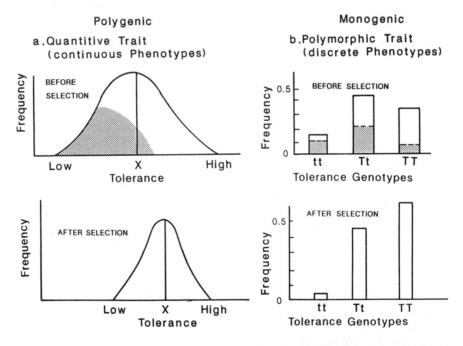

FIGURE 2.   Tolerance acquisition may be driven by directional selection. The vertical axis is the frequency of each tolerance type. The area under the curve or bars represents the population total. Individuals with tolerances in the stippled areas are at a disadvantage relative to the remaining individuals in the population. Distributions can illustrate selection within one generation [i.e., juveniles (before) and adults (after)] or between generations [i.e., parents (before) and offspring (after)]. In panel (a), x indicates the mean tolerance when tolerance is a continuously distributed trait. For tolerance determined by a polymorphic locus with two alleles (panel b), there are three genotypic classes: tt (homozygous sensitive), Tt (heterozygous, partially tolerant), and TT (homozygous tolerant). (Modified from Endler.[82])

the genetic material. Mutation rates are usually 10^{-4} to 10^{-6} mutations per gene per generation and may occur essentially at random. Metal pollution selects on appropriate variation in fitness to produce tolerant populations. Thus selection induced by metal pollution must act on preexisting variation and can uncover variation randomly accumulated over long periods of time. At the population level, variation can also arise from gene flow (migration/dispersal) and genetic recombination of loci influencing multigenic traits. Populations experiencing metal stress are often independent gene pools, and the opportunity for evolution of tolerance will depend on the chance occurrence of genes associated with tolerance. Hence, local populations might be expected to differ in the occurrence of and/or mechanisms associated with metal tolerance.

A large body of evolutionary theory describes factors that are likely to influence population response to metal stress and the evolution of heavy metal tolerance (Table 1). Tolerant phenotypes with high heritabilities and dominance will evolve faster than traits which are recessive or have low heritability. If tolerance is due to one or a few genetic loci with significant effects, then response to selection can be very rapid (see Figure 1); however, if tolerance is controlled by many loci with small effects, then the response would be slower. Response to selection will be rapid if tolerance is dominant in its expression relative to nontolerance. For a dominant gene, even at low frequency when it is most common in the heterozygote, the rate of increase in tolerance in the population can be rapid. Population size is a factor in the evolution of tolerance. Large populations can maintain more additive genetic variation than small populations. However, random genetic drift in small populations is also important. For example, if tolerance were controlled by a single locus with dominant effect, random sampling would be expected to accelerate its approach to fixation (100%) in a small population relative to a large one. Under these circumstances, genetic drift is accelerating the effects of selection. Strong selection will accelerate the evolution of tolerance as long as the selection is not so strong as to exterminate the population or prohibit differential survival. In cases where migration is possible, the evolution of tolerance can be slowed as nontolerant individuals move into the impacted area and mate with tolerant individuals.

Approaches Used To Detect Heavy Metal Tolerance In Aquatic Organisms

The study of tolerance and genetic mechanisms involved in tolerance acquisition is most effectively done when tolerance or a tolerance-related character is readily assayed and objectively defined. For plant studies, the Tolerance Index (ratio of root growth in toxic solution relative to growth in control solution) has been widely used.[25] Assays for aquatic animals include measures of time-to-death, elimination or accumultion rates, body burdens, fin regeneration, larval abnormalities, growth rates, longevity, and fecundity. These attributes are assumed to be components of or related to individual fitness. Under appropriate circumstances, differences in these attributes between individuals and/or populations exposed to metals relative to their nonexposed counterparts can be used as indicators that selection for tolerance has occurred. In addition, for attributes that are readily quantifiable, controlled crosses between tolerant and nontolerant individuals can be used to determine the underlying genetic basis of tolerant phenotypes. Studies reporting metal tolerance in aquatic organisms will be discussed below.

A. Population Comparisons

Many studies involve comparisons of populations from impacted and non-impacted areas. Populations sampled for comparison have included those from above and below effluent sites in streams and from localized areas assumed to be similar in major environmental factors and to differ only (or primarily) in the level of metal contamination. The method usually involved correlation of environmental metal contamination with measures of population tolerance.

Bryan and co-workers found that polychaetes, *Nereis diversicolor,* collected from two estuarine sites with elevated levels of Cu were more tolerant to Cu in laboratory tests than worms from less contaminated areas.[26,27] Cu tolerance seemed to have a genetic basis, as tolerance persisted when animals were maintained in clean conditions,[26] even when moved as juveniles.[21] Also, tolerance was not induced when individuals taken from nonpolluted sites were maintained in Cu-contaminated sediment.[26]

Genetic adaptation to metals has been described for the oligochaete, *Limnodrilus hoffmeisteri,* inhabiting a site with elevated levels of Cd and Ni.[28,29] Worms from the contaminated areas were significantly more tolerant of exposures to a combination of Cd, Co, and Ni than worms from control sites. In addition, in selection experiments involving naive populations, levels of tolerance comparable to those observed in field populations were attained in as few as one to four generations. Calculations of heritability of metal resistance, based on regression of mean offspring survival time in metal-contaminated sediment on midparent survival time in contaminated sediment, indicated the presence of much heritable genetic variation in this population of oligochaetes.

Enhanced tolerance of Cu and Pb has been demonstrated in populations of the isopod *Asellus meridianus* inhabiting rivers receiving mine drainage.[30-32] Tolerant isopods differed from sensitive individuals in uptake and compartmentalization of Cu and Pb. Animals taken from mine-impacted sites had higher LC_{50} values for Pb. This apparent tolerance persisted in the F_2 generation, which was tolerant of Pb concentrations that inhibited isopods collected from nonpolluted sites. Thus this tolerance was assumed to have a genetic basis. Brown[32] described cotolerance in this isopod in which tolerance to Cu seemed to confer tolerance to Pb. In addition, this study found that two populations of *A. meridianus* differed in mechanisms associated with tolerance acquisition to Cu and Pb.

Callahan and Weis[33] demonstrated tolerance of methylmercury in fiddler crabs, *Uca pugnax* and the grass shrimp, *Palaemonetes pugio.* A genetic basis for these tolerances was suggested because tolerance could not be induced by preexposure.

Populations of the killifish, *Fundulus heteroclitus,* have been shown to exhibit tolerance to Hg and methylmercury.[34,35] This tolerance may have a genetic component, as female killifish maintained in nonpolluted conditions retain tolerance.[36] Larval and embryonic stages are very sensitive to the effects of metals which may induce mortality or developmental abnormalities[37] (see this volume, Chapter 6). Susceptibility of embryos to the teratogenic effects of methylmercury was found to be lower for those taken from a polluted site. This embryonic tolerance was attributed to more rapid development and a less permeable chorion.[38]

While these studies make a strong case for the evolution of tolerance in populations inhabiting metal-polluted sites, other studies indicate that populations inhabiting contaminated sites are not necessarily genetically adapted. Bryan[27] reported populations of *Nereis diversicolor* in Cd- or Zn-contaminated sites that were no more tolerant than polychaetes from control sites. In a study of populations inhabiting Ag-contaminated sites, Bryan and Hummerstone[39] found that three populations displayed greater tolerance to Ag than controls, while a fourth population did not. Similarly, flagfish taken from a Zn-contaminated area were not more tolerant to Zn exposures than individuals from noncontaminated sites.[40] Multiple generations of selection for Zn tolerance in this species also did not result in increased tolerance. Bishop and McIntosh[41] found that bluegill, *Lepomis macrochirus,* from contaminated sites were not Cd tolerant.

As discussed by Parsons,[42] adaptations to stressful environments can take place rapidly; however, this may entail trade-offs with other aspects of fitness. Selection regimes and relative fitnesses will be different between contaminated and noncontaminated sites. Organisms can become adapted to the environment in which they are usually found; thus, adaptation to novel environmental conditions is likely to involve lower fitness in the initial environment. Adaptation is limited when tolerant individuals display reduced fitness, i.e., significant loss of growth and/or reproduction. As the contaminated environment becomes increasingly more divergent from noncontaminated sites, the "cost" of adaptation, for example, energy expended for metal-binding proteins or intracellular granule formation, will increase. Energy allocated for adaptation to the stressful environment cannot be so excessive as to prohibit growth and/or reproduction or the population will not persist.

Klerks and Levinton[29] suggest a "cost" of resistance to metals in oligochaetes; worms reared in clean sediment for two generations showed a decrease in resistance. The authors suggest that resistant individuals have a reduced fitness relative to nonresistant individuals in the clean environment. Similarly, embryonic tolerance is not without "cost" in *Fundulus heteroclitus.* Methylmercury-tolerant embryos displayed reduced salinity and $HgCl_2$ tolerance, slower growth, and weakness as adults.[34] Toppin et al.[38] reported that the response of *F. heteroclitus* to heavy metal stress involved changes in life history characteristics: fish from contaminated sites were smaller, reproduced at an earlier age, and had shorter life-spans. These shifts in life history characteristics suggested that the lifetime reproductive success of the impacted population might not be significantly different from that of a control population despite the higher mortality rate. In contrast, Duncan and Klaverkamp[43] and Chapman[44] report poor recruitment of young in white suckers shown to be metal tolerant. Chapman[44] concluded that increased tolerance was sometimes associated with significant sublethal toxicity affecting the condition, growth, and reproduction of fishes. Such sublethal effects may only postpone population extinction. Alternatively, populations displaying sublethal effects may persist long enough for additional tolerance mechanisms to evolve.

Local populations will differ in a large number of characters which may or may not be related to metal tolerance. These differences may reflect genetic differentiation between populations, environmental differences, and/or the interaction of genetics and environment to produce local ecophenotypes. Environments with elevated levels of metals may have other stressful features, such as low productivity, low oxygen levels, turbidity, and mixed pollutants. It may be difficult in such circumstances to assign causative relationships without detailed quantitative genetic laboratory studies.

B. Electrophoretic Variation and Population Tolerance

Ideally, the study of population response to metals would include a description of variability at genetic loci directly responsible for metal tolerance. However, tolerance may be a complex phenomenon, and it may not be possible to identify the controlling loci. Electrophoretically determined genotypic data provide a convenient and reliable tool for the study of population genetic processes and have been used extensively to study populations experiencing metal stress. Protein electrophoresis can be used to identify genetic differences among individuals within populations and to describe differentiation and divergence among populations.

Whether the loci surveyed by electrophoresis are themselves subject to selection or are linked to and "hitchhiking" along with loci selected with metal stress or mark population structure,[45] is usually unknown and is not essential to their application. However, as selection does not act directly on the genotype, the effects of metal stress on electrophoretic variants may be difficult to predict and may vary according to the genetic backgrounds of local populations.[42]

Hughes[46] discussed some possible advantages of population monitoring using electrophoretically determined genetic data: detection of sublethal, as well as lethal, effects; detection of effects even after pollution has stopped; and ease and rapidity of the technique. Electrophoretic surveys of populations should be particularly useful in situations where the pollution history is clearly defined. Chemical monitoring can be used to describe the contaminated environment, and electrophoretic monitoring can be used to describe the impact on populations in the contaminated sites.

One approach is to correlate gene frequencies in natural populations with levels of environmental contamination. Ideally, these studies would involve a number of populations from clean and contaminated sites, adequate sampling from each site (perhaps 40 or more individuals), and a large number of electrophoretically variable loci (> 10). To date, most studies have not involved sampling on this scale. Populations in contaminated sites may diverge from neighboring populations in enzymatic characters. Population differentiation can result from many processes not necessarily related to metal contamination; therefore, these data must be cautiously interpreted, particularly when sampling has been limited. For example, Heber[47] reported differences in isozyme frequencies in *Fundulus heteroclitus* from Hg-impacted and nonimpacted sites; however, these differences were associated with a larger geographical pattern and apparently not related to pollution.

A number of workers[48-54] have suggested that changes in population genetic structure might be used as a biomonitor of metal pollution. This use of allozymes requires that they are sensitive to and vary predictably with the type or level of pollution. One approach used to determine whether gene frequency changes observed at contaminated sites are, in fact, related to the presence of metals and not merely random differences between local populations is to conduct laboratory exposures in conjunction with field studies. This allows a demonstration of the genetic response under controlled conditions.

Nevo and co-workers compared genotypic frequencies in populations of shrimp, *Palaemon elegans*,[52] marine gastropods, *Monodonta turbinata,* and *M. turbiformis*[55] and the acorn barnacle, *Balanus amphitrite*,[51] from polluted and nonpolluted sites in the Mediterranean Sea. Of 15 loci examined in three populations of acorn barnacles, significant differences in allozyme frequencies were detected for 10 loci among the exposed and unexposed populations. Genotypic frequencies varied at the phosphoglucose isomerase (PGI) locus between populations of shrimp from Hg-contaminated and noncontaminated sites.[56] Snails collected from these Hg-contaminated sites had significantly different genotypic frequenices for phosphoglucomutase (PGM) when compared with control sites. Detection of allozyme frequency differences among field populations from contaminated and noncontaminated sites was followed by laboratory exposures to examine the effects of metals on individuals of different electrophoretic genotypes. Variants at enzyme loci in *M. turbinata* and *M. turbiformis* were correlated with survivorship (LC_{50}) during exposure to Cu or Zn, and the more tolerant genotypes were those that were relatively more abundant in the populations residing in the more polluted areas.

Gillespie and Guttman[57] found differences in genotype frequencies at the PGM locus between populations of the central stone roller, *Campostoma anomalum,* upstream and downstream of an impacted region of a stream. The genotypes found to be more abundant in the impacted region of this stream were also found to be most resistant to the effects of Cu exposure in the laboratory.

Not only single-locus effects but also multiple-locus complexes have been examined as indicators of metal pollution. For the marine gastropod, *Cerithium scabridum,* loci with no relationship to metal tolerance when examined singly were found to have significant effects when included in multiple-locus complexes (a 4-locus complex resistant to Hg; a 4-locus complex resistant to Cd; and a 2-locus complex resistant to a mixture of Hg and Cd).[58]

The authors[59,60] expanded on these techniques to determine whether genotype was related to the proness of individual mosquitofish, *Gambusia holbrooki*, to die during exposure to inorganic Hg or arsenate. The naive population of mosquito fish exposed in these studies had significant variability in survival during acute exposures. Time-to-death in both exposures was significantly related to genotype for three of eight loci in the Hg exposure and for two of eight loci in the As exposure.

Using electrophoretic techniques, individuals can be assigned to categories based on the number of loci for which they possess a heterozygous genotype. Numerous studies have reported a positive correlation between this rank-order heterozygosity and measures of fitness (i.e., growth rate,[45,61,62] viability,[63-65] developmental stability[67,68]). Heterozygous individuals may have an optimal metabolic efficiency[62,69] and thus achieve a broader range of physiological tolerance relative to more homozygous individuals. Koehn and Bayne[14] argue that differences in response to environmental stress will be genotype-dependent and can be linked to differences in maintenance metabolism. To the extent that heterozygosity is correlated with low maintenance costs, more heterozygous individuals are expected to display superior performance over a greater range of environmental stress. Nevo et al.[54] describe three species pairs of marine gastropods in which the more genetically heterozygous species of each pair was more resistant to pollution.

The authors[59,60] have also addressed the relationship between multiple locus heterozygosity and viability in populations of *Gambusia holbrooki* acutely exposed to Hg or As in the laboratory. As this population is not subject to heavy metal stress in the field, it was hypothesized that differential survivorship might be associated with a general biochemical or physiological superiority, such as has been attributed to differences in heterozygosity. Proportional hazards models were used to examine the relationship between time-to-death and multiple-locus heterozygosity or single-locus genotypes. For a 10-day Hg exposure, the relationship between multiple-locus heterozygosity and time-to-death was very strong: the estimated median time to death for the most heterozygous class was more than four days longer than the median for the zero heterozygote class. In a second experiment, fish from the same population were exposed to As. For females there was a positive relationship between multiple-locus heterozygosity and time-to-death, but, for males, no effect of multiple-locus heterozygosity and survivorship was observed. However, comparisons of the models generated for multiple-locus heterozygosity and single-locus effects suggest that heterozygosity per se was not related to time-to-death but was a summation of single-locus effects (Figure 3).

The relationship between enzyme genotype and metal tolerance is not well understood. While there is evidence that metals can inhibit enzyme activity *in*

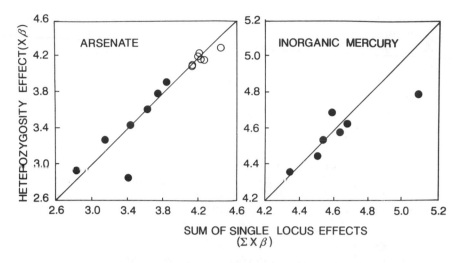

FIGURE 3. Plots of heterozygosity effect against the sum of single-locus effects during inorganic mercury and arsenate exposures in mosquitofish, *Gambusia holbrooki*. The estimates from the models for female (○) and male (●) mosquitofish were presented on the same plot for arsenate. Sexes were pooled (●) for the inorganic mercury models. (See Newman et al.[60] for explanation of model.)

vitro,[70,71] there is little known about how well these results can be extrapolated to living systems. As mentioned previously, the enzyme loci correlated with tolerance may simply be markers for loci coding for a tolerance-imparting factor or factors to which they are physically linked. For example, Wallace[72] demonstrated linkage between the acid phosphatase-1 locus in *Drosophila melanogaster* and a locus imparting tolerance to $CuSO_4$.

At the outset we suggested that it was not essential that electrophoretically detectable polymorphisms be directly affected by heavy metals. Nevo et al.[73] and Lavie and Nevo[55] have suggested that, in some instances, allozymes at loci surveyed electrophoretically may show differential sensitivity to heavy metals. This is consistent with the observation that, when examined in detail, allozymes display differences in kinetic properties, sensitivity to inhibitors, and thermal optima.[64-66,74,75] Thus allozymes least sensitive to metals would be expected to increase in populations inhabiting polluted sites. For allozyme frequency changes in marine snails, Nevo et al.[73] suggested that Cu combines with sulfhydryl groups to interfere with GPI activities and that, for PGM, Zn competes with the enzyme cofactor, Mg. Allozymes of PGM with comparable electrophoretic mobility (isoalleles) were found to differ markedly in their sensitivity to inhibition by Cu.[76] Thus exposure to the metal revealed previously undetected genetic variation when a population of mosquitofish was exposed to Cu. Chagnon and Guttman[76] argue that PGM and GPI may themselves be subjected to selection with metals as these two loci have shown significant effects in marine invertebrates[52,55,77] and freshwater fishes.[57,59,60,76,78] As many loci are routinely surveyed in electrophoretic studies, the recurrence of these loci with significant effects would not be expected and suggests direct differential effects on these loci.

C. Laboratory Exposures

Laboratory studies have been used to describe differences in sensitivity to metals among age classes or life history stages[79] (see also this volume, Chapter 6). Significant differences in sensitivity are often reported, usually with larval or juvenile stages more sensitive than adults. These studies suggest that, in order for populations to reproduce and persist in a contaminated site, the tolerant phenotype should be expressed in the most sensitive life history stage. This most sensitive stage may be a limiting factor in tolerance evolution as illustrated by the effects of tributyltin on marine mollusks. Adults tolerate significant sediment contamination, but larvae and juveniles are more sensitive and recruitment into populations is inhibited (see this volume, Chapter 12).

Laboratory selection experiments can indicate that a population has the potential for the evolution of tolerance. They are necessary to understand patterns of inheritance and specific mechanisms of tolerance. For example, the work of Nevo et al.[54] used both laboratory and field studies to evaluate the association of enzyme genotype with metal tolerance in marine snails. However, the results of laboratory studies may not readily apply to field situations where other population and environmental factors will influence the outcome.

Laboratory studies are necessary to determine if observed tolerances have a genetic basis. Tolerance observed in field-collected animals can be due to acclimation (physiological response) as well as adaptation (genetic response). Populations recently or periodically experiencing contamination might display phenotypic responses to changes in the environment without necessarily constraining the genotype. Genetic responses might be more common in populations experiencing chronic pollution. Retention of tolerance in animals taken from contaminated sites and maintained in clean conditions is evidence (although not conclusive) that the tolerance is genetically determined and not acclimation to prevailing conditions. A second laboratory-reared generation that displayed tolerance would be further evidence of genetic control of tolerance. Matings between stocks of different tolerances can be used to sort out the number of loci involved and whether these act dominantly. Lande[80] provides a method to determine the minimum number of genes affecting a phenotypic trait in a population. These types of experiments require that organisms be maintained and bred under laboratory conditions and that the assay for tolerance be readily quantified. To date, this approach has been much more successfully applied to plants than animals.

DISCUSSION

Despite the variety of metals and aquatic species studied, several generalizations regarding genetic factors and tolerance acquisition can be made. Variability in response of aquatic animal populations exposed to heavy metals has been documented. Such variability in response reflects the natural variability in species, is usually associated with polygenic control, and is not necessarily related to metal tolerance per se.

Many species are excluded from metal-impacted sites and, therefore, must lack the genetically determined variability to respond to selection imposed by metal stress. Populations which persist at impacted sites usually display increased tolerance of the contaminant metal. Tolerance of metals is not a simple condition but may reflect a variety of physiological traits and mechanisms and is, therefore, likely to differ among populations and species. This tolerance may be quite specific, perhaps limited to a single metal[8] or may involve cotolerance.[32] Although studies regarding heritability of tolerance are rare in the aquatic literature, those that have been done indicate a high degree of heritability for the trait.[29] In addition, the rapidity with which populations have acquired tolerance (perhaps only scores of generations) suggests high heritability. Whether tolerance in aquatic animals is controlled by a few genes with major effects or many genes with small effects is not yet known. Klerks and Levinton[29] indicate that metal tolerance of an oligochacte was a polygenic trait. Identification of genetic loci specifically determining tolerance and having large effects has been rare.

An important, and as yet untested, aspect of evolution of metal tolerance is whether fitness variables other than survival of exposure are sufficient for its occurrence. Metals are known to affect developmental stability, growth rate, and reproductive success. The effect of metal exposure on these factors can lead to selection if there is appropriate genetic variation for these traits. Those individuals best able to maintain these functions during an exposure episode would produce more offspring relative to other individuals. In this way the genotypes associated with these individuals would increase in frequency. There would be a correlated response as the entire phenotype is selected and not individual genetic loci.

One point that has been raised concerning genetic factors and tolerance acquisition in populations exposed to metals is the effect on the long-term evolutionary pattern of the population. Populations subjected to metals, especially acute pollution events, often experience high mortality. Such high mortality is likely to change the genetic composition of the remaining population. This change may be the result of selection and the population will display increased tolerance. Tolerance acquisition may be associated with a narrow range of phenotypes and loss of genetic variability. Alternatively, mortality may be nonselective; this can result in random loss of genetic variation due to sampling errors in small populations. Regardless of the underlying cause, one effect of high mortality associated with metal pollution is to reduce the overall genetic variability of exposed populations. As genetic variability is essential to evolution and has been associated with biological fitness, the ability of populations to respond to additional perturbations may be reduced. This includes long-term evolutionary potential as well as human modifications of the environment.

The study of genetic factors in conjunction with studies of animal tolerance of metals will result in a more complete description of the phenomenon. For toxicologists, understanding genetic mechanisms can lead to identification of biochemical and/or physiological mechanisms by which tolerance is achieved. The acquisition of tolerance to metals provides population geneticists and evolutionary biologists with an opportunity to examine evolutionary processes on a tractable time scale.

ACKNOWLEDGMENTS

We thank James M. Novak for critical comments on an earlier version of the manuscript. This work was supported by contract DE-AC09-76SR00-819 at the University of Georgia's Savannah River Ecology Laboratory.

REFERENCES

1. Kettlewell, H. B. D. *The Evolution of Melanism* (Oxford:Clarendon Press, 1973), pp. 51-151.
2. Bishop, J. A. and L. M. Cook. "Industrial Melanism and the Urban Environment," *Adv. Ecol. Res.* 11:373-404 (1980).
3. Whitten, M. J., J. M. Dearn, and J. A. McKenzie. "Field Studies on Insecticide Resistance in the Australian Sheep Blowfly, *Lucilia cuprina*," *Aust. J. Biol. Sci.* 33:725-735 (1980).
4. Partridge, G. G. "Relative Fitness of Genotypes in a Population of *Rattus norvegicus* Polymorphic for Warfarin Resistance," *Heredity* 43:239-246 (1979).
5. Mallet, J. "The Evolution of Insecticide Resistance: Have the Insects Won?" *Trends Ecol. Evol.* 4:336-340 (1989).
6. Drew, A. and C. Reilly. "Observations of Copper Tolerance in the Vegetation of a Zambian Copper Clearing," *J. Ecol.* 69:439-444 (1972).
7. Brooks, R. R. and F. Malaisse. "Metal-Enriched Sites of South Central Africa," in *Heavy Metal Tolerance in Plants: Evolutionary Aspects,* A. J. Shaw, Ed. (Boca Raton, FL:CRC Press, 1990), pp. 53-73.
8. Ernst, W. H. O. "Mine Vegetation in Europe," in *Heavy Metal Tolerance in Plants: Evolutionary Aspects,* A. J. Shaw, Ed. (Boca Raton, FL:CRC Press, 1990), pp. 21-37.
9. Wickland, D. E. "Vegetation of Heavy Metal-Contaminated Soils in North America," in *Heavy Metal Tolerance in Plants: Evolutionary Aspects,* A. J. Shaw, Ed. (Boca Raton, FL:CRC Press, 1990), pp. 39-51.
10. Antonovics, J., A. D. Bradshaw, and R. G. Turner. "Heavy Metal Tolerance in Plants," *Adv. Ecol. Res.* 7:1-85 (1971).
11. Shaw, A. J., Ed. *Heavy Metal Tolerance in Plants: Evolutionary Aspects* (Boca Raton, FL:CRC Press, 1990), p. 355.
12. Klerks, P. L. and J. S. Weis. "Genetic Adaptation to Heavy Metals in Aquatic Organisms: A Review," *Environ. Pollut.* 45:173-205 (1987).
13. Klerks, P. L. "Adaptation to Metals in Animals," in *Heavy Metal Tolerance in Plants: Evolutionary Aspects,* A. J. Shaw, Ed. (Boca Raton, FL:CRC Press, 1990), pp. 311-321.

14. Koehn, R. K. and B. L. Bayne. "Towards a Physiological and Genetical Understanding of the Energetics of the Stress Response," *Biol. J. Linn. Soc.* 37:157-171 (1989).

15. Neuhold, J. M. "The Relationship of Life History Attributes to Toxicant Tolerance in Fishes," *Environ. Toxicol. Chem.* 6:709-716 (1987).

16. Bouquegneau, J. M. "Evidence for the Protective Effect of Metallothioneins Against Inorganic Mercury Injuries to Fish," *Bull. Environ. Contam. Toxicol.* 23:218-219 (1979).

17. Nolan, C. V. and E. J. Duke. "Cadmium Accumulation and Toxicity in *Mytilus edulis.* Involvement of Metallothioneins and Heavy Molecular Weight Protein," *Aquat. Toxicol.* 4:153-164 (1983).

18. Bensen, W. H. and W. J. Birge. "Heavy Metal Tolerance and Metallothionein Induction in Fathead Minnows: Results from Field and Laboratory Investigations," *Environ. Toxicol. Chem.* 4:209-217 (1985).

19. George, S. G. and B. J. S. Pirie. "The Occurrence of Cadmium in Sub-cellular Particles in the Kidney of the Marine Mussel, *Mytilus edulis,* Exposed to Cadmium. The Use of Electron Microprobe Analysis," *Biochim. Biophys. Acta* 580:234-244 (1979).

20. Pynnonen, K., D. A. Holwerda, and D. I. Zandee. "Occurrence of Calcium Concretions in Various Tissues of Freshwater Mussels and Their Capacity for Cadmium Sequestration," *Aquat. Toxicol.* 10:101-114 (1987).

21. Bryan, G. W. and P. E. Gibbs. "Heavy Metals in the Fal Estuary, Cornwall: A Study of Long Term Contamination by Mining Waste and Its Effect on Estuarine Organisms," *Mar. Biol. Assoc. U. K. Occas. Publ.* 2:1-112 (1983).

22. Malueg, K. W., G. S. Schuytema, D. F. Krawczyk, and J. H. Gakstatter. "Laboratory Sediment Toxicity Tests, Sediment Chemistry and Distribution of Benthic Macroinvertebrates in Sediments from the Keweenaw Waterway, Michigan," *Environ. Toxicol. Chem.* 3:233-242 (1984).

23. Hartl, D. L. and A. G. Clark. *Principles of Population Genetics* (Sunderland, MA:Sinauer Associates, 1989), p. 682.

24. Fisher, R. A. *The Genetical Theory of Natural Selection,* 2nd Ed. (New York:Dover Publishers, 1958), p. 291.

25. Phipps, D. A. "Chemistry and Biochemistry of Trace Metals in Biological Systems," in *Effects of Heavy Metal Pollution on Plants,* N. W. Lepp, Ed. (Englewood, NJ:Applied Science Publishers, 1981), p. 1-54.

26. Bryan, G. W. and L. G. Hummerstone. "Adaptation of the Polychaete *Nereis diversicolor* to Sediments Containing High Concentrations of Heavy Metals. I. General Observations and Adaptation to Copper," *J. Mar. Biol. Assoc. U. K.* 51:845-863 (1971).

27. Bryan, G. W. "Some Aspects of Heavy Metal Tolerance in Aquatic Organisms," in *Effects of Pollutants on Aquatic Organisms,* A. P. M. Lockwood, Ed. (Cambridge:Cambridge Univ. Press, 1976).

28. Klerks, P. L. and J. S. Levinton. "Effects of Heavy Metals in a Polluted Aquatic System," in *Ecotoxicology: Problems and Approaches,* S. A. Levin, M. A. Harwell, J. R. Kelly, and K. D. Kimball, Eds. (Berlin:Springer-Verlag, 1987), pp. 41-67.

29. Klerks, P. L. and J. S. Levinton. "Rapid Evolution of Metal Resistance in a Benthic Oligochaete Inhabiting a Metal-Polluted Site," *Biol. Bull.* 176:135-141 (1989).

30. Brown, B. E. "Observations on the Tolerance of the Isopod *Asellus meridianus* Rac. to Copper and Lead," *Water Res.* 10:555-559 (1976).

31. Brown, B. E. "Uptake of Copper and Lead by a Metal-Tolerant Isopod *Asellus meridianus*," *Freshwater Biol.* 7:235-244 (1977).

32. Brown, B. E. "Lead Detoxification by a Copper-Tolerant Isopod," *Nature* 276:388-390 (1978).

33. Callahan, P. and J. S. Weis. "Methylmercury Effects on Regeneration and Ecdysis in Fiddler Crabs *(Uca pugilator, Uca pugnax)* After Short-Term and Chronic Pre-exposure," *Arch. Environ. Contam. Toxicol.* 12:707-714 (1983).

34. Weis, J. S., P. Weis, and M. Heber. "Variation in Response to Methylmercury by Killifish *(Fundulus heteroclitus)* Embryos," in *Aquatic Toxicology and Hazard Assessment: 5th Conf.* J. G. Pearson, R. Foster and W. E. Bishop, Eds., (ASTM S.T.P. 776) (Philadelphia, PA:American Society for Testing and Materials, 1982).

35. Weis, J. S. and P. Weis. "A Rapid Change in Methylmercury Tolerance in a Population of Killifish, *Fundulus heteroclitus,* from a Golf Course," *Mar. Environ. Res.* 13:231-245 (1984).

36. Weis, J. S., P. Weis, M. Renna, and S. Vaidya. "Search for a Physiological Component of Methylmercury Tolerance in the Mummichog, *Fundulus heteroclitus,*" in *Marine Pollution and Physiology: Recent Advances,* F. J. Vernberg, F. Thurberg, A. Calabrese, and W. B. Vernberg, Eds. (Columbia, SC:University of South Carolina Press, 1985), p. 309-326.

37. Weis, J. S. and P. Weis. "Tolerance and Stress in a Polluted Environment," *Bioscience* 39:89-95 (1989).

38. Toppin, S. V., M. Heber, J. S. Weis, and P. Weis. "Changes in Reproductive Biology and Life History in *Fundulus heteroclitus* in a Polluted Environment," in *Pollution Physiology of Estuarine Organisms,* W. Vernberg, A. Calabrese, F. Thurberg, and F. J. Vernberg, Eds. (Columbia, SC:University of South Carolina Press, 1987).

39. Bryan, G. W. and L. G. Hummerstone. "Adaptation of the Polychaete *Nereis diversicolor* to Estuarine Sediments Containing High Concentrations of Zinc and Cadmium," *J. Mar. Biol. Assoc. U. K.* 53:839-857 (1973).

40. Rahel, F. "Selection for Zinc Tolerance in Fish: Results from Laboratory and Wild Populations," *Trans. Am. Fish. Soc.* 110:19-28 (1981).

41. Bishop, W. E. and A. W. McIntosh. "Acute Lethality and Effects of Sublethal Cadmium Exposure on Ventilation Frequency and Cough Rate of Bluegill," *Arch. Environ. Contam. Toxicol.* 10:519-530 (1981).

42. Parsons, P. A. "Environmental Stresses and Conservation of Natural Populations," *Annu. Rev. Ecol. Syst.* 20:29-49 (1989).

43. Duncan, D. A. and J. F. Klaverkamp. "Tolerance and Resistance to Cadmium in White Suckers *(Catostomus commersoni)* Previously Exposed to Cadmium, Zinc or Selenium," *Can. J. Fish. Aquat. Sci.* 40:128-138 (1983).

44. Chapman, G. A. "Acclimation as a Factor Influencing Metal Criteria," in *Aquatic Toxicology and Hazard Assessment* 8th Symp.. R. C. Bahner and D. J. Hansen, Eds., ASTM S.T.P. 891 (Philadelphia, PA:American Society for Testing and Materials, 1985).

45. Zouros, E. and D. W. Foltz. "The Use of Allelic Isozyme Variation for the Study of Heterosis," *Isozymes* 15:1-60 (1987).

46. Hughes, J. "Pollution Impacts and Genetic Variation," *Occas. J. Centre Catchment In-stream Res.* 2:35-39 (1990).

47. Heber, M. A. "Comparative Embryonic Methylmercury Susceptibility and Reproductive Biology of Two Populations of the Mummichog *(Fundulus heteroclitus)."* M. S. Thesis, Rutgers University, Newark, NJ, 1981.

48. Beardmore, J. A. "Genetical Considerations in Monitoring Effects of Pollution," *Rapp. Proces. Verb. Reunions Cons. Perma. Int. Explor. Mer.* 179:258-266 (1980).

49. Battaglia, B., P. M. Bisol. V. U. Fossata, and E. Rodino. "Studies on the Genetic Effect of Pollution in the Sea," *Rapp. Proces. Verb. Reunions Cons. Perma. Int. Explor. Mer.* 179:267-274 (1980).

50. Gyllensten, U. E. and N. Ryman. "Pollution Biomonitoring Programs and the Genetic Structure of Indicator Species," *Ambio* 14:29-31 (1985).

51. Nevo, E., T. Shimony, and M. Libni. "Pollution Selection of Allozyme Polymorphisms in Barnacles," *Experientia* 34:1562-564 (1978).

52. Nevo, E., T. Pearl, A. Beiles, and D. Wool. "Mercury Selection of Allozyme Variation in Shrimps," *Experientia* 37:1152-1154 (1981).

53. Nevo, E., B. Lavie and R. Ben-Shlomo. "Can Allozyme Polymorphism Monitor Marine Environmental Quality?" *J. Etud. Pollut. Cannes, C.I.E.E.S.M.* 6:801-805 (1982).

54. Nevo, E., R. Noy, B. Lavie, A. Beiles, and S. Muchtar. "Genetic Diversity and Resistance to Marine Pollution," *Biol. J. Linn. Soc.* 29:139-144 (1986).

55. Lavie, B. and E. Nevo. "Heavy Metal Selection of Phosphoglucose Isomerase Allozymes in Marine Gastropods," *Mar. Biol.* 71:17-22 (1982).

56. Nevo, E., R. Ben-Shlomo, and B. Lavie. "Mercury Selection of Allozymes in Marine Organisms: Predictions and Verification in Nature," *Proc. Natl. Acad. Sci. U.S.* 81:1258-1259 (1984).

57. Gillespie, R. B. and S. I. Guttman. "Effects of Contaminants on the Frequencies of Allozymes in Populations of the Central Stoneroller," *Environ. Toxicol. Chem.* 8:309-317 (1989).

58. Lavie, B. and E. Nevo. "Multilocus Genetic Resistance and Susceptibility to Mercury and Cadmium Pollution in the Marine Gastropod, *Cerithium scabridum,"* *Aquat. Toxicol.* 13:291-296 (1988).

59. Diamond, S. A., M. C. Newman, M. Mulvey, P. M. Dixon, and D. Martinson. "Allozyme Genotype and Time to Death of Mosquitofish, *Gambusia affinis* (Baird and Girard), During Acute Exposure to Inorganic Mercury," *Environ. Toxicol. Chem.* 8:613-622 (1989).

60. Newman, M. C., S. A. Diamond, M. Mulvey, and P. Dixon. "Allozyme Genotype and Time to Death of Mosquitofish, *Gambusia affinis* (Baird and Girard) During Acute Toxicant Exposure: A Comparison of Arsenate and Inorganic Mercury," *Aquat. Toxicol.* 15:141-156 (1989).

61. Koehn, R. K. and S. E. Shumway. "A Genetic/Physiological Explanation for Differential Growth Rate Among Individuals of the American Oyster, *Crassostrea virginica* (Gmelin)," *Mar. Biol. Let.* 3:35-42 (1982).

62. Koehn, R. K. and P. M. Gaffney. "Genetic Heterozygosity and Growth Rate in *Mytilus edulis,"* *Mar. Biol.* 82:1-7 (1984).

63. Samollow, P. B. and M. E. Soule. "A Case of Stress Related Heterozygote Superiority in Nature," *Evolution* 37:646-649 (1983).

64. Watt, W. B. "Adaptation at Specific Loci. I. Natural Selection on Phosphoglucose Isomerase of *Colias* Butterflies: Biochemical and Population Aspects," *Genetics* 87:177-194 (1977).

65. Watt, W. B. "Adaptation at Specific Loci. II. Demographic and Biochemical Elements in the Maintenance of the *Colias* PGI Polymorphism," *Genetics* 103:691-724 (1983).

66. Watt, W. B., R. C. Cassin, and M. S. Swan. "Adaptation at Specific Loci. III. Field Behavior and Survivorship Differences Among *Colias* PGI Genotypes Are Predictable from *In Vitro* Biochemistry," *Genetics* 103:725-739 (1983).

67. King, D. P. F. "Enzyme Heterozygosity Associated with Anatomical Character Variance and Growth in the Herring *(Clupea harengus* L.)," *Heredity* 54:289-296 (1985).

68. Leary, R. F., F. W. Allendorf, and K. L. Knudsen. "Developmental Stability and Enzyme Heterozygosity in Rainbow Trout," *Nature* 301:71-72 (1983).

69. Garton, D. W., R. K. Koehn, and T. M. Scott. "Multiple-locus Heterozygosity and Physiological Energenetics of Growth in the Coot Clam, *Mulinia lateralis*, from a Natural Population," *Genetics* 108:445-455 (1984).

70. Carty, A. J. and S. F. Malone. "The Chemistry of Mercury in Biological Systems," in *The Biogeochemistry of Mercury in the Environment*, J. O. Nriagu, Ed. (North Holland, Amsterdam:Elsevier/North Holland Biomedical Press, 1975), p. 696.

71. Eichhorn, G. L. "Active Sites of Biological Macromolecules and Their Interaction with Heavy Metals," in *Ecological Toxicology Research: Effects of Heavy Metals and Organohalogen Compounds*, A. D. McIntyre and C. F. Mills, Eds. (New York:Plenum Press, 1975), pp. 123-142.

72. Wallace, B. *"Drosophila melanogaster* Populations Selected for Resistance to NaCl and CuSO$_4$ in Both Allopatry and Sympatry," *J. Hered.* 73:35-42 (1982).

73. Nevo, E., B. Lavie, and R. Ben-Shlomo. "Selection of Allelic Isozyme Polymorphisms in Marine Organisms: Pattern, Theory and Application," *Isozymes* 10:69-92 (1983).

74. Hines, S. A., D. P. Philipp, W. F. Childers, and G. S. Whitt. "Thermal Kinetic Differences Between Allelic Isozymes of Malate Dehydrogenase (Mdh-B locus) of Largemouth Bass, *Micropterus salmoides*," *Biochem. Genet.* 21:1143-1151 (1983).

75. Hall, J. G. "Temperature-Related Kinetic Differentiation of Glucose-Phosphate Isomerase Alleloenzymes Isolated from the Blue Mussel, *Mytilus edulis*," *Biochem. Genet.* 23:705-729 (1985).

76. Chagnon, N. L. and S. I. Guttman. "Biochemical Analysis of Allozyme Copper and Cadmium Tolerance in Fish Using Starch Gel Electrophoresis," *Environ. Toxicol. Chem.* 8:1141-1147 (1989).

77. Lavie, B. and E. Nevo. "Genetic Selection of Homozygote Allozyme Genotypes in Marine Gastropods Exposed to Cadmium Pollution," *Sci. Total Environ.* 57:91-98 (1986).

78. Chagnon, N. L. and S. I. Guttman. "Differential Survivorship of Allozyme Genotypes in Mosquitofish Populations Exposed to Copper or Cadmium," *Environ. Toxicol. Chem.* 8:319-326 (1989).

79. Harrison, F. L., J. P. Knezovich, and D. W. Rice, Jr. "The Toxicity of Copper to the Adult and Early Life Stages of the Freshwater Clam, *Corbicula manilensis*," *Arch. Environ. Contam. Toxicol.* 13:85-92 (1984).

80. Lande, R. "The Minimum Number of Genes Contributing to Quantitative Variation Between and Within Populations," *Genetics* 99:541-553 (1981).

81. Wood, R. J. and J. A. Bishop. "Insecticide Resistance: Populations and Evolution," in *Genetic Consequences of Man-Made Change*. J. A. Bishop and L. M. Cook, Eds. (London:Academic Press, 1981), p. 409.

82. Endler, J. A. *Natural Selection in the Wild* (Princeton, NJ:Princeton University Press, 1986), p. 336.

12

Impact of Low Concentrations of Tributyltin (TBT) on Marine Organisms: A Review

Geoffrey W. Bryan and Peter E. Gibbs

Plymouth Marine Laboratory, Citadel Hill, Plymouth PL1 2PB, United Kingdom

OVERVIEW

The objective of this chapter is to assess evidence indicating that, during the 1980s, hazardous levels of tributyltin (TBT) were reached in estuaries and coastal waters in various parts of the world. Leaching of marine antifoulants containing TBT compounds resulted in seawater TBT concentrations ranging from 10 to 100 ng/L in many estuaries. Furthermore, concentrations of the order of 1000 ng/L were reported in harbors and marinas. Levels of TBT in surface sediments generally exceeded those in the water by 1000 to 10,000 times and, in the biota, bioaccumulation factors (dried tissue/water) ranging from around 2500 to over 500,000 (in the clam *Mya arenaria*) were observed. Degradation of TBT to less toxic metabolites such as dibutyltin occurs in water and sediment; half-times can vary from days to weeks in the water and even longer in sediment, depending on the conditions. The breakdown of TBT has also been observed in organisms ranging from bacteria to birds, but the efficiency of degradation varies markedly between species. Data on bivalve mollusks suggest that lowest TBT bioaccumulation factors occur in species having the most efficient TBT metabolism.

Experimental and field observations have shown that some mollusks are sensitive to seawater TBT levels below 10 ng/L. In neogastropods, imposex is induced in the females of several species at concentrations of no more than a few nanograms per liter. Female *Nucella lapillus* are sterilized at concentrations below 10 ng/L. Shell malformation in the Pacific oyster, *Crassostrea gigas,* occurs over a similar range of concentrations. Examples of deleterious effects of TBT on species exposed to concentrations between 10 and 100 ng/L have also been observed in microalgae, polychaetes, crustaceans, bryozoans, echinoderms, tunicates, and fish. Fish and decapod crustaceans may be among the more tolerant groups since they appear to have a well-developed ability to degrade TBT.

Restrictions on the use of TBT antifoulants have had varying degrees of success, but some commercial oyster fisheries have recovered. Recolonization by some of the most sensitive species, such as *N. lapillus,* is likely to be very slow because of the reservoir of TBT remaining in sediments, the illegal use of TBT, and its continued use on commercial ships.

INTRODUCTION

Antifouling paints are used to prevent the attachment of organisms such as seaweeds, barnacles, tubeworms, and sea squirts to the hulls of ships and small boats. In the past, the effectiveness of these paints was often based on the toxicity of copper, but in the early 1960s, more effective paints containing as much as 20% of tributyltin (TBT) compounds were introduced and have proved very successful.[1] It became evident in the late 1970s that TBT leached from these paints affected organisms other than those targeted. The shells of Pacific oysters, *Crassostrea gigas,* being farmed in France began to exhibit morphological abnormalities; the annual spatfall also declined dramatically.[2] Severity of these effects increased with proximity to yacht marinas, and TBT came under suspicion. Similarly, in the eastern United States, the imposition of male sex characters ('imposex') on female mud snails, *Ilyanassa obsoleta,* was linked to TBT leached from antifouling paints.[3] Additional problems arose during the 1980s when, for example, in Scottish sea lochs antifoulant paints were used on cages employed in the salmon farming industry.[4] Other significant inputs of TBT to estuaries stemmed from contamination of rivers following the use of antifouling paints on freshwater craft,[5] from spillages at wood preservation facilities, and from the disposal of municipal wastewater and sewage sludge.[6]

Reviews on the toxicity of TBT to marine organisms up to 1987 included a high proportion of acute toxicity data.[7-11] However, preliminary evidence was presented showing that extremely low concentrations of TBT were toxic to some organisms. The objectives of this chapter are to demonstrate that some organisms are so sensitive to TBT that its presence in the sea poses a greater hazard to the marine ecosystem than was previously envisaged.

TBT IN THE MARINE ENVIRONMENT
TBT in Saline Waters
Concentrations

A body of evidence indicates that, irrespective of which TBT compounds are added, the same chemical species are produced in seawater.[9] For example, at pH 8, typical of seawater, TBT occurs as the chloride, hydroxide, and carbonato species.[12] In addition, results of experiments showing that the acetate, oxide, fluoride, and chloride of TBT all have similar toxicities to microalgae are consistent with this conclusion.[13] For this reason, concentrations of TBT are usually quoted as those of the ion TBT$^+$.

Measurements of TBT levels in U.K. waters in the mid-1980s revealed a number of fairly general principles:

1. High concentrations, approximately 1000 ng/L, were often found in yacht marinas during the summer and even higher levels were found in localities where boats or ships were cleaned prior to retreatment with TBT paint.[10,14]
2. Concentrations in harbor areas frequented by commercial shipping were low compared with areas occupied by large numbers of small boats.[15]
3. Concentrations in estuaries were generally lowest in winter, increased rapidly as newly-painted boats were launched during the spring, remained high during the summer, and declined gradually in the autumn, often with transient peaks caused by cleaning boats prior to removal.
4. Subsurface concentrations in open water reached several hundred nanograms per liter in estuaries such as the Hamble (Southampton) sheltering large numbers of small boats. In other estuaries, peak concentrations were more commonly in the 50 to 100 ng/L range.[10,15,16]
5. Particularly in stratified estuaries having a salt wedge, concentrations of TBT tended to decrease with increasing depth.[15]
6. In estuaries and harbors, concentrations in the surface microlayer exceeded those just below the surface by factors of 1.9 to 27.[15] This was an important observation, since species inhabiting the surface layer include not only microorganisms but also surface-dwelling egg and larval stages of many commercial fish.[17]

Thus, in the mid 1980s, concentrations of TBT in subsurface waters of U.K. estuaries covered the range <1 ng/L to >1000 ng/L. Values of 10 to 100 ng/L were commonly found in open waters of estuaries and still higher levels in the surface microlayer. Comparable concentrations were reported in other countries, including the United States,[18-20] Australia,[21] and Italy.[22]

Degradation in Seawater

One of the original attractions of TBT as a toxicant was its biodegradability, usually involving stepwise debutylation to dibutyltin (DBT) and monobutyltin (MBT). In the field, concentrations of DBT in seawater are usually somewhat lower than those of TBT, and MBT levels are markedly lower.[10] In a few instances, tetrabutyltin has also been detected[23] and may be a contaminant of some TBT paints. Both bacteria and phytoplankton are important vectors of degradation. The significance of the latter has been stressed by Lee et al.,[24] who demonstrated that TBT degradation in estuarine waters was promoted by light and nitrate supplementation. Half-times for degradation were 3 to 8 days in the light and 7 to 13 days in the dark and tended to be longer at lower temperatures. By contrast, a half-time of 60 days at 5°C has also been reported by Thain et al.,[25] who drew attention to the fact that degradation by microorganisms would have considerable bearing on persistence of TBT in particular waters. Thus a 5000 ng/L spike of TBT in polluted harbor water was degraded with a half-time of 13 days at 20°C, whereas similar spikes in waters from cleaner areas were degraded extremely slowly.[26] Other plant species, including macroalgae and eelgrass, *Zostera marina,* may also play a significant role in the degradation of TBT in some shallow-water areas.[27]

TBT in Sediments
Concentrations
Adsorption of TBT by suspended particles has led to its incorporation into sediments. For example, in Poole Harbour, U.K., TBT concentrations ranged from 49 ng/g dry wt to 1270 ng/g[16] and DBT concentrations were of the same order: sediment:water ratios for TBT were 2400 and 15,000 in the most and least contaminated sediments, respectively. Similarly, in the Severn River and Back Creek area of Chesapeake Bay, sediment concentrations ranged from <50 to 1400 ng/g and sediment:water ratios were 860 to 3800; again, DBT concentrations were of the same order as those of TBT.[18] Experimental studies have demonstrated that, in suspended sediments, the sediment:water ratios are influenced by factors including particulate loading, the size and organic content of the particles, and salinity.[28,29] Kram et al.[29] also studied the desorption of TBT from San Diego Bay sediments; for example, fine rather organic sediments released less TBT (~40%) in 24 h than coarser low-organic sediments (~90%). Observations by Unger et al.[30] showed that TBT adsorbed by sediment particles for 24 h was rapidly desorbed. They suggested that in well-mixed environments, concentrations in water and sediment may be in equilibrium. Furthermore, it was anticipated that in areas where the use of TBT paints was restricted, desorption of sediment-bound TBT and leaching from paint chips would delay the expected decline in seawater concentrations. Field observations in U.K. waters suggest that this may be so.

Degradation in Sediments
Degradation of TBT to DBT and MBT in sediments appears to be far slower than in water and a half-time of 162 days has been reported.[19,31] More recently, Waldock et al.[32] have documented half-times of less than 1 year for aerobic sediment and around 2 years for anaerobic sediment.

Table 1
Mean TBT and DBT Concentrations in Water, Sediment, and Organisms
(Depurated) Collected at Northam Bridge, Itchen Estuary, Southampton, Between
November 1986 and December 1988[a]

	DBT	TBT	%DBT	%TBT	TBT concentration factor (dry wt/water)
Seawater (n = 8)	—	68.2	—	—	—
		± 40.6			
Surface sediment (n = 5)	253	445	35.6	64.4	6,525
	± 92	± 83		± 9.4	
Alga (brown)					
Fucus vesiculosus (n = 5)	67	186	30.5	69.5	2,727
(Bladderwrack)	± 24	± 127		± 20.7	
Polychaete (burrowing)					
Nereis diversicolor (n = 5)	1173	479	72.4	27.6	7,023
(ragworm)	± 461	± 249		± 7.5	
Gastropod mollusk					
Littorina littorea (n = 4)	1377	1009	58.7	41.3	14,795
(common winkle)	± 436	± 428		± 7.2	
Bivalve mollusks (burrowing)					
Petricola pholadiformis (n = 2)	1760	838	67.8	32.2	12,287
(American piddock)	± 18	± 108		± 2.8	
Scrobicularia plana (n = 4)	2278	3375	39.5	60.5	49,487
(clam)	± 642	± 232		± 6.8	
Cerastoderma edule (n = 1)	1368	4128	24.9	75.1	60,528
(cockle)					
Macoma balthica (n = 4)	1020	4587	17.8	82.2	67,258
(clam)	± 669	± 2793		± 6.8	
Mercenaria mercenaria (n = 1)	683	8649	7.3	92.7	126,818
(hard-shelled clam)					
Mya arenaria (n = 4)	5511	36807	11.7	88.3	539,690
(soft-shelled clam)	± 4375	± 9800		± 10.4	

[a] Based on Analyses of Pooled Tissue Samples.

n, Number of samples. Values in ng/g dry wt, except seawater, ng/L.

TBT IN MARINE ORGANISMS
Concentrations in Tissues

Examples of tissue concentrations of TBT and DBT in a range of species collected at a single site in the Itchen Estuary are shown in Table 1. Its mean seawater TBT concentration of 68 ng/L puts this estuary in the moderately contaminated category. In addition, the sediment:water ratio for TBT of 6525 is higher than in other estuaries in the Southampton area, such as the Hamble; this occurrence may be attributed to the additional binding capacity provided by its high organic content of about 7.6%. In the biota, tissue TBT concentrations range from 186 ng/g in the seaweed, *Fucus vesiculosus* (bioaccumulation factor 2727) to 36,800 ng/g in the clam, *Mya arenaria,* a bioaccumulation factor of over one half million. Factors contributing to such variations are considered below.

Bioaccumulation of TBT

Laughlin[33] has proposed that the bioaccumulation of TBT is the result of two mechanisms. The first involves the partitioning of TBT between the surrounding water and lipids within the organism. For example, differences between the acute toxicities of various organotin compounds to zoeae of crabs, *Rhithropanopeus harrisii,* suggested that partitioning behavior, mediated by hydrophobicity, was controlling uptake and, hence, the lethal dose of organotins.[34] TBT, which is more hydrophobic (and lipophilic) than its more water-soluble degradation products dibutyltin and monobutyltin, is thus far more toxic. On the other hand, analyses of species including *Fucus vesiculosus,* the polychaete *Nereis diversicolor,* and the clam *Mya arenaria* from Poole Harbour (U.K.) showed that interspecific differences in lipid content were small, whereas tissue TBT concentrations had a range comparable with that shown in Table 1.[16] Since partitioning cannot account for some of the high bioaccumulation factors observed in the field, Laughlin[33] has proposed that a second bioaccumulation mechanism involves binding to organic ligands. For example, short-term studies with microalgae have shown that, depending on the species, either partitioning or binding control the bioaccumulation of TBT.[35]

Uptake from Solution

Observations on the time-course of TBT uptake in macroorganisms have demonstrated that times taken for the tissues to equilibrate with TBT levels in the environment are often measured in weeks or months rather than days. Experiments with small *Fucus vesiculosus* plants showed that, at a nominal concentration of 500 ng/L of TBT in continuous light, steady-state was reached after about 3 weeks; the bioaccumulation factor was about 1500 but probably higher due to losses from the water (cf. Table 1). Apparently due to degradation in the tissues, almost equal quantities of TBT and DBT were found in the tissues of *F. vesiculosus* during the experiment. In mussels, *Mytilus edulis,* there was no sign of a steady-state after exposure to 64 and 670 ng/L of TBT for more than 6 weeks.[36] On the other hand, at 23 ng/L, maximum tissue TBT concentrations were observed after only 4 to 6 days, suggesting that the mussel is able to regulate its body concentration in some way. Long-term experiments in which dog whelks, *Nucella lapillus,* were exposed to concentrations ranging from about 4 ng/L to about 250 ng/L of TBT showed that initial rates of uptake were roughly proportional to the ambient concentration.[37] Bioaccumulation factors of about 100,000 (dry tissue/water) were reached after 6 to 9 months at seawater concentrations below 50 ng/L, but, at about 250 ng/L, steady-state at a lower bioaccumulation factor of 30,000 was achieved in about 3 months. In the same species, field transplantation experiments showed that equilibration of tissue TBT levels with those of the new environment took about 6 months. Rates of loss of TBT from *N. lapillus* varied, depending on the experimental conditions, but half-times ranged from about 50 to more than 100 days.

Uptake from Food

Although emphasis tends to be placed on the accumulation of TBT from seawater, the amounts found in ingested sediments or other forms of diet may have an important bearing on tissue TBT levels. Experiments in which *N. lapillus* were fed on ^{14}C-TBT-labeled *M. edulis* in seawater containing about 20 ng/L of TBT showed that the compound was absorbed two to three times more rapidly by fed animals than by controls exposed to the water only.[38] Uptake of organotin from the diet was extremely efficient. Due to the degradation of TBT to DBT in the tissues of both fed and unfed *N. lapillus* (particularly in the digestive gland and kidney), concentrations of TBT tended to reach a plateau after about 28 days, whereas total concentrations of ^{14}C, including DBT and other metabolites, were still increasing after 49 days. After 49 days, the diet accounted for about one half the body burden of TBT, and bioaccumulation factors for ^{14}C-labeled TBT in both sexes were similar at about 60,000 for fed and 30,000 for unfed animals. The importance of the dietary route of uptake has also been reported for *Mytilus edulis*[39] and the crab, *Rhithropanopeus harrisii*.[40] In addition, it has been suggested that the relatively high concentrations of TBT found in some burrowing bivalves (cf. Table 1) may reflect the importance of sediment particles as a source of the compound.[16,41] For example, in the deposit-feeding clam, *Scrobicularia plana*, uptake of TBT from sediment appears to be the most important route.[42]

Degradation in Tissues

The ability of organisms to degrade TBT to a variety of metabolites including DBT, hydroxybutyldibutyltin, and monobutyltin, appears to be a widespread phenomenon, although its efficiency varies among species.[43,44] For example, the diatoms *Skeletonema costatum* and *Chaetoceros curvisetus* and the dinoflagellate *Procentrum triestinum* were able to metabolize TBT (25 to 75% remaining after 2 days), whereas in other species, including the green alga *Dunaliella tertiolecta* and the chrysophyte *Isochrysis galbana*, this capacity was very limited.[24] Among the higher plants, eelgrass, *Zostera marina*, decomposes TBT to DBT and then to MBT, which diffuses into the surrounding water.[27]

Of animals that have been studied, decapod crustaceans appear to degrade TBT most efficiently. The blue crab, *Callinectes sapidus*, fed daily for 16 days on a diet of TBT-contaminated shrimp, *Palaemonetes pugio* (1890 ng/g wet wt), reached a peak TBT concentration of 120 ng/g wet wt after 4 days which declined thereafter to 45 ng/g at 16 days; concentrations of DBT and MBT peaked after 8 and 12 days, respectively.[45] In the same species, a study of the role of the hepatopancreas in TBT metabolism indicated that TBT absorbed from the diet was partitioned to the lipids stored in the R cells and subsequently metabolized in the F cells and eliminated.[46] A capacity to degrade TBT absorbed from the water or from the diet was also observed in the shrimp, *Penaeus aztecus*.[46]

Among fish, examples of degradation have been reported in the sheepshead minnow, *Cyprinodon variegatus*, [47] and the spot, *Leiostomus xanthurus*.[44] An experimental study of the rainbow trout, *Salmo gairdneri,* exposed for 15 days to 1000 ng/L of TBT indicated that TBT became distributed between three main tissue compartments: (1) liver, gall bladder/bile, and kidney; (2) peritoneal fat; and (3) other tissues.[48] The presence of the highest proportions of metabolites in liver and gall bladder/bile suggested that TBT is metabolized in the liver and excreted by the gall bladder-fecal route. The presence of high concentrations of TBT and metabolites in the kidney was thought to reflect translocation from the liver by the blood, rather than implying a significant level of urinary excretion. Data for TBT and DBT in the tissues of flounder, *Platichthys flesus,* from Sutton Harbour, Plymouth (Table 2) are consistent with the proposal that the excretion of TBT metabolites occurs mainly in the bile rather than in the urine.

Evidence in mollusks suggests that the capacity for degrading TBT varies

Table 2
TBT and DBT Concentrations in Flounder, *Platichthyus flesus,* Tissues from Sutton Harbour, Plymouth, U.K. Compared with Other Species and Sediment (Nov./Dec. 1987)

	DBT[b]	TBT[b]	Total[b]	%TBT	TBT concentration factor (dry wt/water)
Platichthys flesus[a]					
Liver (*n* = 3)[c]	10,528	5132	15,660	32.1	16,937
	± 5218	± 3886	± 6964	± 16.0	
Gall bladder and bile (*n* = 3)	2810	1760	4570	44.0	5,809
	± 2095	± 244	± 2125	± 19.5	
Gut (*n* = 9)	5160	3087	8247	38.1	10,188
	± 1884	± 1002	± 2373	± 10.1	
Kidney (*n* = 3)	5762	2789	8551	31.9	9,234
	± 1107	± 1166	± 2088	± 7.0	
Urine (*n* = 1)	73	12	8.5	14.1	39.6
Gill filaments (*n* = 3)	3857	2212	6069	36.7	7,300
	± 1344	± 987	± 1552	± 14.3	
Muscle (*n* = 6)	1709	863	2572	35.5	2,848
(ventral and dorsal)	± 689	± 543	± 343	± 24.5	
Skin (*n* = 6)	1505	1251	2756	47.3	4,129
(ventral and dorsal)	± 706	± 623	± 358	± 28.3	
Ovary (*n* = 3)	1177	767	1944	40.1	2,531
	± 402	± 269	± 428	± 13.4	
Fucus vesiculosus	96	699	795	87.9	2,307
(brown seaweed)					
Nereis diversicolor	3098	4626	7724	59.9	15,267
(ragworm)					
Littorina saxatilis	2298	635	2933	21.7	2,096
(rough periwinkle)					
Surface sediment	1256	1884	3140	60.0	6,218
Water (mean for Oct.-Dec., 1986)	55	303	358	85.6	—
(Waldock et al[10]) (*n* = 3)	± 57	± 318	± 374		

[a] *Platichthys* females: 610, 410, and 380 g.
[b] Values in ng/g dry wt except Urine, ng/g wet wt and Water, ng/L.
[c] *n*, Number of samples.

considerably from species to species. TBT was metabolized only slowly by *Crassostrea virginica*[43] in comparison with fish and crustaceans. Experiments referred to earlier, in which ^{14}C-TBT-labeled *Mytilus edulis* were fed to *N. lapillus,* showed that TBT was metabolized far more rapidly by the latter than by the former.[38] Interspecific variability in the capacity to degrade TBT may, in part at least, explain the differences between the data in Table 1. Thus, animals having low concentrations of TBT contain a higher proportion of DBT which, in turn, may reflect their greater capacity to degrade TBT. On this basis, degradation appears to be most efficient in the polychaete *Nereis diversicolor,* followed by the bivalve *Petricola pholadiformis,* and then by the gastropod *Littorina littorea.* TBT degradation appears to be relatively inefficient in the bivalves *Mercenaria mercenaria* and *Mya arenaria.* However, the contrast in TBT levels between these two species suggests an additional factor is responsible for the very high concentration in *M. arenaria.* In this connection, it was observed that insecticides, including DDT and aldrin, were absorbed and lost far more readily by *M. arenaria* than *M. mercenaria.*[49]

Biomagnification?

Results in Table 2 show that the concentration of TBT in *N. diversicolor,* a prey species of *P. flesus,* exceeds that in the whole body of the flounder, and hence, there is no evidence for biomagnification up this step in the food chain. Similar conclusions were reached for the phytoplankton-*Mytilus edulis* and the *M. edulis-Nucella lapillus* food chains.[39,38] In addition, data on oystercatchers, *Haematopus ostralegus,* from the Exe Estuary (U.K.) and some typical dietary species (Table 3)[50] suggest that the intake of dietary TBT by the bird is counteracted by its degradation in the liver.

TOXICITY TO MARINE ORGANISMS (EXCLUDING MOLLUSKS)
Algae

While some macroalgae, including *Enteromorpha* sp. and *Ectocarpus* sp., are numbered among the principal fouling agents on ships' bottoms,[51] and *Fucus* sp. appears to thrive in TBT-polluted estuaries, some planktonic algae are quite sensitive. TBT suppressed growth in the centric diatoms *Skeletonema costatum* and *Thalassiosira pseudonana,* the EC_{50} concentrations at low cell densities being about 350 and 1150 ng/L, respectively.[13] By comparison, the EC_{50} concentrations for stannous chloride, dibutyltin dichloride, and tetrabutyltin in *S. costatum* were 325, 40, and 17 µg/L, respectively. Furthermore, it was found that, after 12 serial transfers to increasing concentrations of TBT, these species were unable to improve their resistance to the compound. This is a surprising discovery since *S. costatum* possesses the ability to degrade TBT to DBT and other products.[24] Other experiments with phytoplankton[52] demonstrated that *S. costatum* would not grow at a nominal concentration of 100 ng/L. In contrast, growth in *Dunaliella tertiolecta* and *Pavlova lutheri* was only slightly affected at 100 ng/L and was not totally prevented at 1000 ng/L. Surprisingly, *D. ter-*

Table 3
TBT and DBT Concentrations in Tissues of Oystercatchers and Dietary Species (Pooled Samples Not Depurated) from Lympstone, Exe Estuary (Feb./March, 1987)

	DBT[a]	TBT[a]	Total[a]	%TBT
Haematopus ostralegus (oystercatcher)	891	93	984	10.8
Liver (*n* = 13)[b]	± 636	± 54	± 670	± 5.6
Muscle (*n* = 13)[b]	61	149	210	65.3
	± 39	± 132	± 105	± 27.8
Nereis diversicolor (ragworm)	373	455	828	55.0
Littorina littorea (common winkle)	261	362	623	58.1
Cerastoderma edule (cockle)	273	494	767	64.4
Mytilus edulis (mussel)	269	694	963	72.1
Scrobicularia plana (clam)	724	1488	2212	67.3

Source: From Osborn and Leach.[50]

[a] Values in ng/g dry wt.
[b] *n*, Number of samples.

tiolecta appears to have a very limited ability to degrade TBT[24] and thus tolerance to TBT is not necessarily a function of an ability to degrade the compound.

Invertebrates

Experimental systems for the study of TBT toxicity in marine invertebrates have ranged between two extremes. On the one hand, experiments with larvae are often carried out in small containers, using nominal seawater concentrations of TBT which are changed daily. Losses of TBT from solution in this type of system can be quite dramatic, and mean exposure concentrations over 24 h may be only 50% of the nominal levels (e.g., Bryan et al.;[38] Laughlin et al.[53]). On the other hand, various flowing systems have been devised in which organisms are exposed to relatively constant, measured concentrations of TBT (e.g., Hall et al.,[54] Bushong et al.,[55] Bryan et al.[56]). Sometimes a considerable effort has been made to create environmentally realistic systems, as, for example, that employed by Henderson.[57] In his set-up, flowing seawater from Pearl Harbor, Hawaii, was piped to eighteen 155-L tanks in which panels treated with anti-fouling paint provided the TBT, and water concentrations were measured at intervals. Experiments were of three types: (1) TBT-free panels were allowed to develop a naturally recruited fouling community for 14 weeks in Pearl Harbor and then placed into the tanks for 5 weeks to adapt prior to TBT exposure. Species on panels were considered together with organisms colonizing the tank walls, and changes in colonization were estimated from photographs; (2) settlement panels for the study of colonization during TBT exposure were placed in the contaminated tanks; (3) other species, including oysters, were also introduced into the tanks. After 2 months of TBT treatment, the painted panels were removed and 2 months were allowed for recovery. Data from this experimental system are mentioned frequently in the following discussion of the effects of TBT on invertebrates.

Coelenterates

Experiments by Henderson[57] on species from Pearl Harbor showed that, after 26 days exposure to measured concentrations of 40, 100, 500, 1800, and 2500 ng/L of TBT, the anemone *Aiptasia pulchella* had disappeared at the three higher concentrations, but another anemone, *Haliplanella luciae,* appeared unaffected even at the highest levels. Laboratory experiments with the colonial hydroid *Laomedea (Campanularia) flexuosa* demonstrated that TBT levels ranging from 10 to 100 ng/L (nominal) stimulated colonial growth by up to 20%. However, growth was inhibited at higher levels, particularly between 500 and 1000 ng/g.[58,59] Since growth is regulated, it was proposed that increased colonial growth (hormesis) is the result of overcorrection of the inhibition caused by low levels of toxicant. In the field, studies of hard-bottom communities at TBT-contaminated sites (mean 208 ng/L) and less contaminated sites (mean 63 ng/L) in San Diego Bay revealed virtually no hydroids.[60] Available evidence suggests that some hydroids are certainly sensitive to TBT levels between 100 and 500 ng/L.

Polychaete Annelids

Little information exists on the sensitivity of polychaete worms to TBT. However, Henderson[57] observed that the sabellid *Sabellastarte sanctijosephi* exhibited increased mortality during a period in clean seawater following exposure to TBT levels of 40 and 100 ng/L for 60 days. Some individuals shed their tentacle crowns in the 100 ng/L treatment both during and after exposure. By contrast, in the same experimental system, the serpulids *Hydroides elegans, Pileolaria militaris,* and *P. pseudomilitaris* colonized settlement panels exposed to 2500 ng/L of TBT. Species of *Hydroides* were also characteristic of sites in San Diego Bay having a high density of boats and TBT concentrations of 89 to 400 ng/L.[60] In addition, studies on the survival and morphology of embryos and larvae of the lugworm, *Arenicola cristata,* showed that they were unaffected by exposure for 168 h to a seawater TBT concentration of 2000 ng/L.[61] *Nereis diversicolor* is another fairly tolerant species; it occurred in Sutton Harbour, Plymouth, where seawater concentrations sometimes exceeded 1000 ng/L in the summer[10] and the sediments contained nearly 2000 ng/g of TBT (Table 2). In a recent field study, Matthiessen and Thain[62] contaminated a sterilized sediment artificially to levels of 100, 1000, and 10000 ng/g dry wt with TBT paint particles (of the type eroded when yachts are scrubbed down). The sediment was then transferred to an uncontaminated site and recolonization studied over the next 164 days. No clear effects on colonization were seen in most species, but populations of the polychaete *Scoloplos armiger* (and the burrowing amphipod crustacean *Urothoe poseidonis*) exhibited a dose-related decline in numbers, suggesting that they are moderately sensitive to TBT.

Crustaceans

Evidence obtained to date suggests that copepods are the most sensitive of crustaceans. Early experimental work with *Acartia tonsa* produced a 144-h LC_{50} concentration of 550 ng/L.[63] However, later experiments with the same species showed that egg production was impaired by 18, 19, and 37% at concentrations of only 10, 50, and 100 ng/L, respectively,[64] and significantly lower survival was observed at 23 ng/L.[65] Similarly, in *Eurytemora affinis*, the 72-h LC_{50} concentration was 600 ng/L,[55] whereas Hall[66] reported that only 21% of the copepods survived exposure to 85 ng/L for 6 days. Mysids, *Acanthomysis sculpta*, were also quite sensitive since the release of viable juveniles from the brood pouches of gravid females was reduced at concentrations above 140 ng/L.[67] In the amphipod, *Gammarus oceanicus*, larval survival was reduced by exposure to 300 ng/L of TBT for 8 weeks.[68] Experiments with an unidentified species of *Gammarus*[54] showed that, after 24 days in concentrations ranging from 29 to 579 ng/L, survival was not significantly affected, although growth was inhibited at the highest concentration. Crabs appear to be relatively resistant to TBT. For larvae of *Rhithropanopeus harrisii*, LC_{50} concentrations calculated for the duration of zoeal development were 13,000 ng/L for a California population but 33,600 ng/L for a population from Florida.[69] Corresponding values for DBT in the two populations were 807 and 1660 μg/L, respectively.

Dissolved TBT caused a range of deformities during the regeneration of autotomized limbs in the fiddler crab, *Uca pugilator*.[70,71] The most common malformation was a curling of the dactyl so that it curved away from the pollex, making the claw ineffective. Males were more sensitive than females because of the greater sensitivity of the regenerating major claw. Some effects were seen at a nominal concentration of 500 ng/L, although the effective concentration was probably much lower. Furthermore, treatment with 500 ng/L for 1 to 3 weeks also resulted in an acceleration of the righting reflex in females, an indication of hyperactivity.[72] It is considered that while this latter effect reflects the neurotoxicity of TBT, the limb deformities reflect its behavior as a teratogen.

Bryozoans

There is little evidence for the toxicity of TBT to bryozoans, but Henderson[57] observed that after 63 days exposure to 100 ng/L of TBT, 50% of encrusting bryozoans, *Schizoporella errata*, had died.

Echinoderms

A study of the influence of TBT on arm regeneration in the brittle star, *Ophioderma brevispina*, revealed some evidence of inhibition at 10 ng/L and significant inhibition at 100 ng/L.[73] It was hypothesized that TBT acts via the nervous system, although direct action on the tissues at the point of breakage could not be excluded.

Chordates
Tunicates

In his experiments on Pearl Harbor organisms, Henderson[57] observed that all colonial tunicates, *Botrylloides* spp., died following 63 days exposure to 100 ng/L of TBT. Mortality in another colonial tunicate, *Didemnum candidum,* was 100% at 500 ng/L, but settlement on unfouled surfaces was observed at 100 ng/L. Solitary species, including *Ascidia nigra* and *A. interrupta,* withstood exposure to 2500 ng/L and were observed to settle at 1800 ng/L.

Fish

Pioneering work on the toxicity of TBT to the sheepshead minnow, *Cyprinodon variegatus,* by Ward et al.[47] gave a 21-day LC_{50} concentration of 960 ng/L. In addition, long-term experiments demonstrated that fecundity was not significantly affected at 450 ng/L, nor was the hatching success of juveniles from embryos produced by fish exposed to 450 ng/L for up to 147 days. Ultimately, some mortality occurred in these F_1 juveniles at both 240 and 450 ng/L, although growth of the surviving F_1 juveniles 28 days after hatching did not appear to be affected at 450 ng/L. In another study, exposure of the developing embryos of the Californian grunion, *Leuresthes tenuis,* to concentrations ranging from 150 to 1720 ng/L revealed no adverse effects on hatching success or growth.[74] Surprisingly, exposure of eggs and sperm to 10,000 ng/L during fertilization did not significantly reduce hatching success. On the other hand, 13-day-old larval striped bass, *Morone saxatilis,* exhibited reduced growth and increased mortality after 6 days exposure to 766 ng/L of TBT at a salinity of 3‰.[75] In addition, morphometric observations showed that the body depth of larvae was significantly reduced at only 67 ng/L. Observations on ionic regulation and gill ATPase activity in juveniles of the same species following 14 days exposure to 100 ng/L TBT revealed a significant increase in Na^+, K^+-ATPase but no effects on Mg^{2+}-ATPase or on ionic concentrations in the blood serum.[76] The most sensitive fish of five species subjected to acute toxicity experiments were larval inland silversides, *Menidia beryllina* (96-h LC_{50} = 3000 ng/L) and juvenile Atlantic menhaden, *Brevoortia tyrannus* (96-h LC_{50} = 4500 ng/L).[55] Chronic exposure of the same species to 93 and 490 ng/L of TBT for 28 days had no effect on survival, but both concentrations inhibited the growth of larval *M. beryllina.*[54] A morphometric study of the gills of mummichog, *Fundulus heteroclitus,* following 6 weeks exposure to TBT concentrations varying from 105 to 2000 ng/L of TBT revealed no pathological changes.[77] However, with a 96-h LC_{50} of 23,800 ng/L this is a particularly resistant species.[55]

TBT TOXICITY TO MOLLUSKS
Bivalves
Oysters

Crassostrea gigas were successfully introduced into French waters in 1968 but, in the mid-1970s, oysters having thickened shells became increasingly evident and the annual spatfall declined dramatically.[78] In Arcachon Bay on the west coast of France, the most serious problems were encountered in the vicinity of boat moorings and TBT leached from antifouling paints came under suspicion.[2,78] For convenience, the shell thickening and reproductive aspects will be discussed separately.

Shell Thickening. The link between shell thickening in *C. gigas* and TBT was first reported by Thomas[79] following an investigation into the use of TBT-containing preservatives on wooden oyster trays. Thickening results from the growth of additional layers of shell which are separated by chambers filled with gelatinous material; in the worst cases, ball-shaped oysters are produced containing very little soft tissue.[78] The effect is reversible, in that thickened oysters will revert to normal growth when transferred to clean water.[2] Problems with shell thickening in *C. gigas* were not confined to France. In England, where the species does not breed but is grown-on from cultivated spat, serious thickening occurred in several regions during the 1970s and 1980s.[80] As an index of thickening, the shell length:shell thickness ratio for the upper valve is commonly employed. Spat grown for 49 days in seawater containing a nominal concentration of 2 ng/L had an index of 23, comparable with controls; however, at 20 ng/L, the index was only 8 and at 100 ng/L had fallen to 5.[25] By interpolation, it was concluded that a concentration below 8 ng/L was necessary for the production of marketable *C. gigas* (index 12) and field observations in England and Wales tended to confirm this. In France, Alzieu et al.[81] and Chagot et al.[82] reported that a measured TBT concentration as low as 2 ng/L can initiate shell chambering in adults and concluded that this would explain the persistence of shell deformation in areas where low TBT concentrations were prevalent. Based on a concentration factor of 16,000 (wet tissue/water),[10] these data would suggest that tissue concentrations associated with the initiation of shell malformation lie in the region of 30 ng/g wet wt, with more severe effects occurring at concentrations over 100 ng/g. Reports of shell thickening in *C. gigas* in relation to TBT pollution have also come from Scotland,[83] Ireland,[84] Australia,[21] New Zealand,[85] and the United States.[86,87]

Although the shell-thickening response in *C. gigas* has been attributed to other factors, including high concentrations of suspended sediment, TBT has certainly been the main causal agent.[80] Dramatic improvements in *C. gigas* fisheries were observed when restrictions on the use of TBT paints on boats of less than 25-m length were introduced in France (1982) and in the U.K. (1987). For example, at Cap Ferret in Arcachon Bay, France, the proportion of *C. gigas* having chambers in both shell valves fell from a range of 62 to 91% during 1980-1982 to 0 to 11% during 1983-1987.[81]

An explanation of shell thickening has been proposed by Machado et al.[88] who observed that exposing the freshwater mussel, *Anodonta cygnea,* to 200 ng/L of TBT for 8 days caused secretion of a calcified organic matrix over the inner surface of the shell. Electrophysiological studies on the outer (shell side) epithelium of the mantle lining the shell showed that the short-circuit current was inhibited by exposure to TBT. It was hypothesized that this reflects inhibition of the physiological pump transporting protons (H^+) from the outer mantle epithelium toward the shell, coupled with the inhibition of bicarbonate transport into the hemolymph. As a consequence, diffusion of Ca and bicarbonate across the outer mantle epithelium increases and results in enhanced deposition of shell calcium carbonate. While it might be expected that such a mechanism would produce shell malformation in many species of mollusks, even in oysters, sensitivity of shell formation to TBT varies enormously. For example, thickening has been reported in *Crassostrea gigas* and *C. angulata* but not in *C. virginica*.[2,89,90] In Australia, a deformity known as shell curl has been described in *Saccostrea commercialis*.[21] Shell curl, coupled with low growth, appeared to occur in waters containing over 70 ng/L of TBT when the tissue concentration generally exceeded 200 ng/g wet wt. This finding would suggest that *S. commercialis* is less sensitive than *C. gigas*. Similarly, in New Zealand, King et al.[85] observed shell thickening in *C. gigas* but not in *Saccostrea glomerata* from the same site; however, mention was made of shell thickening in *S. glomerata* from a marina site. No appreciable shell thickening has been encountered in the European oyster, *Ostrea edulis*,[2] or in the New Zealand species, *Ostrea helfordii*.[85]

Other Effects of TBT on Oysters. The declining annual spatfall of *C. gigas* observed in Arcachon Bay, France, during the mid-1970s led to studies on the potential effects of TBT on larval and postlarval oysters. For example, experiments by His et al.[91] showed that, while exposure of *C. gigas* larvae to water containing 50 ng/L of TBT for 10 days resulted in high mortality and slow growth, exposure to 200 ng/L for 12 days caused total mortality. Studies by Lawler and Aldrich[92] demonstrated that in *C. gigas* spat (0.15 g wt) oxygen consumption and feeding rate were significantly reduced at concentrations down to 50 ng/L, growth based on shell length was inhibited down to 20 ng/L, and the ability to compensate for hypoxia was reduced even at 10 ng/L. On the other hand, no effects on growth were recorded when juvenile (0.21 g) *C. gigas* were exposed to 157 ng/L of TBT for 56 days.[93] By comparison, concentrations of 200 ng/L were observed in the Arcachon marina in 1982 and had probably been higher in previous years.[94] The ban on usage of TBT paints in this area early in 1982 led to a dramatic increase in *C. gigas* spatfall from zero in 1980 and 1981 to good in 1982 and excellent in 1983.[2]

Crassostrea virginica may be less sensitive to TBT than *C. gigas*, since spat exposed for 5 weeks to nominal concentrations of 20, 200 and 2000 ng/L grew normally at the two lower levels.[95] Also no effect on growth was observed in juvenile *C. virginica* exposed for 56 days to 156 ng/L.[93] However, in adults of the same species, Henderson[57] observed a very significant loss of condition following exposure to 100 ng/L for 57 days.

Thain and Waldock[89] concluded that TBT was probably a major contributory factor in the decline of the fishery for the European oyster, *Ostrea edulis*, on the east coast of England during the 1970s. In laboratory experiments, recently metamorphosed (3-mm) spat exhibited signs of reduced growth at 20 ng/L and a 50% reduction at 60 ng/L. On the other hand, older spat exposed to 240 ng/L for 7 days grew normally.[89] The same authors also observed that no larvae were produced when adult *O. edulis* were exposed to water concentrations of 240 and 2620 ng/L for 74 days during which they achieved tissue concentrations of 400 and 1230 ng/g wet wt, respectively. On the other hand, 4.8×10^6 larvae were produced by the "controls" in which tissue TBT levels reached only 190 ng/g. Histological examination indicated that TBT may have retarded the hormonally determined change in sex from male to female that normally occurs during the gametogenic cycle.[95] No comparable effects were seen when experiments of a similar type were carried out with *C. virginica*.[96]

Effects on Other Bivalves

As far as we are aware, there is little evidence for the shell-thickening response in bivalves other than the examples reported above. For example, unpublished work on shell length:shell weight relationships in *Scrobicularia plana* from TBT-contaminated sites in U.K. estuaries (e.g., Itchen; Table 1) showed no evidence of abnormalities. However, it is suspected that TBT may be responsible for a decline in the abundance of the clam over the past decade, apparently due to recruitment failure.[41,42] Other problems with shellfish arose in Scotland and Ireland where the use of TBT-treated cages on salmon farms contaminated the water in many remote areas.[83] There is strong circumstantial evidence that spatfall failures encountered by shellfish growers in the vicinity of salmon farms were caused by TBT. For example, in Mulroy Bay, Ireland, where observations on the settlement of bivalves were carried out from 1979 to 1986, the mean number of scallops, *Pecten maximus*, per settlement panel fell from 2000 in 1979 to zero during 1983-1985 but recovered in 1986.[84] These trends in settlement success coincided remarkably well with the increased use of TBT-treated nets in nearby salmon farms from 1981 to 1985, when usage ceased. In 1984, adult *P. maximus* contained 1600 to 2100 ng/g wet wt of TBT which, assuming a bioaccumulation factor of 35,000 (more than double that for oysters),[10,97] gives a water concentration of about 50 ng/L. Similar trends in settlement to those in scallops were seen in the bean mussel, *Musculus marmoratus*, and the flame shell, *Lima hians*, although recovery of the latter was not observed in 1986. Settlement of *Mytilus edulis* occurred every year, suggesting that they or their larvae are less sensitive to TBT than these other species. Even so, Beaumont and Budd[98] reported that the growth of *M. edulis* veliger larvae was inhibited at only 100 ng/L and mortality also increased, the 15-day LC_{50} concentration being 100 ng/L. Older mussels appear less sensitive, since growth in juveniles was inhibited by 400 ng/L of TBT but not by lower concentrations (Strømgren and Bongard).[99] On the other hand, Salazar and Salazar[100] concluded that, although at measured concentrations of 200 ng/L or greater the growth of mussels was inhibited in the field, at levels below 100 ng/L other environmental factors affecting growth tended to mask any obvious influence of TBT. Widdows and Donkin[101] reported that, under field conditions, a body burden of about 1000 ng/g (dry wt) of TBT in *M. edulis* reduced the capacity for growth. Similar tissue concentrations were associated with reduced growth capacity in mussels *Arca zebra* transplanted to sites on Bermuda where seawater concentrations of 20 to 50 ng/L were measured.[102]

Studies on the effects of TBT on the embryos and larvae of *Mercenaria mercenaria* by Laughlin et al.[53] revealed two main effects. First, the mean valve length of veligers and postlarvae declined with exposure and was significantly reduced at a nominal concentration of only 10 ng/L. Second, at concentrations of 100 ng/L or more, veligers did not reach the pediveliger stage within 14 days. It was concluded that a link exists between growth and metamorphosis, with TBT acting on the former to inhibit the latter.

Neogastropods (= Stenoglossans)
Ilyanassa obsoleta

The imposition of male characters, including a penis and vas deferens, on female mud snails, *Ilyanassa obsoleta*, was linked to TBT contamination by Smith.[3,103] Both the degree of penis development and the frequency of "imposex," as this phenomenon was termed by Smith,[104] declined with increasing distance from harbors and marinas and both measurements were proposed as indices of TBT contamination.[3] A field study of the relationship between imposex in *I. obsoleta* and measurements of environmental concentrations of TBT led Bryan et al.[105] to conclude that the syndrome is probably initiated at a seawater concentration of around 2 ng/L and a tissue concentration of the order of 20 ng/g dry wt. In an area where the seawater concentration was around 20 ng/L and the tissue concentration was 630 ng/g, all females exhibited imposex (frequency = 100%), and penis development approached that of males. As far as we are aware, there is no unequivocal evidence that populations of *I. obsoleta* have declined as a result of imposex, although a decrease in the proportion of females was observed with increasing frequency of imposex.[105] In the event of TBT causing reduced fecundity in *I. obsoleta*, its possession of a planktonic larva would probably ensure a supply of recruits from less polluted areas, thereby tending to mask the effect.

Nucella lapillus

Imposex is not confined to *I. obsoleta*. and it may be significant that this phenomenon came to light in several other species worldwide almost simultaneously. These include *Nucella lapillus*,[106] *Ocenebra erinacea*,[107] and *Thais emarginata*.[108] Subsequently, imposex has been reported in more than 40 species of neogastropods[109-111] although definite links with TBT contamination have been established in only a few of them. During the mid-1980s, studies on imposex in *N. lapillus* on the rocky coast around southwest England showed that its intensity increased with proximity to harbors and with increasing tissue TBT concentrations; furthermore, the most affected populations were observed to be declining.[56] The intensity of imposex was expressed as the relative penis size (RPS) index [(mean length of female penis^3/mean length of male penis^3) × 100]. A value of 50% shows that the bulk of the female penis is one half that of the male. Studies of the morphology of imposex in *N. lapillus*[112] showed that the formation of a vas deferens and a penis on the female could be divided into six stages, of which the first four covered development from the initial appearance of a short length of vas deferens near the genital papilla to the formation of a continuous vas deferens running from the genital papilla to a moderately sized penis.[112] Up to this stage, the female is able to lay egg capsules normally. At stage 5, however, the vas deferens overgrows the genital papilla and renders the female sterile by preventing the release of egg capsules. In addition, a build-up of aborted capsules in the capsule gland often occurs (stage 6); this mass of material can rupture the oviduct and is assumed to be a cause of the more rapid disappearance of females than males in declining populations. The vas deferens sequence (VDS) index is the average stage of imposex for females in a population; values exceeding 4 show that the population contains sterile females (stages 5 and 6). The RPS and VDS indices of imposex in *N. lapillus* have been employed successfully to monitor the distribution of TBT contamination from shipping and salmon farms around the coasts of the U.K.[56,113,114]

Laboratory experiments in which *N. lapillus* were reared from egg capsules to maturity at 2 years of age showed that one third of females exposed to a seawater TBT concentration of 4.0 ± 1.9 ng/L were sterile. These animals had an RPS index of 48% and the tissues contained 585 ng/g of TBT, giving a bioaccumulation factor of 146,000 (dried tissue/water). At a concentration of 9.4 ± 3.9 ng/L, all females were sterile and the RPS index was 96.6%. Tissues contained 1390 ng/g of TBT, giving a bioaccumulation factor of 148,000. Exposure to 50 ± 15 ng/L led to further masculinization, as oogenesis was supplanted by spermatogenesis.[115] Experiments in which female *N. lapillus* were injected with small amounts of TBT and other organotin and tributyl compounds (e.g., tributylsilane) showed that TBT was by far the most effective inducer of imposex, although limited effects were produced by tetrabutyltin and tripropyltin.[116,117]

Field observations on relationships between the development of imposex in *Nucella lapillus* and seawater and tissue concentrations of TBT agree remarkably well with experimental results and point to the initiation of imposex at dissolved concentrations below 1 ng/L and a tissue concentration of less than 100 ng/g.[112,115] This great sensitivity is reflected by the fact that during the 1980s very few populations lacking some signs of imposex were found around the U.K. It has been argued that this observation is indicative of the natural occurrence of imposex or evidence that it is caused by other contaminants. Against the natural occurrence hypothesis is the fact that imposex in *N. lapillus* was not noticed before the late 1960s, and examination of preserved material showed that several populations having high intensities of imposex in the 1980s were free of imposex in the past.[56,118] Although it would be surprising if imposex could only be induced by TBT, similarities between data on imposex from southern England, where the main source of TBT is relatively polluted harbors, and data from remote areas of Scotland, where virtually the only source of TBT and other pollution is salmon farms, suggests that, for practical purposes, TBT is the only cause.[112,113]

In the U.K., populations of *N. lapillus* are most heavily affected on shores along the south coast of England where there are large numbers of harbors and marinas. Most populations exhibit moderate to high RPS indices and the majority include sterile females; some are completely sterile and others have already disappeared (Figure 1). Since *N. lapillus* develop directly within the attached egg capsules and hatch as miniature adults, populations tend to be self-sustaining and are particularly vulnerable to suppression of reproduction. Subtle effects of TBT on populations are not immediately evident to the casual observer since the decline in numbers of egg capsules and juveniles tends to be masked by the continued presence of adults which may live for 10 years.

The ecological consequences of reduced predation on barnacles and mussels as dog whelk numbers decline may be relatively limited, although it is thought that on semiexposed shores the natural cycle of barnacle or fucoid seaweed domination is likely to be modified.[114] The decline of *N. lapillus* has also been implicated in the disappearance of the hermit crab, *Clibanarius erythrops,* for which the shell of *N. lapillus* is a favored home.[119]

No evidence of shell deformation has been found in *N. lapillus*. Crothers[120] reported that, at sites near Milford Haven in South Wales, where populations had declined markedly or disappeared by 1988, there was no detectable change in mean shell shape compared with data from 1966/1971.

Over the 3 years since the 1987 ban on the use of TBT paint on small boats, there has been a variable decline in seawater concentrations in different areas: for example, in Plymouth Sound the decrease was about 50%. Since imposex in adult *N. lapillus* appears irreversible, the first evidence of recovery has come from observations (1990) that intensities of imposex are declining in juveniles aged about 1 year in populations from relatively uncontaminated sites. At more heavily contaminated sites, little evidence of improvement is apparent, and it cannot be expected until ambient TBT concentrations fall to about 4 ng/L. This is the concentration at which about 30% of females were sterilized in laboratory studies. In areas where populations have disappeared or all females are sterile, recolonization will probably take many years.

Other Species of Nucella

In *Nucella lima* from Auke Bay, Alaska, RPS and VDS indices of imposex increased along a gradient of TBT, and treatment of the shells with TBT paint was shown to promote imposex.[121] Evidence of reduced reproduction at some sites and disappearance of the species from others was also obtained. Field observations on *Nucella lamellosa* in British Columbia indicated that it probably responds to TBT in the same way as *N. lapillus*.[122,123] On the other hand, although

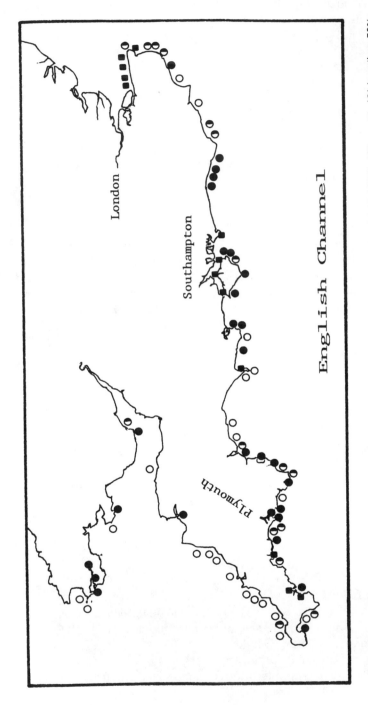

FIGURE 1. *Nucella lapillus*: Incidence of sterility in female dog whelks around southern Britain (1986-1989). (○) zero%; (◑) less than 50%; (●) over 50%; (■) population absent where previously recorded.

Nucella emarginata and *Nucella canaliculata* both exhibited imposex, there was no evidence for sterilization through blockage of the oviduct.

Ocenebra erinacea

Research on *Ocenebra erinacea* in France showed that imposex was widespread in the 1970s and a link with TBT pollution was established.[124,125] In the U.K., comparisons between *O. erinacea* and *N. lapillus* from the same sites showed that, in the vicinities of harbors, both species had similar RPS indices.[126] On the other hand, at lower levels of contamination, RPS indices in *O. erinacea* were zero when values in *N. lapillus* were around 15. Based on this comparison and on TBT measurements at field sites, it is thought that imposex in *O. erinacea* is probably initiated at 2 to 3 ng/L or at a tissue concentration of around 200 ng/g. Abnormalities of the oviduct caused by imposex in *O. erinacea* differ from those in *N. lapillus* but still appear to prevent copulation and egg capsule production in the worst affected females. Females in this condition were observed to contain around 700 ng/g of TBT and had been exposed to seawater concentrations of the order of 5 to 10 ng/L. Definite evidence that TBT has caused *O. erinacea* populations to decline is largely wanting on U.K. shores, but the decline of the species in Arcachon Bay, France, is attributable to TBT pollution (see also section on Oysters).

Thais haemastoma

Observations on *Thais haemastoma* from the Azores, the Canary Islands and from Spain have demonstrated that imposex occurs in populations from the vicinity of harbors and that blockage of the oviduct causes sterilization of females.[127] Unlike species of *Nucella*, *T. haemastoma* has a planktonic larva, thus allowing recruitment from outside in areas where all females may be sterile.

Lepsiella scobina

Field observations on this New Zealand thaid whelk revealed that the development of imposex resembled that in *N. lapillus*.[128] The VDS index in *L. scobina* was related to tissue TBT concentration and increased from zero at the least contaminated site to over 5 at sites in Auckland Harbor, where some sterile females were present. Two other thaids, *Thais orbita* and *Haustrum haustorium*, were similarly affected and, in the harbor, many females were sterile.

Hinia (= Nassarius) reticulatus

This species occurs sublittorally and sometimes intertidally at sites in southwest England also occupied by *N. lapillus* and *O. erinacea*. Comparisons between the intensities of penis development in *H. reticulatus* and in the other two species, coupled with measurements of seawater TBT, suggest that imposex is initiated in *H. reticulatus* at concentrations of 2 to 3 ng/L. So far, no evidence of sterilization of the females has been observed in *H. reticulatus* and, even so, its possession of a planktonic larva would help to sustain recruitment. However, reduced numbers have been reported in marina areas in Sweden (B. Tallmark, pers. commun. in Laughlin and Linden).[129]

In conclusion, it seems likely that imposex is induced by TBT in the majority of neogastropods. Sensitivity is generally high and penes are developed on females at concentrations ranging from less than one to a few nanograms per liter of TBT. Other modifications of the female reproductive system caused by imposex differ in detail from species to species, and even within the same genus. Thus, of the *Nucella* species studied, sterilization of females by imposex has been observed in *N. lapillus, N. lima,* and *N. lamellosa* but not in *N. emarginata* or *N. canaliculata.* Of the other species, evidence of sterility has been reported in female *Ocenebra erinacea, Thais haemastoma, Thais orbita, Lepsiella scobina,* and *Haustrum haustorium* but not in *Ilyanassa obsoleta* and *Hinia* (= *Nassarius*) *reticulatus.* It is fairly certain that imposex in neogastropods is caused when the endocrine system is disrupted by TBT and, in *O. erinacea,* TBT has been shown to activate the neuroendocrine mechanism responsible for penis differentiation.[125] In addition, Spooner and Goad[130] found that levels of testosterone were increased in *N. lapillus* following a small injection of TBT and that injections of testosterone caused an increase in penis length. Lee[131] has drawn attention to the probability that some organic pollutants and reproductive hormones are metabolized by the same mixed-function oxidase system. Thus, if the pollutant and hormone have common metabolic pathways, the presence of the pollutant may change the rate of hormone production or clearance.

DISCUSSION
TBT in the Marine Environment
The highest subsurface seawater concentrations of TBT to which organisms are likely to be exposed are found in harbors and marinas and are of the order of 1000 ng/L. In the open waters of estuaries or bays, concentrations are more likely to lie between 10 to 100 ng/L, although several hundred nanograms per liter have been found in estuaries harboring exceptionally large numbers of small boats. Since outflow from an estuary often hugs the coast, concentrations of a few nanograms per liter are sometimes found in shallow coastal waters tens of kilometers from the original source. These low levels seem to persist despite evidence for the rapid degradation of TBT observed in some laboratory studies. In the sea-surface microlayer, concentrations may be an order of magnitude higher than in the subsurface water.[15,18] In the surface layer is the neuston, which includes not only a wide range of microorganisms but also the eggs and larvae of macroorganisms, including fish. The degree of risk is debatable and depends on the extent to which the bioavailability of TBT is suppressed by binding to detritus or partitioning to lipids within the microlayer.[132,133] However, hexane-extractable components of the microlayer, presumably including TBT, have been shown to be toxic to developing embryos of the Baltic herring, *Clupea harengus.*[134]

Concentrations of TBT in the upper layers of sediment are usually one thousand to several thousand times higher than those in the water. The highest concentrations in harbors or near marinas often exceed 1000 ng/g but, in estuaries, the highest levels are usually measured in hundreds of nanograms per gram. Degradation of TBT in sediments appears to be much slower than in the overlying water, with half-times of the order of months or even years, rather than days. The reservoir of TBT provided by the sediments has probably contributed to the relatively slow decline of TBT levels in the water and biota of some U.K. estuaries following restrictions on the usage of TBT paints on small boats introduced in 1987.

The capacity for accumulating TBT varies widely between different species. Even at a single site, concentration factors ranged from a few thousand (dry tissue:water) in *Fucus vesiculosus* to more than one half million in *Mya arenaria* (Table 1). Exceptional TBT concentrations of the order of 50,000 ng/g dry wt have been observed in this species,[16] but, in most organisms, concentrations ten times lower would be regarded as very high. In animals, absorption from the diet or from ingested sediment appears to be a major route for TBT accumulation and may explain some interspecific differences between tissue concentrations. Possibly an even more important factor is the varying capacity of different species to excrete TBT or degrade it to less toxic metabolites such as DBT and MBT. Since so few detailed studies of degradation have been carried out, it is difficult to generalize with much confidence. Examples of TBT degradation have been reported in phytoplanktonic algae, macroalgae, and eelgrass, *Zostera marina;* plants may also play a significant role in the degradation of TBT in seawater. On the other hand, some species of phytoplankton appear to have little capacity for degradation.[24] Of the animals examined, decapod crustaceans, fish, and a single species of bird, the oystercatcher, all appear to possess degradation systems in which the hepatopancreas or liver plays an important role. Of the mollusks, *Mytilus edulis* and *Crassostrea virginica* metabolize TBT slowly, whereas TBT degradation in *Nucella lapillus* is more efficient.[38,44] Based on the assumption that organisms having a high proportion of DBT in the tissues are metabolizing TBT most rapidly, data for six bivalves from the same site suggested that *Mya arenaria* and *Mercenaria mercenaria,* with the highest TBT concentrations, were less efficient at metabolizing the compound than *Petricola pholadiformis,* in which the proportion of DBT was high and the concentration of TBT was low (Table 1). Also, by comparison with most bivalves, the gastropod *Littorina littorea* and the polychaete *Nereis diversicolor* appeared to metabolize TBT efficiently. The high interspecific variability observed both between tissue concentrations of TBT and the ability to metabolize the compound within groups such as algae or mollusks makes it difficult to draw generalizations about the capabilities of different groups to cope with the compound. Since, in most cases, the intensities of chronic effects of TBT are probably related to tissue TBT burdens, organisms most at risk might be expected to be those having a high affinity for the compound, whether in the water or the diet, coupled with an inability to degrade it. The bivalve *Mya arenaria* appears to be one such example, but there may be others within other groups.[16]

TBT Toxicity

Before discussing the present data on the low-level effects of TBT, some conclusions drawn from earlier work will be considered. Experiments carried out by Henderson[57] (see section on Coelenterates) on the settlement of organisms from the waters of Pearl Harbor onto panels in TBT-contaminated tanks showed that, although no difference was observed between the number of faunal species and species diversity in the control and 40 ng/L treatments, overall settlement was reduced at 200 ng/L. Henderson's general conclusion was that sponges, coelenterates, bryozoans, and mollusks were the most sensitive groups. Least affected species included macroalgae, serpulid polychaetes, solitary ascidians, and decapod crustaceans. Similar conclusions can be drawn from the work of Lenihan et al.[60] on the distribution of hard-bottom species in embayments in San Diego Bay having high and low intensities of small boats. Areas with the most boats and the highest TBT concentrations (mean 208 ng/L) were characterized by the presence of filamentous algae, serpulid polychaetes, and the solitary ascidian *Ciona intestinalis*. In embayments with few boats (mean TBT concentration, 63 ng/L), these species were replaced by other tunicates, sponges, bryozoans, and mussels. In both cases, species living among the sessile organisms included errant polychaetes and crustaceans but no gastropods or fish. Based on evidence available to mid-1987, Rexrode[8] concluded that mollusks were the group containing the most susceptible species, followed by crustaceans, algae, and fish. Similar conclusions were drawn by Waldock et al.[10] except that crustaceans and algae were transposed. However, it was stressed that, until more long-term studies were carried out, such generalizations are prone to inaccuracy. One of the predictable aspects of TBT toxicity seems to be its unpredictability.

In agreement with earlier evidence, recent work tends to confirm the high sensitivity of some species of mollusks to TBT, especially neogastropods (Table 4). Initiation of imposex (the imposition of male sex characters), including a penis and vas deferens, on female snails, occurs in several species at measured concentrations ranging from less than 1 ng/L in *Nucella lapillus* to perhaps 3 ng/L in *Ocenebra erinacea*. The sensitivity of *N. lapillus* is such that, at present, there appear to be very few populations around the U.K. in which evidence of the syndrome is absent. The slightly lower sensitivities of the other species means that imposex is absent in populations at coastal sites where there are no nearby sources of TBT. Sterilization through blockage of the oviduct by vas deferens tissue in *N. lapillus* was initiated at around 4 ng/L and, at 10 ng/L, all females were sterile. These values were determined by growing snails from egg capsules to adults over a period of 2 years and were supported by observations in the field.[115] Significantly, while earlier long-term experiments with adult females confirmed the sensitivity of penis development to low levels of TBT, it was necessary to expose them to about 250 ng/L for 12 months to produce any evidence of sterility.[37] This observation underlines the need to study the effects of TBT throughout the life cycle of an organism as far as possible. This is especially the case when it is considered that TBT appears to be disturbing the balance of reproductive hormones in neogastropods. Indeed, the sterilization of other species in other groups by less obvious forms of imposex is not difficult to envisage. For example, in *Ostrea edulis,* total inhibition of larval production at 240 ng/L of TBT may have been due to hormonal retardation of the sex change from male to female that normally occurs during the gametogenic cycle.[95]

In other bivalves such as scallops, *Pecten maximus* and *Lima hians,* it is not clear whether spatfall failure observed in TBT-polluted areas was due to the sensitivity of embryos or larvae, or whether reproduction was inhibited in the adults.[84] Even if reproduction is not affected, the embryos and larvae of some species are certainly sensitive. For example, deleterious effects of TBT on embryos and early larvae were reported in *Crassostrea gigas* and *Mercenaria mercenaria* at concentrations from 10 ng/L upward (Table 4). The effects that contribute to the failure of larval settlement include growth inhibition, delayed metamorphosis, and increased mortality.

Among other groups, evidence of deleterious effects at concentrations between 10 and 100 ng/L is relatively sparse, although this may partially reflect the need for more long-term experimental work, or possibly the lack of appropriate measurements. For example, although female *Nucella lapillus* reared from egg capsules to maturity over a period of 2 years were sterilized at concentrations below 10 ng/L, there was no evidence that parameters including growth, mortality, and ultimate size changed appreciably, even at 250 ng/L. As might be expected, some planktonic species are quite sensitive to TBT. Of the phytoplanktonic organisms studied, *Skeletonema costatum* did not grow at a nominal concentration of 100 ng/L which, in reality, was probably much lower. Other species including *Dunaliella tertiolecta* were clearly more resistant, although curiously, unlike *S. costatum,* this species has little ability to degrade TBT.[24] What seems fairly predictable is that TBT contamination in estuaries is very likely to influence the species composition of the phytoplankton. Whether TBT is in any way involved in the development of blooms of toxic phytoplankton has, as far as we are aware, not been reported. Among zooplanktonic organisms, deleterious effects of TBT on *Acartia tonsa* and *Eurytemora affinis* were observed at concentrations between 10 and 100 ng/L (Table 5). Possibly the processes responsible for the degradation of the compound in copepods are less well developed than they appear to be in decapod crustaceans such as *Callinectes sapidus.*[45] In this group, effects of TBT on the regeneration of limbs in *Uca pugilator* were observed at the relatively high concentration of 500 ng/L (Table 5). By comparison, arm regeneration in the brittle star, *Ophioderma brevispina,* was inhibited at levels ranging from 10 to 100 ng/L. Evidence of effects at concentrations below 100 ng/L has also been observed in a polychaete and can be assumed to occur in a bryozoan and a tunicate, since a high mortality was reported at 100 ng/L (Table 5). In fish, there are examples of slight effects on larval growth at concentrations below 100 ng/L.

Table 4
Summary of Low-Level Effects of TBT on Mollusks Described in Text[a]

Conc. (ng/L)	Crassostrea gigas	Crassostrea virginica	Saccostrea commercialis	Ostrea edulis	Mytilus edulis	Mercenaria mercenaria	Nucella lapillus	Ocenebra erinacea	Ilyanassa obsoleta	Hinia (= Nassarius) reticulatus
1	(2) Initiation of shell thickening[82]	(8) Upper limit of shell thickening for salable oyster[25]					(<1) Initiation of imposex[115] (4) One third of females sterile[115]	(2-3) Initiation of imposex[126] (5-10) Sterilization of females[126]	(2) Initiation of imposex[105]	(2-3) Initiation of imposex
10	(10) Reduced ability to compensate for hypoxia in spat[92] (20 +) Reduced growth of spat[92]			(20-60) Reduced growth of small spat[89]		(10) Reduced larval growth[53]	(10) All females sterile[115] (25 +) Oogenesis in females supplanted by spermatogenesis[115]		(20) 100% frequency of imposex in females; high penis intensity[105]	

(50)
High mortality; low growth of larvae[91]

(50 +)
Reduced oxygen consumption and feeding rate in spat[92]

100

(200)
Total mortality of larvae[91]

(70)
Shell curl; low growth[21]

(100)
Loss of condition in adult[57]

(240)
Inhibition of larval production in adult[95]

(100)
15-day LC_{50} for larvae[98]

(200)
Inhibition of growth in small mussels[100]

(100)
Delayed metamorphosis[53]

a Concentrations are given in parentheses. References numbers are also shown.

Table 5
Summary of Low-Level Effects of TBT Discussed in Text[a]

Conc. (ng/L)	Microalgae	Coelenterates	Polychaetes	Crustaceans	Bryozoans	Echinoderms	Tunicates	Fish
10		(10-100) *Laomeda flexuosa* (hormesis)[58]		(10-100) *Acartia tonsa* (reduced egg production;[64] mortality)[65]		(10-100) *Ophioderma brevispina* inhibition of arm regeneration)[73]		(67) *Morone saxatilis* (reduced body depth of larval fish)[75]
50			(40-100) *Sabellastarte sanctijosephi* (increased mortality)[57]	(85) *Eurytemora affinis* (high mortality over 6 days)[66]				(93-490) *Menidia beryllina* (reduced larval growth)[54]
100	(100) *Skeletonema costatum* (zero growth)[52] (100-1000) *Dunaliella tertiolecta* and *Pavlova lutheri* (growth inhibition)[52]			(140+) *Acanthomysis sculpta* (reduced release of viable juveniles)[67] (300) *Gammarus oceanicus* (reduced larval survival over 8 weeks)[68]	(100) *Schizoporella errata* (63-day LC50)[57]		(100) *Botrylloides spp* (100% mortality)[57]	(100) *Morone saxatilis* (increased gill Na$^+$ K$^+$-ATPase in juvenile)[76] (240) *Cyprinodon variegatus* (juvenile mortality)[47]

500

(500)
Aiptasia pulchella (100% mortality)[57]

(500-1000)
Laomeda flexuosa (growth inhibition)[58]

(500)
Uca pugilator (hypersensitivity in females; deformity of regenerated limbs in both sexes)[70-72]

(550)
Acartia tonsa (144-h LC_{50})[63]
(600)
Eurytemora affinis (72-h LC_{50})[55]

(500)
Didemnum candidum (100% mortality)[57]

(766)
Morone saxatilis (reduced growth and increased mortality in larvae)[75]

a Concentrations are shown in parentheses. Reference numbers are also indicated.

Evidence of a well developed capacity for degrading TBT has been reported in several species, including *Cyprinodon variegata*.[47] Life cycle tests carried out during the same project indicated that the fecundity of adults was not impaired up to a concentration of 450 ng/L, although mortality was observed in some of the juveniles produced during exposure to 240 ng/L.

Based on the limited amount of evidence, groups of organisms can be placed in an approximate order, depending on the concentrations affecting their most sensitive species. Mollusks are the most sensitive since some neogastropods and some oysters respond to levels below 10 ng/L. Organisms sensitive to the 10 to 100 ng/L range include copepod crustaceans, an echinoderm, a polychaete, and a tunicate. Some species of phytoplankton and fish may also fall within this range although, possibly due to a well developed ability to metabolize TBT, fish and decapod crustaceans appear relatively resistant. However, tolerance to TBT is no guarantee of protection to an organism if major components of its diet are sensitive to the compound. For example, a decline in the recruitment of several species of bivalve observed in heavily TBT-contaminated U.K. estuaries may have influenced the distributions of some wading birds.

Following restrictions on the usage of TBT, evidence from France and Ireland showed that settlement of bivalves, including *Crassostrea gigas* and *Pecten maximus,* improved dramatically.[78,84] On the other hand, recruitment of deposit-feeding bivalves such as *Scrobicularia plana* in TBT-contaminated estuaries was not evident 2 years after the ban and is attributable to residual contamination in sediments. Due to this reservoir of TBT in sediments, the illegal use of TBT paints and their continued use on commercial ships, the recovery of the most sensitive species, such as *Nucella lapillus,* seems unlikely to occur in the near future in the vicinity of some harbors. For example, along the French coast, widespread imposex was observed in *N. lapillus* and *Ocenebra erinacea* 7 years after restrictions were introduced.[117,126]

In conclusion, there seems little doubt that restrictions on the usage of TBT antifoulants are necessary, since seawater concentrations in many areas were approaching, and, in some areas, far exceeded 100 ng/L. Its insidious effects on reproduction in some mollusks at levels of a few nanograms per liter (almost certainly caused by the disruption of hormonal balance), coupled with its toxicity to embryonic and larval organisms at concentrations of 10 to 100 ng/L, had led to accelerating degradation of the ecosystem, not only in many estuarine areas but also on open coasts. Furthermore, it is probable that the effects of TBT discussed in this review are far from being the only consequences of releasing TBT into the sea.

REFERENCES

1. Anderson, C. D. and R. Dalley. "Use of Organotins in Antifouling Paints," in *Oceans '86 Proceedings, Vol 4. International Organotin Symposium* (New York:Institute of Electrical and Electronics Engineers, 1986), pp. 1108-1113.
2. Alzieu, C. and J. E. Portman. "The Effect of Tributyltin on the Culture of *C. gigas* and Other Species," *Proc. Annu. Shellfish Conf.* 15:87-101 (1984).
3. Smith, B. S. "Tributyltin Compounds Induce Male Characteristics on Female Mud Snails *Nassarius obsoletus = Ilyanassa obsoleta*," *J. Appl. Toxicol.* 1:141-144 (1981).
4. Balls, P. W. "Tributyltin (TBT) in the Waters of a Scottish Sea Loch Arising from the Use of Antifoulant Treated Netting by Salmon Farms," *Aquaculture* 65:227-237 (1987).
5. Waite, M. E., K. E. Evans, J. E. Thain, and M. J. Waldock. "Organotin Concentrations in the Rivers Dure and Yare, Norfolk Broads, England," *Appl. Organometal. Chem.* 3:383-391 (1989).
6. Fent, K. "Organotin Speciation in Municipal Wastewater and Sewage Sludge: Ecotoxicological Consequences," *Mar. Environ. Res.* 28:477-483 (1989).
7. Thompson, J. A. J., M. G. Sheffer, R. C. Pierce, Y. K. Chan, J. J. Cooney, and R. J. Maquire. "Organotin Compounds in the Environment: Scientific Criteria for Assessing Their Effects on Environmental Quality," Natl. Res. Coun. Can. Ottawa, Publ. No. NRCC 22494 (1985), p. 284.
8. Rexrode, M. "Ecotoxicity of Tributyltin," in *Oceans '87 Proceedings, Vol. 4. International Organotin Symposium* (New York:Institute of Electrical and Electronics Engineers, 1987), pp. 1443-1455.
9. Maguire, R. J. "Environmental Aspects of Tributyltin," *Appl. Organometal. Chem.* 1:475-498 (1987).
10. Waldock, M. J., J. E. Thain, and M. E. Waite. "The Distribution and Potential Toxic Effects of TBT in UK Estuaries During 1986," *Appl. Organometal. Chem.* 1:287-301 (1987).
11. Hall, L. W. and A. E. Pinkney. "Acute and Sublethal Effects of Organotin Compounds on Aquatic Biota," *Crit. Rev. Toxicol.* 14:159-209 (1984).
12. Laughlin, R. B., H. E. Guard, and W. M. Coleman. "Tributyltin in Seawater: Speciation and Octanol Water Partition Coefficient," *Environ. Sci. Technol.* 20:201-204 (1986).
13. Walsh, G. E., L. L. McLaughlin, E. M. Lores, M. K. Louie, and C. H. Deans. "Effects of Organotins on Growth and Survival of Two Marine Diatoms, *Skeletonema costatum* and *Thalassiosira pseudonana*," *Chemosphere* 14:383-392 (1985).
14. Waldock, M. J., M. E. Waite, and J. E. Thain. "Inputs of TBT to the Marine Environment from Shipping Activity in the U.K.," *Environ. Technol. Lett.* 9:999-1010 (1988).
15. Cleary, J. J. and A. R. D. Stebbing. "Organotin in the Surface Microlayer and Subsurface Waters of Southwest England," *Mar. Pollut. Bull.* 18:238-246 (1987).
16. Langston, W. J., G. R. Burt, and Zhou Mingjiang. "Tin and Organotin in Water, Sediment and Benthic Organisms of Poole Harbour," *Mar. Pollut. Bull.* 18:634-639 (1987).
17. Hardy, J., S. Kiesser, L. Antrim, A. Stubin, R. Kocan, and J. Strand. "The Sea-Surface Microlayer of Puget Sound: Part 1. Toxic Effects on Fish Eggs and Larvae," *Mar. Environ. Res.* 23:227-249 (1987).

18. Matthias, C. L., S. J. Bushong, L. W. Hall, J. M. Bellama, and F. E. Brinckman. "Simultaneous Butyltin Determination in the Microlayer, Water Column and Sediment of a Northern Chesapeake Bay Marina and Receiving System," *Appl. Organometal. Chem.* 2:547-552 (1988).

19. Seligman, P. F., J. G. Grovhoug, A. O. Valkirs, P. M. Stang, R. Fransham, M. O. Stallard, B. Davidson, and R. F. Lee. "Distribution and Fate of Tributyltin in the United States Marine Environment," *Appl. Organometal. Chem.* 3:31-47 (1989).

20. Stang, P. M., D. R. Bower, and P. F. Seligman. "Stratification and Tributyltin Variability in San Diego Bay," *Appl. Organometal. Chem.* 3:411-416 (1989).

21. Batley, G. E., K. J. Mann, C. I. Brockbank, and A. Maltz. "Tributyltin in Sydney Harbour and Georges River Waters," *Aust. J. Mar. Freshwater Res.* 40:39-48 (1989).

22. Bacci, E. and C. Gaggi. "Organotin Compounds in Harbour and Marine Waters from the Northern Tyrrhenian Sea," *Mar. Pollut. Bull.* 20:290-292 (1989).

23. Hall, L. W., M. J. Lenkevich, W. S. Hall, A. E. Pinkney, and S. J. Bushong. "Evaluation of Butyltin Compounds in Maryland Waters of Chesapeake Bay," *Mar. Pollut. Bull.* 18:78-83 (1987).

24. Lee, R. F., A. O. Valkirs, and P. F. Seligman. "Importance of Microalgae in the Degradation of Tributyltin in Estuarine Waters," *Environ. Sci. Technol.* 23:1515-1518 (1989).

25. Thain, J., M. J. Waldock, and M. E. Waite. "Toxicity and Degradation Studies of Tributyltin (TBT) and Dibutyltin (DBT) in the Aquatic Environment," in *Oceans '87 Proceedings, Vol 4. International Organotin Symposium* (New York:Institute of Electrical and Electronics Engineers, 1987), pp. 1398-1404.

26. Hattori, Y., A. Kobayashi, K. Nonaka, A. Sugimae, and M. Nakamoto. "Degradation of Tributyl Tin and Dibutyl Tin Compounds in Environmental Waters," *Water Sci. Technol.* 20:71-76 (1988).

27. Francois, R., F. T. Short, and J. H. Weber. "Accumulation and Persistence of Tributyltin in Eelgrass *(Zostera marina* L.)," *Environ. Sci. Technol.* 23:191-196 (1989).

28. Harris, J. R. W. and J. J. Cleary. "Particle-Water Partitioning and Organotin Dispersal in an Estuary," in *Oceans '87 Proceedings, Vol 4. International Organotin Symposium* (New York:Institute of Electrical and Electronics Engineers, 1987), pp. 1370-1374.

29. Kram, M. L., P. M. Stang, and P. F. Seligman. "Adsorption and Desorption of Tributyltin in Sediments of San Diego Bay and Pearl Harbor," *Appl. Organometal. Chem.* 3:523-536 (1989).

30. Unger, M. A., W. G. MacIntyre, and R. J. Huggett. "Sorption Behaviour of Tributyltin on Estuarine and Freshwater Sediments," *Environ. Toxicol. Chem.* 7:907-915 (1988).

31. Stang, P. M. and P. F. Seligman. "Distribution and Fate of Butyltin Compounds in the Sediment of San Diego Bay," in *Oceans '86 Proceedings, Vol 4. International Organotin Symposium* (New York:Institute of Electrical and Electronics Engineers, 1986), pp. 1256-1261.

32. Waldock, M. J., J. E. Thain, D. Smith, and S. Milton. "The Degradation of TBT in Estuarine Sediments," Proceedings of the 3rd International Organotin Symposium, Monaco (1990), pp. 46-48.

33. Laughlin, R. B. "Bioaccumulation of Tributyltin: The Link Between Environment and Organisms," in *Oceans '86 Proceedings, Vol 4. International Organotin Symposium* (New York:Institute of Electrical and Electronics Engineers, 1986), pp. 1206-1209.

34. Laughlin, R. B., R. B. Johannesen, W. French, H. Guard, and F. E. Brinckman. "Structure-Activity Relationships for Organotin Compounds," *Environ. Toxicol. Chem.* 4:343-351 (1985).

35. Chiles, T. C., P. D. Pendoley, and R. B. Laughlin. "Mechanisms of Tri-n-butyltin Bioaccumulation by Marine Phytoplankton," *Can. J. Fish. Aquat. Sci.* 46:859-862 (1989).

36. Laughlin, R. B. and W. French. "Concentration Dependence of Bis(tributyl)tin Oxide Accumulation in the Mussel, *Mytilus edulis,*" *Environ. Toxicol. Chem.* 7:1021-1026 (1988).

37. Bryan, G. W., P. E. Gibbs, G. R. Burt, and L. G. Hummerstone. "The Effects of Tributyltin (TBT) Accumulation on Adult Dog-Whelks, *Nucella lapillus:* Long-Term Field and Laboratory Experiments," *J. Mar. Biol. Assoc. U.K.* 67:525-544 (1987).

38. Bryan, G. W., P. E. Gibbs, L. G. Hummerstone, and G. R. Burt. "Uptake and Transformation of 14C-Labelled Tributyltin Chloride by the Dog-Whelk, *Nucella lapillus:* Importance of Absorption from the Diet," *Mar. Environ. Res.* 28:241-245 (1989).

39. Laughlin, R. B., W. French, and H. E. Guard. "Accumulation of Bis(tributyltin) Oxide by the Marine Mussel *Mytilus edulis,*" *Environ. Sci. Technol.* 20:884-890 (1986).

40. Evans, D. W. and R. B. Laughlin. "Accumulation of Bis(tributyltin) Oxide by the Mud Crab, *Rhithropanopeus harrisii,*" *Chemosphere* 13:213-219 (1984).

41. Langston, W. J., G. W. Bryan, G. R. Burt, and P. E. Gibbs. "Assessing the Impact of Tin and TBT in Estuaries and Coastal Regions," *Funct. Ecol.* 4:433-443 (1990).

42. Langston, W. J. "Bioavailability and Effects of TBT in Deposit-Feeding Clams, *Scrobicularia plana,*" Proceedings of the 3rd International Organotin Symposium, Monaco (1990), pp. 110-113.

43. Lee, R. F. "Metabolism of Tributyltin Oxide by Crabs, Oysters and Fish," *Mar. Environ. Res.* 17:145-148 (1985).

44. Lee, R. F. "Metabolism of Bis(tributyltin) Oxide by Estuarine Animals," in *Oceans '86 Proceedings, Vol 4. International Organotin Symposium* (New York:Institute of Electrical and Electronics Engineers, 1986), pp. 1182-1188.

45. Rice, S. D., J. W. Short, and W. B. Stickle. "Uptake and Catabolism of Tributyltin by Blue Crabs Fed TBT-Contaminated Prey," *Mar. Environ. Res.* 27:137-145 (1989).

46. Lee, R. F. "Metabolism and Accumulation of Xenobiotics Within Hepatopancreas Cells of the Blue Crab, *Callinectes sapidus,*" *Mar. Environ. Res.* 28:93-97 (1989).

47. Ward, G. S., G. C. Cramm, P. R. Parrish, H. Trachman, and A. Slesinger. "Bioaccumulation and Chronic Toxicity of Bis(tributyltin)oxide (TBTO): Tests with a Saltwater Fish," *Aquatic Toxicology and Hazard Assessment, 4th Conf.* D. R. Branson and K. L. Dickson, Eds., ASTM S.T.P. (American Society for Testing and Materials, 1981), pp. 183-200.

48. Martin, R. C., D. G. Dixon, R. J. Maguire, P. V. Hodson, and R. J. Tkacz. "Acute Toxicity, Uptake, Depuration and Tissue Distribution of Tri-n-butyltin in Rainbow Trout *Salmo gairdneri*," *Aquat. Toxicol.* 15:37-52 (1989).

49. Butler, P. A. "Influence of Pesticides on Marine Ecosystems," *Proc. R. Soc. London B.* 177:321-329 (1971).

50. Osborn, D. and D. V. Leach. "Organotin in Birds: A pilot Study." Final Rep. Dept. of the Environment, Institute of Terrestrial Ecology (1987), p. 15.

51. Evans, L. V. "Marine Algae and Fouling: A Review with Particular Reference to Ship-Fouling," *Bot. Mar.* 24:161-171 (1981).

52. Beaumont, A. R. and P. B. Newman. "Low Levels of Tributyltin Reduce Growth of Marine Micro-algae," *Mar. Pollut. Bull.* 17:457-461 (1986).

53. Laughlin, R. B., R. Gustafson, and P. Pendoley. "Chronic Embryo-Larval Toxicity of Tributyltin (TBT) and the Hard Shell Clam *Mercenaria mercenaria*," *Mar. Ecol. Prog. Ser.* 48:29-36 (1988).

54. Hall, L. W., S. J. Bushong, M. C. Ziegenfuss, and W. E. Johnson. "Chronic Toxicity of Tributyltin to Chesapeake Bay Biota," *Water Air Soil Pollut.* 39:365-376 (1988).

55. Bushong, S. J., L. W. Hall, W. S. Hall, W. E. Johnson, and R. L. Herman. "Acute Toxicity of Tributyltin to Selected Chesapeake Bay Fish and Invertebrates," *Water Res.* 22:1027-1032 (1988).

56. Bryan, G. W., P. E. Gibbs, L. G. Hummerstone, and G. R. Burt. "The Decline of the Gastropod *Nucella lapillus* Around South-West England: Evidence for the Effect of Tributyltin from Antifouling Paints," *J. Mar. Biol. Assoc. U.K.* 66:611-640 (1986).

57. Henderson, R. S. "Effects of Tributyltin Antifouling Paints Leachates on Pearl Harbor Organisms," Naval Ocean Systems Centre, San Diego, California, Tech. Rep. No. 1079 (1985) p. 25.

58. Stebbing, A. R. D. "Hormesis-Stimulation of Colony Growth in *Campanularia flexuosa* (Hydrozoa) by Copper, Cadmium and Other Toxicants," *Aquat. Toxicol.* 1:227-238 (1981).

59. Stebbing, A. R. D. and B. E. Brown. "Marine Ecotoxicological Tests with Coelenterates," in *Ecotoxicological Testing for the Marine Environment, Vol. 1,* G. Persoone, E. Jaspers, and C. Claus, Eds. (Bredene, Belgium:State University, Ghent and Inst. Mar. Scient. Res., 1984), pp. 307-339.

60. Lenihan, H. S., J. S. Oliver, and M. A. Stephenson. "Changes in Hard Bottom Communities Related to Boat Moorings and Tributyltin in San Diego Bay: A Natural Experiment," *Mar. Ecol. Prog. Ser.* 60:147-159 (1990).

61. Walsh, G. E., M. K. Louie, L. L. McLaughlin, and E. M. Lores. "Lugworm *(Arenicola cristata)* Larvae in Toxicity Tests: Survival and Development When Exposed to Organotins," *Environ. Toxicol. Chem.* 5:749-754 (1986).

62. Matthiessen, P., and J. E. Thain. "A Method for Studying the Impact of Polluted Marine Sediment on Intertidal Colonising Organisms; Tests with Diesel-Based Drilling Mud and Tributyltin Antifouling Paint," *Hydrobiologia* 188/189:477-485 (1989).

63. U'ren, S. "Acute Toxicity of Bis(tributyltin)oxide to a Marine Copepod," *Mar. Pollut. Bull.* 14:303-306 (1983).

64. Johansen, K. and F. Møhlenberg. "Impairment of Egg Production in *Acartia tonsa* Exposed to Tributyltin Oxide," *Ophelia* 27:137-141 (1987).

65. Bushong, S. J., M. C. Ziegenfuss, M. A. Unger, and L. W. Hall. "Chronic Tributyltin Toxicity Experiments with the Chesapeake Bay Copepod *Acartia tonsa*," *Environ. Toxicol. Chem.* 9:359-366 (1990).

66. Hall, L. W. "Tributyltin Environmental Studies in Chesapeake Bay," *Mar. Pollut. Bull.* 19:431-438 (1988).

67. Davidson, B. M., A. O. Valkirs, and P. F. Seligman. "Acute and Chronic Effects of Tributyltin on the Mysid *Acanthomysis sculpta* (Crustacea, Mysidacea)," in *Oceans '86 Proceedings, Vol 4. International Organotin Symposium* (New York:Institute of Electrical and Electronics Engineers, 1986), pp. 1219-1225.

68. Laughlin, R., K. Nordlund, and O. Linden. "Long-Term Effects of Tributyltin on the Baltic Amphipod, *Gammarus oceanicus*," *Mar. Environ. Res.* 12:243-271 (1984).

69. Laughlin, R. B. and W. French. "Population-Related Toxicity Responses to Two Butyltin Compounds by Zoeae of the Mud Crab *Rhithropanopeus harrisii*," *Mar. Biol.* 102:397-401 (1989).

70. Weis, J. S., J. Gottlieb, and J. Kwiatkowski. "Tributyltin Retards Regeneration and Produces Deformities of Limbs in the Fiddler Crab, *Uca pugilator*," *Arch. Environ. Contam. Toxicol.* 16:321-326 (1987).

71. Weis, J. S. and K. Kim. "Tributyltin is a Teratogen in Producing Deformities in Limbs of the Fiddler Crab, *Uca pugilator*," *Arch. Environ. Contam. Toxicol.* 17:583-587 (1988).

72. Weis, J. S. and J. Perlmutter. "Effects of Tributyltin on Activity and Burrowing Behaviour of the Fiddler Crab *Uca pugilator*," *Estuaries* 10:342-346 (1987).

73. Walsh, G. E., L. L. McLaughlin, M. K. Louie, C. H. Deans, and E. M. Lores. "Inhibition of Arm Regeneration by *Ophioderma brevispina* (Echinodermata, Ophiuroidea) by Tributyltin Oxide and Triphenyltin Oxide," *Ecotoxicol. Environ. Safety* 12:95-100 (1986).

74. Newton, F., A. Thum, B. Davidson, A. Valkirs, and P. Seligman. "Effects on the Growth and Survival of Eggs and Embryos of the California Grunion *(Leuresthes tenius)* Exposed to Trace Levels of Tributyltin," U.S. Naval Ocean Systems Center, San Diego, California, Tech Rep. No. 1040 (1985), p. 17.

75. Pinkney, A. E., L. L. Matteson, and D. A. Wright. "Effects of Tributyltin on Survival, Growth, Morphometry and RNA-DNA Ratio of Larval Striped Bass, *Morone saxatilis*," in *Oceans '88 Proceedings, Vol 4. International Organotin Symposium* (New York:Institute of Electrical and Electronics Engineers, 1988), pp. 987-991.

76. Pinkney, A. E., D. A. Wright, M. A. Jepson, and D. W. Towle. "Effects of Tributyltin Compounds on Ionic Regulation and Gill ATPase Activity in Estuarine Fish," *Comp. Biochem. Physiol.* 92C:125-129 (1989).

77. Pinkney, A. E., D. A. Wright, and G. M. Hughes. "A Morphometric Study of the Effects of Tributyltin Compounds on the Gills of the Mummichog, *Fundulus heteroclitus*," *J. Fish Biol.* 34:665-677 (1989).

78. Alzieu, C. "TBT Detrimental Effects on Oyster Culture in France — Evolution Since Antifouling Paint Regulation," in *Oceans '86 Proceedings, Vol 4. International Organotin Symposium* (New York:Institute of Electrical and Electronics Engineers, 1986), pp. 1130-1134.

79. Thomas, M. L. H. "Experiments in the Control of Shipworm *Teredo* sp. Using TBTO," *Fish. Res. Bd. Can. Tech. Rep. No. 21*, p. 27 (1967).

80. Waldock, M. J. and J. E. Thain. "Shell Thickening in *Crassostrea gigas:* Organotin Antifouling or Sediment Induced," *Mar. Pollut. Bull.* 14:411-415 (1983).

81. Alzieu, C., J. Sanjuan, P. Michel, M. Borel, and J. P. Dreno. "Monitoring and Assessment of Butyltins in Atlantic Coastal Waters," *Mar. Pollut. Bull.* 20:22-26 (1989).

82. Chagot, D., C. Alzieu, J. Sanjuan, and H. Grizel. "Sublethal and Histopathological Effects of Trace Levels of Tributyltin Fluoride on Adult Oysters *Crassostrea gigas," Aquat. Living Resour.* 3:121-130 (1990).

83. Davies, I. M., J. Drinkwater, J. C. McKie, and P. Balls. "Effects of the Use of Tributyltin Antifoulants in Mariculture," in *Oceans '87 Proceedings, Vol 4. International Organotin Symposium* (New York:Institute of Electrical and Electronics Engineers, 1987), pp. 1477-1481.

84. Minchin, D., C. B. Duggan, and W. King. "Possible Effects of Organotins on Scallop Recruitment," *Mar. Pollut. Bull.* 18:604-608 (1987).

85. King, N., M. Miller, and S. de Mora. "Tributyltin Levels for Sea Water, Sediment and Selected Marine Species in Coastal Northland and Auckland, New Zealand," *N.Z. J. Mar. Freshwater Res.* 23:287-294 (1989).

86. Smith, D. R., M. D. Stephenson, J. Goetzi, G. Ichikawa, and M. Martin. "The Use of Transplanted Juvenile Oysters to Monitor the Toxic Effects of Tributyltin in California Waters," in *Oceans '87 Proceedings, Vol 4. International Organotin Symposium* (New York:Institute of Electrical and Electronics Engineers, 1987), pp. 1511-1516.

87. Wolniakowski, K., M. D. Stephenson, and G. S. Ichikawa. "Tributyltin Concentrations and Pacific Oyster Deformation in Coos Bay, Oregon," in *Oceans, '87 Proceedings, Vol. 4. International Organotin Symposium* (New York:Institute of Electrical and Electronics Engineers, 1987), pp. 1438-1442.

88. Machado, J., J. Coimbra, and C. Sá. "Shell Thickening in *Anodonta cygnea* by TBTO Treatments," *Comp. Biochem. Physiol.* 29C:77-80 (1989).

89. Thain, J. E. and M. J. Waldock. "The Impact of Tributyl Tin (TBT) Antifouling Paints on Molluscan Fisheries," *Water Sci. Technol.* 18:193-202 (1986).

90. Marcus, J. M., G. I. Scott, and D. D. Heizer. "The Use of Oyster Shell Thickness and Condition Index Measurements as Physiological Indicators of No Heavy Metal Pollution Around Three Coastal Marinas," *J. Shellfish Res.* 8:87-94 (1989).

91. His, E., D. Maurer and R. Robert. "Estimation de La Teneur en Acétate de Tributylétain dans L'Eau de Mer, par Une Méthod Biologique," *J. Moll. Stud. Suppl.* 12A:60-68 (1983).

92. Lawler, I. F. and J. C. Aldrich. "Sublethal Effects of Bis(tri-n-butyltin)oxide on *Crassostrea gigas* Spat," *Mar. Pollut. Bull.* 18:274-278 (1987).

93. Salazar, S. M., B. M. Davidson, M. H. Salazar, P. M. Stang, and K. J. Meyers-Schulte. "Effects of TBT on Marine Organisms: Field Assessment of a New Site-Specific Bioassay System," in *Oceans '87 Proceedings, Vol 4. International Organotin Symposium* (New York:Institute of Electrical and Electronics Engineers, 1987), pp. 1461-1470.

94. Alzieu, C., J. Sanjuan, J. P. Deltreil, and M. Borel. "Tin Contamination in Arcachon Bay: Effects on Oyster Shell Abnormalities," *Mar. Pollut. Bull.* 15:494-498 (1986).

95. Thain, J. E. "Toxicity of TBT to Bivalves: Effect on Reproduction, Growth and Survival," in *Oceans '86 Proceedings, Vol 4. International Organotin Symposium* (New York:Institute of Electrical and Electronics Engineers, 1986), pp. 1306-1313.

96. Roberts, M. H., M. E. Bender, P. F. DeLisle, H. C. Sutton, and R. L. Williams. "Sex Ratio and Gamete Production in American Oysters Exposed to Tributyltin in the Laboratory," in *Oceans '87 Proceedings, Vol 4. International Organotin Symposium* (New York:Institute of Electrical and Electronics Engineers, 1987), pp. 1471-1476.

97. Davies, I. M., J. C. McKie, and J. D. Paul. "Accumulation of Tin and Tributyltin from Anti-Fouling Paint by Cultivated Scallops *(Pecten maximus)* and Pacific Oysters (Crassostrea gigas)," *Aquaculture* 55:103-114 (1986).

98. Beaumont, A. R. and M. D. Budd. "High Mortality of the Larvae of the Common Mussel at Low Concentrations of Tributyltin," *Mar. Pollut. Bull.* 15:402-405 (1984).

99. Strømgren, T. and T. Bongard. "The Effect of Tributyltin Oxide on Growth of *Mytilus edulis,"* *Mar. Pollut. Bull.* 18:30-31 (1987).

100. Salazar, M. H. and S. M. Salazar. "Tributyltin and Mussel Growth in San Diego Bay," in *Oceans '88 Proceedings, Vol 4. International Organotin Symposium* (New York:Institute of Electrical and Electronics Engineers, 1988), pp. 1188-1197.

101. Widdows, J. and P. Donkin. "The Application of Combined Tissue Residue Chemistry and Physiological Measurements of Mussels *(Mytilus edulis)* for the Assessment of Environmental Pollution," *Hydrobiologia* 188/189:455-461 (1989).

102. Widdows, J., K. A. Burns, N. R. Menon, D. S. Page, and S. Soria. "Measurements of Physiological Energetics (Scope for Growth) and Chemical Contaminants in Mussels *(Arca zebra)* Transplanted Along a Contamination Gradient in Bermuda," *J. Exp. Mar. Biol. Ecol.* 138:99-117 (1990).

103. Smith, B. S. "Male Characteristics in Female Mud Snails Caused by Antifouling Bottom Paints," *J. Appl. Toxicol.* 1:22-25 (1981).

104. Smith, B. S. "Sexuality in the American Mud Snail, *Nassarius obsoletus* (Say)," *Proc. Malacol. Soc. London* 39:377-378 (1971).

105. Bryan, G. W., P. E. Gibbs, R. J. Huggett, L. A. Curtis, D. S. Bailey, and D. M. Dauer "The Effects of Tributyltin Pollution on the Mud Snail, *Ilyanassa obsoleta,* from the York River and Sarah Creek, Chesapeake Bay," *Mar. Pollut. Bull.* 10:458-462 (1989).

106. Blaber, S. J. M. "The Occurrence of a Penis-Like Outgrowth Behind the Right Tentacle in Spent Females of *Nucella lapillus* (L.)," *Proc. Malacol. Soc. London* 39:231-233 (1970).

107. Poli, G., B. Salvat, and W. Strieff. "Aspect Particulier de la Sexualité Chez *Ocenebra erinacea,"* *Haliotis* 1:29-30 (1971).

108. Houston, R. S. "Reproductive Biology of *Thais emarginata* (Deshayes, 1839) and *Thais canaliculata* (Duclos, 1832)," *Veliger* 13:348-357 (1971).

109. Jenner, M. G. "Pseudohermaphroditism in *Ilyanassa obsoleta* (Mollusca: Neogastropoda)," *Science* 205:1407-1409 (1979).

110. Smith, B. S. "Reproductive Anomalies in the Stenoglossan Snails Related to Pollution from Marinas," *J. Appl. Toxicol.* 1:15-21 (1981).

111. Ellis, D. V. and L. A. Pattisina. "Widespread Neogastropod Imposex: A Biological Indication of Global TBT Contamination," *Mar. Pollut. Bull.* 21:248-253 (1990).

112. Gibbs, P. E., G. W. Bryan, P. L. Pascoe, and G. R. Burt. "The Use of the Dog-Whelk, *Nucella lapillus,* as an Indicator of Tributyltin (TBT) Contamination," *J. Mar. Biol. Assoc. U. K.* 67:507-523 (1987).

113. Bailey, S. K. and I. M. Davies. "The Effects of Tributyltin on Dog-Whelks *(Nucella lapillus)* from Scottish Coastal Waters," *J. Mar. Biol. Assoc. U. K.* 69:335-354 (1989).

114. Spence, S. K., G. W. Bryan, P. E. Gibbs, D. Masters, L. Morris, and S. J. Hawkins. "Effects of TBT Contamination on *Nucella* Populations," *Funct. Ecol.* 4:425-432 (1990).

115. Gibbs, P. E., P. L. Pascoe, and G. R. Burt. "Sex Change in the Female Dog-Whelk, *Nucella lapillus,* Induced by Tributyltin from Antifouling Paints," *J. Mar. Biol. Assoc. U. K.,* 68:715-731 (1988).

116. Bryan, G. W., P. E. Gibbs, and G. R. Burt. "A Comparison of the Effectiveness of Tri-*n*-butyltin Chloride and Five Other Organotin Compounds in Promoting the Development of Imposex in the Dog-Whelk, *Nucella lapillus," J. Mar. Biol. Assoc. U. K.* 68:733-744 (1988).

117. Gibbs, P. E. and G. W. Bryan. "TBT-Induced Imposex in the Dog-Whelk, *Nucella lapillus:* Aspects of its Regional Distribution," in Proceedings of the 3rd International Organotin Symposium, Monaco (1990) pp. 98-104.

118. Bailey, S. K. and I. M. Davies. "Tributyltin Contamination in the Firth of Forth (1975-87)," *Sci. Total Environ.* 67:185-192 (1988).

119. Southward, A. J. and E. C. Southward. "Disappearance of the Warm-Water Hermit Crab *Clibanarius erythropus* from South-West Britain," *J. Mar. Biol. Assoc. U. K.* 68:409-412 (1988).

120. Crothers, J. H. "Has the Population Decline Due to TBT Pollution Affected Shell-Shape Variations in the Dog-Whelk, *Nucella lapillus* (L.)?," *J. Molluscan Stud.* 55:461-467 (1989).

121. Short, J. W., S. D. Rice, C. C. Brodersen, and W. B. Stickle. "Occurrence of Tri-*n*-Butyltin-Caused Imposex in the North Pacific Marine Snail *Nucella lima* in Auke Bay, Alaska," *Mar. Biol.* 102:291-297 (1989).

122. Bright, D. A. and D. V. Ellis. "A Comparative Survey of Imposex in N. E. Pacific Neogastropods (Prosobranchia) Related to Tributyltin Contamination, and Choice of A Suitable Bio-indicator," *Can. J. Zool.* 68:1915-1924 (1990).

123. Saavedra Alvarez, M. M. and D. V. Ellis. "Widespread Neogastropod Imposex in the Northeast Pacific: Implications for TBT Contamination Surveys," *Mar. Pollut. Bull.* 21:244-247 (1990).

124. Féral, C. "Analyse Expérimentale de la Morphogénèse et du Cycle du Penis Chez Les Femelles d' *Ocenebra erinacea,* Espèce Gonochorique," *Haliotis* 6:267-271 (1976).

125. Féral, C. and S. Le Gall. "The Influence of a Pollutant Factor (Tributyltin) on the Neuroendocrine Mechanism Responsible for the Occurrence of a Penis in the Females of *Ocenebra erinacea,*" in *Molluscan Neuro-endocrinology: Proceedings of the International Minisymposium on Molluscan Endocrinology,* J. Lever and H. H. Boer, Eds. (Amsterdam:North Holland 1982), pp. 173-175.

126. Gibbs, P. E., G. W. Bryan, P. L. Pascoe, and G. R. Burt. "Reproductive Abnormalities in Female *Ocenebra erinacea* (Gastropoda) Resulting from Tributyltin-Induced Imposex," *J. Mar. Biol. Assoc. U. K.* 70:639-656 (1990).

127. Spence, S. K., S. J. Hawkins, and R. S. Santos. "The Mollusc *Thais haemastoma* — An Exhibitor of "Imposex" and Potential Biological Indicator of Tributyltin Pollution," *Mar. Ecol. Naples,* 11:147-156 (1990).

128. Stewart, C., S. J. de Mora, M. R. L. Jones, and M. C. Miller. "Degree of Imposex and Tri(n-butyl)tin Body Burdens in Three Indigenous Species of Marine Gastropods in Auckland, New Zealand," *Sci. Total Environ.* (1991), in press.

129. Laughlin, R. B. and O. Linden. "Tributyltin-Contemporary Environmental Issues," *Ambio* 16:252-256 (1987).

130. Spooner, N. and L. J. Goad. "Steroids and Imposex in the Dog Whelk *Nucella lapillus,*" in Proceedings of the 3rd International Organotin Symposium, Monaco (1990) pp. 88-92.

131. Lee, R. F. "Possible Linkages Between Mixed Function Oxygenase Systems, Steroid Metabolism, Reproduction, Moulting and Pollution in Aquatic Animals," in *Toxic Contaminants and Ecosystem Health: A Great Lakes Focus,* Adv. Environ. Sci. Tech. Vol. 21 M. S. Evans, Ed. (New York:John Wiley & Sons, 1988), pp. 201-213.

132. Cardwell, R. D. and A. W. Sheldon. "A Risk Assessment Concerning the Fate and Effects of Tributyltins in the Aquatic Environment," in *Oceans '86 Proceedings, Vol 4. International Organotin Symposium* (New York:Institute of Electrical and Electronics Engineers, 1986), pp. 1117-1129.

133. Gucinski, H. "The Effect of Sea Surface Microlayer Enrichment on TBT Transport," in *Oceans '86 Proceedings, Vol. 4. International Organotin Symposium,* New York:Institute of Electrical and Electronics Engineers, 1986), pp. 1266-1274.

134. Kocan, R. M., H. von Westernhagen, H. L. Landolt, and G. Furstenberg. "Toxicity of Sea-Surface Microlayer: Effects of Hexane Extract on Baltic Herring *(Clupea harengus)* and Atlantic Cod *(Gadus morhua)* Embryos," *Mar. Environ. Res.* 23:291-305 (1987).

13

Community Responses of Stream Organisms to Heavy Metals: A Review of Observational and Experimental Approaches

William H. Clements

Department of Fishery and Wildlife Biology, Colorado State University, Fort Collins, Colorado 80523

OVERVIEW

Laboratory studies investigating acute and chronic toxicity of metals, modes of toxicity, tolerance mechanisms and environmental factors that influence metals toxicity have greatly increased our knowledge of how these contaminants impact aquatic organisms. By comparison, our understanding of how metals affect natural populations, communities, and ecosystems in the field is greatly limited. Field investigations on the impact of heavy metals generally fall into two categories: descriptive and experimental. Descriptive approaches, such as routine biomonitoring of community structure and measurement of levels of metals in organisms, are the most commonly employed techniques for assessing the biological integrity of streams. Biomonitoring approaches are often limited by the lack of appropriate spatial and/or temporal controls, therefore making it difficult to attribute observed changes to specific causes. Experimental approaches, including the use of stream mesocosms and introduction of contaminants into natural systems, provide the strongest evidence for a causal relationship between metals contamination and changes in benthic communities. The purpose of this chapter is to examine community responses of benthic organisms (periphyton and macroinvertebrates) to heavy metals in streams. The advantages and limitations of descriptive (biomonitoring) and experimental (microcosms, mesocosms, field experiments) approaches for assessing the impact of metals on these communities in the field will be discussed. The application of outdoor stream mesocosms for predicting effects of metals on benthic communities and for developing an index of metals impact will be described. Finally, the potential for food chain transfer of metals from benthic invertebrates to fish will be examined.

INTRODUCTION

Over the past several decades, a wealth of data have been generated on the effects of heavy metals on aquatic organisms and the fate of these contaminants in aquatic systems. Numerous laboratory investigations have contributed significantly to our understanding of acute and chronic toxicity, sublethal effects, modes of action, tolerance mechanisms, and various environmental factors that influence metals toxicity for a diverse assortment of organisms. Perhaps more than any other group of contaminants, our knowledge of heavy metals toxicity has increased dramatically in recent years. Several reviews describing the fate and effects of heavy metals on aquatic organisms have been published.[1-8]

Most research on heavy metals toxicity has been conducted in the laboratory and therefore our understanding of how metals impact natural populations, communities, and ecosystems is greatly limited. The relative lack of information on effects of these contaminants on higher levels of biological organization is certainly not unique to metals. Kimball and Levin[9] surveyed 699 papers published in ecotoxicology-related journals between 1980 to 1982 and reported that only 12% of these papers examined population- or ecosystem-level effects.

Several potential reasons for the reluctance of aquatic toxicologists to employ more environmentally realistic procedures have been suggested. Historically, there has been relatively little overlap between the fields of aquatic toxicology and ecology.[10] Odum[11] discusses the philosophical differences between the reductionist approaches employed by traditional toxicologists and the holistic approaches used by population, community, and ecosystem ecologists. The driving force of traditional aquatic toxicology has been regulatory pressure for measurement of effects and establishment of criteria. As a result, a high premium has been placed on standardization and there has been less opportunity for conducting basic research in this field.[12] Single-species toxicity tests have been employed to generate LC_{50} values, no observed effect concentrations (NOEC), and maximum allowable toxicant concentrations (MATC), for a large number of aquatic organisms under a variety of experimental conditions. Because of their simplicity, low cost, and potential for standardization, single-species toxicity tests play a major role in establishment of water quality criteria for metals.

These tests continue to dominate the field of aquatic toxicology, despite recent developments in the emerging field of ecotoxicology that suggest single-species tests may not be protective of higher levels of biological organization.[13,14] Unfortunately, there is little evidence to support the hypothesis that end points employed in these tests, such as mortality, growth, and reproduction, are protective of natural communities.[15] Field validation of laboratory findings occurs infrequently. Furthermore, examples of indirect effects of contaminants on community- and ecosystem-level processes, transformation of chemicals in the environment, and movement of contaminants through food chains are relatively common in the literature and suggest that an ecosystem perspective is necessary.[9]

A number of investigators have noted the limitations of single-species toxicity tests and have suggested that these procedures be supplemented with more environmentally realistic techniques to assess impact on populations, communities, and ecosystems.[9,11,15,16] Although a variety of approaches have been employed to investigate effects of heavy metals in the field, these approaches generally fall into two categories: descriptive and experimental. Descriptive approaches include routine biomonitoring, whereas experimental approaches include the use of microcosms, mesocosms, and experimental introduction of metals into natural systems.

This chapter will examine descriptive and experimental approaches for investigating community-level responses of aquatic organisms to heavy metals. It will be limited to lotic systems and emphasis will be placed on periphyton and macroinvertebrate communities. Because these organisms readily bioaccumulate metals and therefore represent a potential pathway to higher trophic levels, the importance of food chain transfer of heavy metals from benthic invertebrates to fish will be discussed.

Biomonitoring Approaches

Clearly, one of the oldest and most widely used approaches for assessing the integrity of streams is biomonitoring. Based on the assumptions that the structure of biological communities present in a stream is a reflection of relative health and that we can define a "healthy" community, numerous studies have been conducted to establish a relationship between species composition and the degree of impact. In one of the earlier field investigations, Carpenter[17,18] reported results of a series of studies on the effects of Pb mining on stream fauna. More recently, distribution and abundance of periphyton,[19,20] protozoans,[21] macroinvertebrates,[16,22-27] and fish[28,29] in streams have been employed in biomonitoring studies.

Concentrations of heavy metals in aquatic organisms are also suggested as indicators of contamination.[30] Many aquatic organisms accumulate metals to concentrations much greater than those found in overlying water. Since concentrations of metals in water are highly variable and often below detection, a predictive relationship between heavy metals in water and organisms may be useful for monitoring heavy metal contamination. However, Moriarty et al.[31]

have noted several problems with this approach and suggested that it may be more appropriate to analyze abiotic samples than plant and animal tissues for detecting metals contamination.

Limitations of Biomonitoring

Stream biomonitoring approaches for evaluating the impact of contaminants typically involve comparison of upstream reference sites to downstream impacted and recovery sites. Ideally, these locations should be similar in all respects except for the presence of contaminants.[32] However, because of natural changes in structural and functional parameters along longitudinal stream gradients,[33] as well as variation in other parameters such as substrate composition, current velocity, stream gradient, and riparian vegetation, it is often difficult to locate comparable up- and downstream sites. Effects due to the presence of contaminants are therefore confounded by microhabitat and longitudinal variation in structural and functional characteristics. Crossey and La Point[34] note the problems associated with determining changes in complex systems due to spatial and temporal variability, particularly if contaminant effects are subtle. In systems receiving impacts from several sources, determining specific causes for changes in downstream communities is greatly complicated.

Green[32] describes an optimal sampling design for conducting biomonitoring studies, and notes the importance of obtaining pre- and postimpact data from both reference and impacted sites. Spatial *and* temporal controls are necessary in this design to test the null hypothesis of no change due to impact, where a significant interaction between location and time indicates impact has occurred.[32] Hurlbert[35] criticizes this design and concludes that inferential statistics are not appropriate in most biomonitoring studies because of the problem of temporal and spatial pseudoreplication. Green's[32] design, argues Hurlbert, allows for the determination of significant differences among locations, but these differences cannot be attributed to a particular discharge. Recently, Stewart-Oaten et al.[36] have suggested a way to salvage this design by collecting multiple samples, pseudoreplicated in time, before and after impact. Inferential statistics are then employed to test for a change in the difference between control and impacted sites. While this approach may solve the problem of attributing observed changes to perturbation, generalizations among streams will be limited. In other words, because treatment and control streams are not replicated, it is not statistically valid to generalize among streams. This problem, however, may not be a severe limitation of biomonitoring studies since impact assessments are often site-specific and, therefore, it is usually not necessary to extend results to other systems.[36]

Regardless of the statistical validity of the optimal impact designs described above, the fact remains that preimpact data are rarely available in most biomonitoring studies. Consequently, investigators simply monitor changes in communities during or after impact has occurred and assume a causal relationship. From a regulatory perspective, if a stream receives discharges from several different sources, the demonstration of causality is often critical. In the absence of preimpact data, an alternative to descriptive surveys of stream communities is experimentation. Experimental approaches involve the use of microcosms, mesocosms, or introduction of contaminants into natural systems. Results obtained from such experimental studies provide the strongest evidence for causal relationships between concentrations of heavy metals and changes in community structure of lotic systems.[24]

COMMUNITY RESPONSES OF ALGAE AND PERIPHYTON
Descriptive Approaches

Responses of algae and periphyton communities to heavy metals in streams have received considerable attention. Much of the early research on metals pollution emphasized the use of freshwater algae as indicators of contamination. Numerous investigators have characterized algal communities of streams and related abundance of specific taxa to the presence of heavy metals. Whitton[37] reviewed these earlier studies on the effects of heavy metals from mining activities in West Wales. Several reports cited in this review indicate that *Cladophora* sp. is particularly sensitive to heavy metals whereas *Stigeoclonium tenue* is quite tolerant. Armitage[38] also reported increased abundance of *S. tenue* in areas impacted by Zn from mining. In a more recent review of phycology and heavy metal pollution, Rai et al.[39] notes that *Cladophora* sp. is widely employed as an indicator of heavy metals due to its filamentous habit, large size, abundance, widespread distribution, and capacity to concentrate metals.

Because of several anomalies reported in the literature (e.g., *Cladophora* occurring at metal-contaminated sites), it is unlikely that any particular species will be a useful indicator of metal pollution.[37] Furthermore, normally sensitive taxa may be found at metal-impacted sites in the field because of their ability to adapt to heavy metals.[19] For example, Say et al.[40] reported that filamentous green algae, *Hormidium* sp., was abundant at field sites where Zn levels were greater than 20.0 mg/L. Laboratory experiments with *Hormidium* spp. collected from impacted sites indicated that these organisms were more tolerant of heavy metals than the same taxa collected from control streams.

A more appropriate use of algae and periphyton as indicators of heavy metals is to examine structure and composition of these communities. Reduced species diversity and richness are the most frequently reported consequences of heavy metals impact; however, these may not be the most sensitive indicators. Say and Whitton[20] reported that change in species composition was a more evident response to a Zn gradient than the number of species. A general trend in responses of periphyton communities to metals reported in several investigations is decreased abundance of diatoms, with a concomitant increase in filamentous green or blue-green algae.[40-42] Foster[19] compared the algal flora of the River Hayle and the River Gannel, which drain Cu and Pb mining regions, respectively. Similar responses of algal communities were observed in each river, suggesting that metal concentration was a better predictor of community structure than which metal was present.[19] Principal components analysis of these data was employed to characterize sites based on a 64-species matrix. Algal associations at low, low-moderate, moderate, moderate-high, and high metal levels were extracted from this analysis. In general, polluted sites in each stream were characterized by a *Microspora* community, whereas *Spirogyra* and *Mougeotia* dominated control sites. Biomonitoring studies on the impacts of heavy metals on periphyton have historically focused on structural measurements rather than functional changes in these communities. The relationship between these two classes of measurements is discussed by Cairns and Pratt.[43] In general, structural measurements involve counts (abundance, number of taxa) of organisms comprising a system whereas functional measurements are rate processes (primary productivity, detritus processing). The usefulness of structural and functional measurements for demonstrating effects of heavy metals on periphyton communities is discussed by Crossey and La Point.[34] In a stream that received several metals (Cd, Cu, Pb, and Zn), community respiration, gross primary productivity, chlorophyll a content, and ash free dry mass were greater at impacted and recovery sites than at reference stations. Impacted sites were dominated by green and blue-green algae (*Ulothrix* sp. and *Chroococcus* sp.) whereas reference sites were dominated by diatoms. Crossey and La Point[34] observed that while these structural measurements of periphyton were sensitive to heavy metals, functional parameters could better indicate the ecological consequences of these changes. However, due to the functional redundancy of ecosystems, greater variability of functional parameters, and the difficulty measuring these parameters compared to structural variables, the usefulness of functional variables for detecting effects of heavy metals is limited. Recently, Schindler[44] concluded that variables reflecting ecosystem function (primary production, nutrient cycling, respiration) were relatively poor indicators of early stress and that structural measurements were more useful.

Bioaccumulation Studies

Concentrations of heavy metals in periphyton are often employed as indicators of the presence of heavy metals in streams. Because of their role in energy transfer in aquatic systems, these organisms represent an important link in the transfer of heavy metals to higher trophic levels. Kelly[2] summarized results of several investigations demonstrating a relationship between metal concentrations in water and plants (Figure 1). Because of the ability of plants to bioconcentrate metals, levels in these organisms may be several orders of magnitude greater than concentrations in water. Consequently, analysis of metals in periphyton may be more practical than determining levels in water, particularly if ambient concentrations approximate limits of detection. Foster[19] reported linear relationships between concentrations of several metals in algae and water and concluded that, at least for some metals, the levels in algae were better indicators of site conditions than water concentrations. The use of algae and periphyton as indicators of heavy metals will require knowledge of the relationship between concentrations in these organisms and water, in addition to the environmental factors that may influence this relationship.[45]

CONCENTRATION IN PLANTS (MG/KG)

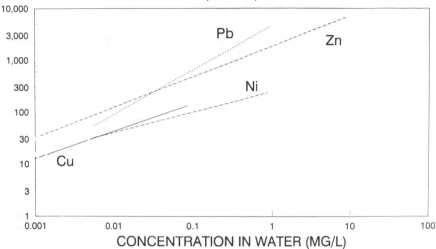

FIGURE 1. Uptake of metals by plants. Regression lines were obtained from data summarized by Kelly.[49]

Although it is generally assumed that levels of metals in algae and periphyton result from biological uptake, results of bioaccumulation studies with these organisms should be interpreted cautiously when field samples are not adequately characterized. Bioaccumulation of metals by *aufwuchs*, procedurally defined as the biotic and abiotic material accumulating on submerged surfaces, has been reviewed by Newman and McIntosh.[46] These investigators noted problems associated with bioaccumulation studies where the *aufwuchs* are poorly characterized. Since both biotic and abiotic components of *aufwuchs* concentrate metals, levels measured do not necessarily reflect bioaccumulation. Furthermore, metals associated with the abiotic portion of *aufwuchs* may not be available for transfer to higher trophic levels.[46]

Experimental Investigations

As noted above, because of the lack of preimpact data and the difficulty locating reference sites, field investigations into the impact of metals on periphyton are limited.[47] Experimental studies are a useful alternative to biomonitoring and allow investigators to establish cause and effect relationships between heavy metal concentrations and community responses.

Effects of metals on both structural and functional responses of periphyton have been investigated experimentally. Kaufman[48] exposed stream mesocosms to Cu (30 µg/L) and reported reductions in diatom species numbers, diversity, biomass, ATP, and chlorophyll a, relative to a control stream. Kaufman[48] also observed an inverse relationship between community age and resistance to Cu, and hypothesized that initial tolerance to Cu was a result of early colonization by opportunistic species. Results of this study are difficult to interpret since only nominal Cu levels are reported and since each of the four experimental streams was subjected to a different treatment without replication. A more rigorous statistical analysis of periphyton responses to Zn in experimental streams was conducted by Genter et al.[49] Several Zn treatments (ambient, 0.05, 0.50,and 1.00 mg/L) could be identified based on dominant periphyton communities present in these streams. In general, diatoms, which dominated control streams, were replaced by filamentous green algae (0.05 and 0.50 mg/L treatments) and blue-green algae (1.00 mg/L treatment) at higher Zn levels. Genter et al.[50] also examined responses of epilithic communities (heterotrophic and autotrophic microbial communities) to Zn exposure and grazing by snails in experimental streams. Both snails and Zn individually altered community structure and function, however combined effects could not be inferred from individual effects. Interaction between snail and Zn treatments on algal abundance was observed, illustrating the importance of examining community-level responses to metals.[50]

Sigmon et al.[51] exposed outdoor experimental streams to Hg and reported significant reductions in algal numbers, biomass, diversity, and species composition at 0.01 μg/L. Stream microcosms were also employed to examine subacute effects of Cd on *aufwuchs* communities.[52] Although high levels of Cd were associated with *aufwuchs,* less than 0.3% of the total Cd passing through the microcosms was accumulated. Most of the Cd retained in the system resulted from precipitation of $CdCO_3$ on the surface of the *aufwuchs* community and was not due to biological uptake.[52] These researchers also presented evidence that this $CdCO_3$ "frosting" was readily available to higher trophic levels.

Experimental addition of Cu (ambient, 2.5, 5.0, and 10.0 μg/L) to an oligotrophic stream was employed to document structural and functional changes in periphyton communities.[47,53] Significant alterations in structural and functional variables were observed at the lowest Cu level, suggesting that these communities are extremely sensitive to metals. At 2.5 μg/L, autotrophic and heterotrophic productivity were inhibited by 57 to 81% and 28 to 63%, respectively.[53] Abundance of most taxa was significantly reduced at 5 to 10 μg/L, although *Lyngbya* spp., a blue-green alga, was reduced at 2.5 μg/L. These investigators also reported that Cu reduced competition for attachment sites among tolerant algal taxa.

In summary, periphyton communities are useful indicators of heavy metal contamination. Changes in the number of species and, more importantly, the composition of these communities, are usually correlated with water quality. These organisms readily concentrate heavy metals and thus levels of metals in periphyton may be employed to estimate concentrations in water. The use of periphyton to monitor changes in water quality is, however, limited by taxonomic difficulties. For example, Patrick[41] notes that a typical diatom community in an unpolluted habitat may consist of 300 to 400 species occupying a relativley small area. The taxonomic expertise required to identify algae and periphyton to species may prohibit the use of these organisms for routine biomonitoring in streams.

COMMUNITY RESPONSES OF BENTHIC MACROINVERTEBRATES
Descriptive Approaches

The distribution and abundance of benthic macroinvertebrates have routinely been employed as indicators of water quality in streams. The relationship between benthic community structure and degree of heavy metal pollution was examined in the early 1920s by Carpenter[17] who reported reduced number of taxa and an "impoverishment of the fauna" at sites downstream from Pb mining operations. Carpenter[17] also noted the recovery of these streams following cessation of mining and speculated that restoration of the benthic fauna was an important initial step. The classic book by Hynes[54] on the biology of polluted waters and the work of Hart and Fuller[55] on pollution ecology both demonstrate the importance of freshwater invertebrates as indicators of water quality. More recently, Wiederholm[56] reviewed responses of aquatic insects to pollution and Voshell et al.[57] discussed field techniques for determining effects of toxicants on stream benthic invertebrates.

Because of their influence on various functional parameters in streams, such as primary productivity, detritus processing, and energy flow, aquatic macroinvertebrates are an important component of lotic ecosystems. These organisms are often quite abundant in streams, have a relatively short generation time, and represent several functional feeding groups. Finally, because of their close association with the substrate, tendency to bioaccumulate toxic materials, and their importance in aquatic food chains, benthic invertebrates are useful for monitoring the transport of contaminants in lotic systems.

Considerable research has been devoted to describing responses of stream benthic communities to heavy metals (Table 1). To be useful as an indicator of heavy metals impact, these responses should be predictable, allowing some degree of generalization among locations. Winner et al.[58] reported similar responses of benthic communities to heavy metals in two Ohio streams. These researchers suggested that changes in benthic communities were predictable and proposed using benthic community structure as an index of heavy metal pollution. Similarly, in a study of the relationship among metal concentrations, water quality criteria, and benthic community structure, La Point et al.[24] noted that benthic communities responded in a "predictable and indicative manner, which overall may be more sensitive than any single species tests." Clements et al.[16] observed similar community responses of benthic invertebrates to metals in natural stream and in outdoor experimental streams, suggesting that these responses were highl predictable.

Biomonitoring approaches using stream invertebrates for measuring impact of metals typically show reduced abundance and number of species at downstream sites (Table 1). Winner et al.[22] compared responses of several community indices (abundance, number of species, Margalef Index, Shannon Diversity) to Cu and concluded that the number of species was the most sensitive index examined. In studies conducted at the Clinch River, a system impacted by metals from a coal-fired electric generating plant, the author found that macroinvertebrate abun-

Table 1
Summary of Community Responses of Benthic Macroinvertebrates to Heavy Metals from Descriptive and Experimental Investigations

Location	Metal(s)	Summary of Results	References
I. Descriptive Studies			
1. Rivers Rheidol and Ystwyth, Aberystwyth, Wales	Pb	Reduced number of species downstream; faunal type related to level of Pb pollution; recovery observed following cessation of mining activities	17, 18
2. Clinton River, Michigan	Cu, Cr	Reduced abundance, taxa; *Cricotopus bicinctus* (Chironomidae:Orthocladiinae) accounted for >95% total organisms at impacted sites	70
3. Nent System, North England	Zn	High Zn sites dominated by Orthocladiinae and Lumbricidae (Oligochaeta); moderate Zn sites dominated by *Amphinemura* and *Leuctra* (Plecoptera); significant negative correlation between [Zn] and number of taxa	38
4. Elam's Run, Ohio	Cu, Cr	Increased abundance of *Tubifex tubifex* and *Limnodrilus hoffmeisteri* (Oligochaeta) at impacted sites; percentage Chironomidae (*Cricotopus bicinctus* and *C. infuscatus*) correlated with levels of metals; number of chironomid species per genus decreased with metal concentrations	58, 71
5. Grua Stream, Italy	Cr	Ephemeroptera eliminated from all impacted sites; Ephemeroptera were good indicators of Cr impact	23
6. Kolbacksan and Ybbarpsan Rivers, Sweden	Cr, Hg, Pb, Ni	Hydropsychidae (Trichoptera) highly tolerant of heavy metals; percentage anomalies in hydropsychid capture nets increased at metal-impacted sites and inversely correlated with Shannon diversity	60
7. Adair Run, Virginia	As, Cd, Cu, Cr, Se, Zn	Reduced diversity, abundance of Ephemeroptera; increased abundance of *Psephenus herricki* (Coleoptera: Psephenidae)	62
8. Prickley Pear Creek, Montana	Cu, Zn	Reduced abundance and number of taxa	24
9. Slate River, Colorado	As, Zn	Reduced abundance and number of taxa; recovery zone dominated by Orthocladiinae	24
10. Yoshino, Mazawa, and Senasaka Rivers, Japan	Cd, Cu, Pb, Zn	Impacted sites dominated by *Orthocladius*, *Cricotpus*, and *Eukiefferiella* (Orthocladiinae); *Tanytarsus* spp. (Chironominae) only collected from unpolluted sites	25
11. Clinch River, Virginia	Cu, Zn	Reduced number of taxa, individuals; Shannon diversity not affected; Ephemeroptera was most sensitive group; linear relationship between percentage reduction in Ephemeroptera and [Cu] employed to calculate site specific water quality criterion	64

12. Silver Bow Creek, Montana	Cu, Fe, Zn	Reduced abundance, number of taxa, diversity; Ephemeroptera, Plecoptera eliminated from all sites; Hydropsychidae dominated downstream sites; Chironomids (Orthocladiinae and Diamesinae) accounted for >90% total individuals at heavily impacted sites	59
13. Arkansas River, Colorado	Cd, Cu, Pb, Zn	Reduced diversity at all sites; *Alloperla* sp., *Isoperla* sp. (Plecoptera), and *Hydropsyche* sp. showed tolerance to heavy metals	61
14. Clinch River, Virginia	Cu, Zn	Reduced abundance and number of taxa; greatest effects on Ephemeroptera and Tanytarsini chironomids; impacted sites dominated by Orthocladiinae and Hydropsychidae; significant increases in Hydropsychidae at recovery sites	16, 26

II. Experimental Studies
A. Microcosms and Mesocosms

1. Outdoor concrete channels	Hg	No effect on Shannon diversity or richness; no effect on community composition	51
2. Stream microcosms	Cd	Increased abundance of Chironomidae and Plecoptera in treated streams	52
3. Laboratory streams	Cu	Reduced abundance and number of taxa in dosed streams during winter, spring, and summer experiments; Shannon diversity was not sensitive to Cu; *Baetis* sp. (Ephemeroptera) reduced by 85-98%; percentage composition of Orthocladiinae increased in dosed streams	63
4. Laboratory streams	Cu	Vulnerability of net-spinning caddisflies to stonefly predation significantly greater in dosed streams than controls; vulnerability to predation was a more sensitive indicator of Cu stress than mortality	77
5. Comparison of laboratory and field streams	Cu	Reduced abundance and number of taxa; reduced numbers of Ephemeroptera; increased percentage composition of Hydropsychidae and *Stenelmis* sp. in dosed streams; effects of Cu greater in laboratory streams	65
6. Field streams	Cu, Zn	Responses of benthic invertebrates highly predictable and similar to responses observed in the field; Orthocladiinae was significantly greater in treated streams than controls	16
7. Field streams	Cu	Community responses of benthic invertebrates to Cu were affected by water quality	26

B. Field Manipulations

1. Shayler Run, Ohio	Cu	Reduced abundance and number of taxa; Trichoptera increased in recovery zones; percentage chironomids correlated with [Cu]; number of species was the most sensitive index examined	22, 58

Table 1 (continued)
Summary of Community Responses of Benthic Macroinvertebrates to Heavy Metals from Descriptive and Experimental Investigations

Location	Metal(s)	Summary of Results	References
2. Convict Creek, California	Cu	Composition and density of invertebrate drift varied in control and treated sections; increased drift density of Ephemeroptera in treated sections resulted in reduced benthic abundance; drift and abundance of *Lepidostoma* spp. and *Symphitopsyche oslari* (Trichoptera) and Elmidae (Coleoptera) not affected	74
3. Convict Creek, California	Cu	Total biomass and density not affected; differences in sensitivity among taxa related to trophic preferences. *Rhyacophila* spp. was more abundant in treated section than in control section. Correspondence Analysis showed effects on community structure at the lowest Cu concentrations (5-10 µg/L)	75

dance was a more sensitive indicator than number of taxa, based on percentage reduction relative to control sites (Figure 2A). Macroinvertebrate abundance was, however, more variable than the number of taxa, thus making it more difficult to delineate zones of impact and recovery using this index (Figure 2B).

Although reduced species diversity is often observed in streams receiving heavy metals[59-62] this index is not always a useful indicator of impact.[22,63,64] Unlike organic enrichment, in which increased abundance of a few tolerant species decreases the evenness component of Shannon's Diversity, heavy metals often decrease abundance of all taxa without a substantial reduction in evenness.

Changes in percentage composition of dominant macroinvertebrate taxa are frequently reported in biomonitoring studies (Table 1). The most commonly observed changes in benthic communities exposed to metals include reduced percentage composition of Ephemeroptera with concomitant increases in abundance of Chironomidae (Diptera) and/or Hydropsychidae (Trichoptera). Other groups showing moderate tolerance to heavy metals include Oligochaeta,[38,58] Coleoptera,[62,65] and Plecoptera.[38,52,61] The sensitivity of mayflies to heavy metals is well established in both field biomonitoring studies and laboratory experiments. These organisms are frequently the last group to recover in streams impacted by metals. Several explanations for observed differences in sensitivity among aquatic insects have been proposed, including feeding habits,[66,67] body size,[68] presence of body coverings,[69] and the presence of external, platelike gills.[26]

Many investigators have noted the tolerance of chironomids, particularly Orthocladiinae, to metals.[16,25,26,58,59,70,71] These organisms may account for >95% of total abundance at metal-impacted sites.[16,26] The effects of metals on other groups of Chironomidae are quite variable. For example, Tanytarsini chironomids were reported to be as sensitive as mayflies to Cu and Zn.[16,26] Similar findings were reported by Yasuno et al.[25] and are supported by laboratory toxicity tests with these organisms.[72] These results indicate the importance of identifying chironomids beyond the level of family.

Net-spinning caddisflies (Hydropsychidae), which show moderate tolerance to heavy metals, are frequently abundant at impacted streams and may show significant increases at downstream recovery sites compared to upstream reference sites.[16,26] Petersen and Petersen[60] reported high tolerance of Hydropsychidae to heavy metals and noted that anomalies in capture nets of these organisms could be employed as an indicator of metals impact.

Experimental Studies

Although field biomonitoring studies have greatly increased our understanding of community responses to heavy metals, changes in benthic community structure at impacted sites relative to reference areas cannot definitively be associated with heavy metals.[73] Consequently, experimental studies are necessary to establish direct cause and effect relationships between metal concentrations and benthic community structure. Experimental investigations of the impact of heavy metals on benthic community structure have been conducted in laboratory streams,

PERCENT REDUCTION

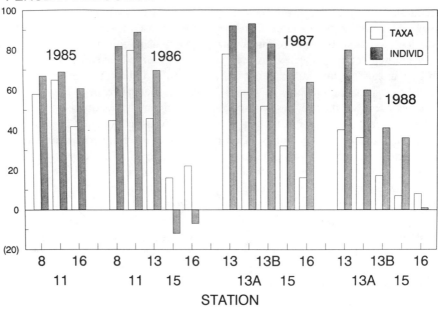

FIGURE 2(A). Comparison of the sensitivity of number of taxa and number of individuals to heavy metals. Data were obtained from biomonitoring studies conducted at the Clinch River, Virginia, 1985 to 1988. Station 8 was located approximately 100 m downstream from a coal-fired electric generating facility. Remaining stations were located 0.5 to 4.5 km downstream.

outdoor mesocosms, and natural streams (Table 1). Experimental introduction of Cu into Shayler Run, Ohio (120 μg/L) resulted in decreased abundance and number of taxa and increased percentage composition of Orthocladiinae and Trichoptera.[58] These researchers suggested that percentage composition of chironomids in streams may be a useful index of heavy metal pollution. Benthic communities exposed to Cu in Convict Creek, California, displayed increased drift and alterations in community structure at 5 to 10 μg/L.[66,74,75] Detrended correspondence analysis was employed to summarize spatial distributions of aquatic insects along a Cu gradient established in the stream. Sensitivity to Cu was expressed by changes in both relative abundance of dominant taxa and functional feeding groups in dosed sections.[66,75] These investigators suggested that differences in sensitivity to Cu were related to trophic habits, as herbivores and detritivores were more sensitive than predators.

The studies cited above demonstrate the usefulness of experimental introduction of metals into natural streams and generally support findings of biomonitoring studies. However, because of logistical, economic, and legal considerations, this approach is quite limited and cannot be employed on a routine basis. Furthermore, because of the difficulty obtaining replicate streams, the problem of pseudoreplication[35] will limit generalizations from these studies.

CV (s.d./mean x 100%)

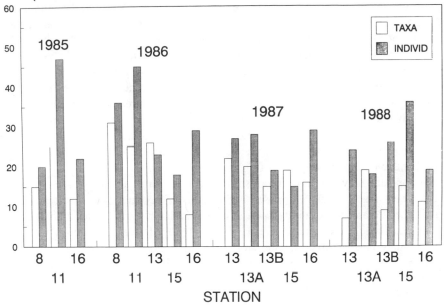

FIGURE 2(B). Coefficients of variation for number of taxa and individuals collected from the Clinch River, 1985 to 1988.

The use of stream microcosms and mesocosms[11] offers a practical alternative to experimental introduction of metals into natural systems. This approach, however, also has limitations and may sacrifice ecological realism for additional replicates.[44] Responses observed in these test systems are usually scale-dependent[76] and may not reflect changes in natural systems. Furthermore, because large predators cannot be included in most microcosm or mesocosm studies, the indirect effects of contaminants on predator-prey interactions cannot be assessed.

Despite these limitations, microcosm and mesocosm studies are an important intermediate step between laboratory bioassays and experiments in natural systems. Experimental streams have been employed in several studies to investigate the effects of heavy metals on benthic communities.[16,26,51,52,63,65,77] Benthic communities may be established in these streams by either allowing organisms to colonize naturally through replacement water or by field collections. Although the former approach may be more realistic, the length of time required to establish representative communities in these streams may be prohibitive. In addition, differences in initial community composition among streams due to variability in colonization may complicate these studies. Selby et al.[52] noted that random factors resulted in considerable variation in benthic invertebrate populations among experimental streams prior to exposure. Clements et al.[16] described an experimental procedure in which benthic communities were established on introduced substrates in the field (Figure 3) and then transferred to outdoor stream mesocosms (Figure 4). The advantage of this approach is that initial community composition is similar among streams, thus allowing precise evaluation of the effects of metals on community structure. The main disadvantage of this approach is that introduced substrates are often selective for certain taxa.

The flexibility of obtaining benthic communities from different sources has several interesting applications. If communities are obtained from reference sites in a stream receiving metals, responses in experimental streams may be compared to responses at impacted sites in the field.[16] Similarly, responses of benthic communities obtained from several different source streams may be compared. For example, resistance to heavy metals could be examined by comparing effects on benthic communities with different histories of previous exposure. Finally, this approach may also have important applications for establishing water quality criteria. Pontasch et al.[78] proposed deriving benthic communities from regional reference streams to examine responses of standardized communities to toxicants.

The approach described above has been employed to estimate effects of Cu on benthic communities in laboratory streams[63] and to compare responses in laboratory and outdoor experimental streams.[65] Results of these experiments indicate that benthic communities are more sensitive to metals in the laboratory, possibly due to limited food availability. The influence of water quality on community responses to Cu was examined in outdoor experimental streams receiving diluent water from two different systems.[26] Experimental streams have

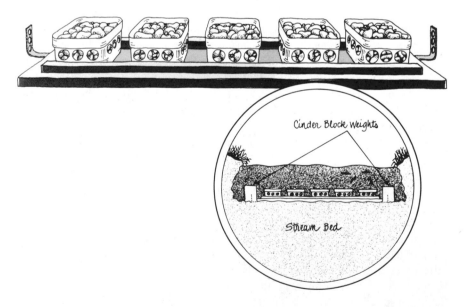

FIGURE 3.   Schematic of substrate-filled trays used for collecting benthic invertebrate communities. From Clements, W. H. et al., *Hydrobiologia* 173 (1989). By permission.

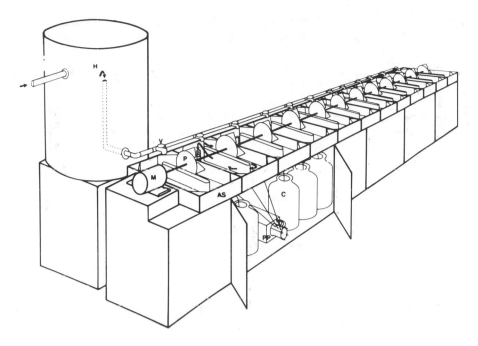

FIGURE 4.   Schematic of outdoor stream mesocosms. As, artificial stream; H, headbox; M, motor; P, paddlewheel; PP, peristaltic pump; C, carboy; V, valve. From Clements, W. H. et al., *Aquat. Toxicol.* 14 (1989). By permission.

recently been employed to examine the influence of heavy metals on predator-prey interactions.[77] Net-spinning caddisflies (*Hydropsyche morosa* and *Chimarra* sp.) were relatively tolerant of Cu exposure; however vulnerability of these organisms to predation by stoneflies (*Paragnetina media*) was significantly greater in streams dosed with Cu than in controls (Figure 5). Hydropsychid caddisflies were also significantly more abundant in the stomach contents of stoneflies collected from dosed streams.[77] Results of these experiments indicate that vulnerability of net-spinning caddisflies to predation was a more sensitive indicator of heavy metals stress than mortality.

Responses of benthic communities to Cu in outdoor experimental streams were compared to responses at impacted sites in the Clinch River, Virginia.[16] Ephemeroptera and Tanytarsini chironomids were eliminated from treated streams and from impacted sites in the field. Highly impacted field sites and high Cu-Zn experimental streams were dominated by Orthocladiini chironomids, whereas moderatly impacted field sites and low Cu-Zn streams were dominated by Hydropsychidae. The similarity of these experimental and observational results suggests that benthic community structure is a predictable indicator of heavy metal contamination. The observed sensitivity of Ephemeroptera and the tolerance of Orthocladiini and Hydropsychidae to heavy metals in experimental streams support findings reported in previous biomonitoring studies[58-62,64,70,71] suggesting that the distribution and abundance of these organisms may be employed as indicators of heavy metal impact.

Index of Heavy Metals Stress

The responses of benthic macroinvertebrates to contaminants in streams are quite diverse, and several biotic indices have been developed based on observed variation among taxa.[79] In general, these indices include two components, relative abundance of dominant taxa and their tolerance values. Unfortunately, most of these biotic indices have been developed for organic pollution, and their usefulness for other classes of contaminants (e.g., heavy metals, acidification, pesticides) is uncertain. Despite evidence of considerable variation among benthic invertebrates in responses to heavy metals, there have been few attempts to develop tolerance values specific to these contaminants. Research described above indicates a general increase in sensitivity to heavy metals from chironomids to caddisflies to stoneflies to mayflies. Within each of these broad taxonomic groups, however, there may be considerable variation in effects of heavy metals. For example, in experimental stream mesocosms receiving Cu, Clements et al.[16,26] noted that Orthocladiini and Tanytarsini chironomids were, respectively, the most and least tolerant organisms examined.

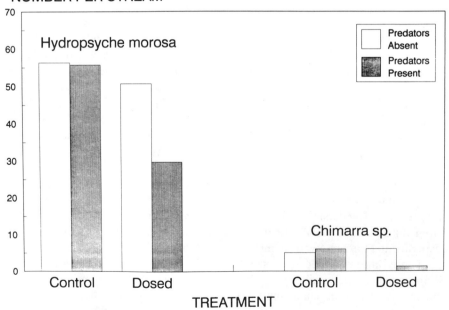

FIGURE 5.   Effects of predation and Cu exposure (6 µg/L) on two species of net-spinning caddisflies in stream mesocosms.

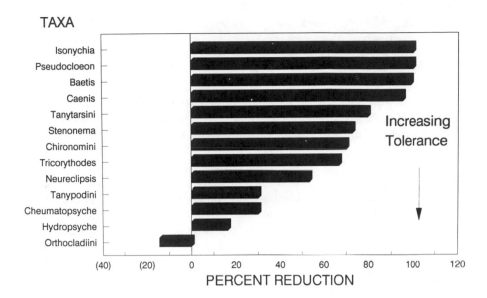

FIGURE 6. Relative sensitivity of benthic macroinvertebrates to Cu. Data shown are the percentage reduction of dominant taxa following 10 days exposure to 25 µg/L.

Establishment of tolerance values for benthic invertebrates is difficult and highly subjective. Clearly a more objective approach would be to estimate relative sensitivity to contaminants experimentally, for example, in stream mesocosms. An index based on abundance and relative sensitivity to metals could be developed and then validated in the field. Since tolerance values could be generated independently from field data, such an index would be useful for objectively delineating zones of impact and recovery in streams receiving metals.

Experiments were conducted in outdoor stream mesocosms during August, 1987 to examine responses of benthic communities to Cu (Figure 6). After 10 days exposure, percentage reduction of the 13 dominant taxa in treated streams (25 µg Cu/L) compared to controls was employed to estimate relative sensitivity. Mayflies (*Isonychia bicolor, Pseudocloeon* sp., *Baetis brunneicolor, Caenis* sp., *Stenonema* sp., and *Tricorythodes* sp.) were highly sensitive to Cu exposure. Percentage reduction of these organisms ranged from 67 to 100%. Tanytarsini chironomids were also highly sensitive to Cu, as these organisms were reduced by 80% in treated streams. Net-spinning caddisflies (*Cheumatopsyche* sp. and *Hydropsyche* sp.) showed intermediate tolerance to Cu. Orthocladiini chironomids were quite tolerant to heavy metals and were the only group to show increased abundance in treated streams compared to controls.

An index of metal impact (M.I.) was developed based on these estimates of relative sensitivity and abundance of dominant taxa in the field. The index is given as

$$M.I. = \Sigma n_i s_i / N$$

where M.I. = metals index
n_i = abundance of ith species
s_i = sensitivity (percentage reduction in experimental streams) of ith species
N = total individuals

M.I. values were calculated from samples collected in 1987 at the Clinch River, a system receiving Cu from a coal-fired electric generating plant (Figure 7). Values ranged from −11.5 at station 13, located 1.0 km downstream from the effluent, to 66.5 at upstream reference station 2. Little evidence of recovery was observed at station 16, the furthest downstream site. M.I. values at this station were significantly lower than at the upstream reference site, reflecting the presence of highly tolerant Orthocladiini chironomids and moderately tolerant Hydropsychidae. These organisms comprised 58 to 85% of the total individuals at all downstream sites. The number of taxa was also significantly reduced at downstream sites; however, this variable was less sensitive to Cu exposure. Results of multiple range tests showed the number of taxa recovered at station 16. In contrast, M.I. values were significantly reduced at all downstream sites.

This proposed index, which integrates experimentally derived estimates of Cu

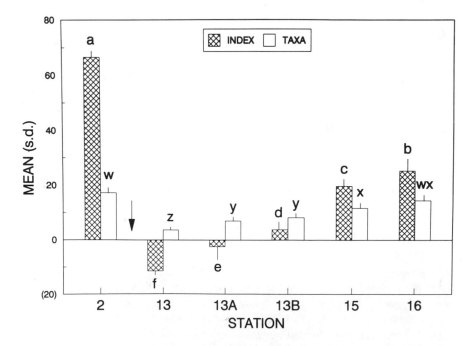

FIGURE 7.   Comparison of the index of metals impact (M.I.) and number of taxa at the Clinch River (1987). Stations with the same letter were not significantly different based on Ryan's Q Multiple Range Tests (a-e for index; w-z for taxa). Arrow indicates location of effluent. s.d., Standard deviation.

sensitivity with estimates of relative abundance, is an effective indicator of water quality. The index may also be useful for quantitatively assessing the impact of metals on streams and objectively delineating zones of impact and recovery. Furthermore, by using regional estimates of species-specific sensitivity to metals, such an index may provide evidence of the degree of impact in systems where appropriate reference sites are not available.

Summary of Benthic Community Responses

The distribution and abundance of benthic invertebrates have been employed in both descriptive and experimental studies for documenting the impact of metals on streams. Responses of these communities are highly sensitive to metals and therefore useful indicators of water quality. Because of limitations of biomonitoring studies, experimental investigations can provide the best direct evidence of cause and effect relationships between metal concentrations and benthic community structure. Experimental introduction of metals into natural systems has corroborated findings of biomonitoring studies, but because of obvious limitations cannot be routinely employed to assess the impact of these contaminants. It is recommended that experimental streams are the most practical alternative to these costly and logistically impractical field manipulations. Experimental streams may be employed to predict changes in field communities exposed to metals, examine the influence of environmental variables on community responses, and investigate the indirect effects of metals on predator-prey interactions. Finally, relative sensitivity values for dominant taxa, estimated experimentally, can be employed to develop an objective index of heavy metals impact. Since the index is developed independently of field data, it can be field-validated and employed to delineate zones of impact and recovery.

FOOD CHAIN TRANSFER OF HEAVY METALS

Benthic invertebrates readily accumulate metals from their surrounding environment.[80-82] Concentrations of metals in these organisms, which may be several orders of magnitude greater than ambient levels, are often highly correlated with levels in water, sediments, and/or food.[30,67,68,83-85] As with periphyton samples, the concentration of metals in aquatic insects may be a better indicator of water quality than direct measurement because of high variability of ambient concentrations as well as analytical limitations at very low metal levels.[86] Several researchers have noted the ability of benthic invertebrates to actively regulate uptake, storage, and elimination of metals.[68,83,87] This ability may be particularly well developed in previously exposed populations of benthic invertebrates,[68,83,87] thus limiting the usefulness of this approach for biomonitoring.

Accumulation of heavy metals in benthic invertebrates occurs by uptake from water,[68,83,84] sediments,[88] interstitial water,[89] and food.[67] Sediments, which are considered an ultimate sink for contaminants in aquatic systems, may contribute significantly to concentrations of metals in benthic invertebrates either by absorption/adsorption from interstitial water or by direct ingestion. Reynoldson[90] reviewed interactions between sediment contaminants and benthic organisms and noted the effects of sediments on these organisms as well as the processes by which benthic invertebrates influenced movement of sediment contaminants back into the water column.

The relative contribution of food and water to total body burdens of metals in benthic invertebrates has been the subject of considerable research. Although for certain metals the general consensus is that uptake from food is less efficient than from water, few studies have attempted to assess the relative importance of these two pathways simultaneously.[91] Recent experimental work suggests that water and interstitial water are the primary routes of exposure for many benthic invertebrates.[68,89,91,92] Surface adsorption also contributes significantly to total body burdens of metals in benthic invertebrates,[85,93] and may account for the observed inverse relationship between size and metal concentrations in aquatic insects.[85] Krantzberg and Stokes[93] reported similar concentrations of certain metals in live and dead *Chironomus* sp. exposed to contaminated sediments, and concluded that surface adsorption accounted for most of the metals associated with these organisms.

Although water and interstitial water are the primary routes of exposure for many benthic invertebrates, uptake from food may also be significant.[52,67,80,82] The relative importance of food as a source of metals in benthic invertebrates will depend on species-specific feeding habits. Smock[67] reported that feeding habits influenced concentrations of metals in benthic invertebrates, with highest concentrations found in organisms that ingest sediments and lowest levels in predators. Similarly, Selby et al.[52] found higher concentrations of metals in grazers and collectors than in predators. Thus, while food may contribute to total body burden of metals in benthic invertebrates, comparisons of predators with other functional feeding groups indicate little tendency for increased concentrations with trophic level.[80,91,94]

Benthic invertebrates are often a major component in the diet of many fish species. Because of their close association with sediments, and the ability of certain species to tolerate and accumulate metals, benthic invertebrates represent an important link in the transfer of contaminants to higher trophic levels. The relative importance of food and water as sources of metals in fish is, however, controversial. As with benthic invertebrates, most evidence derived from laboratory studies with fish indicates that uptake of metals from water is the predominant route of exposure.[30,95] However, several recent studies have suggested that dietary accumulation of metals, particularly Cd, may contribute significantly to total body burdens in fish.[94,96-100] Hatakeyama and Yasuno[100] reported that 90% of Cd accumulation by the guppy, *Poecilia reticulata,* was derived from feeding on contaminated chironomids. Stone loach (*Noemacheilus barbatulus*) readily accumulated Cd from feeding on contaminated tubificid worms.[98] Harrison and Klaverkamp[99] reported that uptake of Cd from food was more important than from water in both rainbow trout (*Oncorhynchus mykiss*) and lake whitefish (*Coregonus clupeaformis*).

The importance of food as a source of metals has also been reported in field studies. Dallinger and Kautzky[96] demonstrated that *O. mykiss,* which fed predominantly metal-enriched isopods (*Asellus aquaticus*), accumulated metals primarily from the diet when levels in the water were low. Van Hassel et al.[94] reported that, despite low concentrations of metals in a stream receiving highway runoff, concentrations in fish were similar to those observed during laboratory exposure to much higher levels. High concentrations of metals in sediments and benthic invertebrates accounted for the observed levels in fish, indicating that feeding habits as well as the degree of association with sediments influence accumulation of heavy metals.[94,101]

Although most metals, with the important exceptions of Hg and Se, show little tendency to biomagnify up food chains, concentrations in fish can reach harmful levels due to increased consumption of highly contaminated prey. The tolerance of many benthic invertebrates for heavy metals is well documented,[68,73,83,87] and several potential mechanisms have been suggested which have important implications for food chain transfer of these contaminants. For example, some tolerant organisms have the ability to accumulate and sequester high concentrations of metals (see this volume, Chapters 3 and 4). This mechanism has been reported in several groups of benthic organisms including polychaetes,[87] isopods,[83] and oligochaetes.[73] Tolerant benthic invertebrates will dominate metal-impacted sites in the field and abundance of these organisms may actually increase relative to reference sites.[16,26] Many fish species are opportunistic predators and able to switch feeding habits as prey availability changes. Several investigators have shown that feeding habits of fish at impacted sites may be modified to include tolerant prey.[102-104] In streams polluted by mining effluents, Jefree and Williams[102] reported that fish switched from pollution-sensitive to pollution-tolerant prey types. As a result of reduced prey diversity and increased consumption of contaminated prey, dietary accumulation of metals may contribute significantly to total body burdens. This phenomenon, which has been termed the "food chain effect,"[97] may account for high concentrations of metals observed in fish.

In summary, when concentrations of heavy metals in water are high, the contribution of food to total body burdens in fish will be relatively insignificant because of the greater rate and efficiency of transport across the gills. When concentrations in water are low, food chain transfer of metals may be the primary route of exposure. The relative magnitude of food chain transfer will therefore depend on specific feeding habits of fish, concentrations of metals in prey, the ability of prey species to tolerate and accumulate metals, the relative abundance of tolerant prey, and the tendency of fish to switch to tolerant prey.

REFERENCES

1. Forstner, U. and G. T. W. Whittman. *Metal Pollution in the Aquatic Environment* (Berlin:Springer-Verlag, 1979), p. 486.
2. Kelly, M. *Mining and The Freshwater Environment* (London, England:Elsevier Applied Science, 1988), p. 231.
3. Moore, J. W. and S. Ramamoorthy. *Heavy Metals in Natural Waters: Applied Monitoring and Impact Assessment* (New York:Springer-Verlag, 1984), p. 268.
4. Nriagu, J. O. *The Biogeochemistry of Lead in the Environment. Part B. Biological Effects* (Amsterdam:Elsevier/North Holland Biomedical Press, 1978), p. 397.
5. Nriagu, J. O. *Copper in the Environment. Part II: Health Effects* (New York:John Wiley & Sons, 1979), p. 489.
6. Nriagu, J. O. *The Biogeochemistry of Mercury in the Environment. Part B. Biological Effects* (Amsterdam:Elsevier/North Holland Biomedical Press, 1979), p. 696.
7. Nriagu, J. O. *Zinc in the Environment. Part II: Health Effects* (New York:John Wiley & Sons, 1980), p. 480.
8. Nriagu, J. O. and J. B. Sprague. *Cadmium in the Aquatic Environment* (New York:John Wiley & Sons, 1987), p. 908.
9. Kimball, K. D. and S. A. Levin. "Limitations of Laboratory Bioassays: The Need for Ecosystem-Level Testing," *BioScience* 35:165-171 (1985).
10. Brungs, W. A. "Review of Multispecies Toxicity Testing," *BioScience* 36:677-678 (1986).
11. Odum, E. P. "The Mesocosm," *Bioscience* 34:558-562 (1984).
12. Macek, K. J. "Aquatic Toxicology: Fact or Fiction?," *Environ. Health. Perspect.* 34:159-163 (1980).
13. Cairns, J. Jr. "Putting the Eco in Ecotoxicology," *Reg. Toxicol. Pharmacol.* 8:226-238 (1988).
14. Cairns, J. Jr. "Will The Real Ecotoxicologist Please Stand Up?," *Environ. Toxicol. Chem.* 8:843-844 (1989).
15. Cairns, J. Jr. "Are Single Species Toxicity Tests Alone Adequate for Estimating Environmental Hazard?," *Hydrobiologia* 100:47-57 (1983).
16. Clements, W. H., D. S. Cherry, and J. Cairns, Jr. "The Impact of Heavy Metals on Macroinvertebrate Communities: A Comparison of Observational and Experimental Results," *Can. J. Fish. Aquat. Sci.* 45:2017-2025 (1988).
17. Carpenter, K. E. "A Study of The Fauna of Rivers Polluted by Lead Mining in The Aberystwyth District of Cardiganshire," *Ann. Appl. Biol.* 11:1-23 (1924).
18. Carpenter, K. E. "On the Biological Factors Involved in the Destruction of River-Fisheries by Pollution Due to Lead-Mining," *Ann. Appl. Biol.* 12:1-13 (1925).
19. Foster, P. L. "Species Associations and Metal Contents of Algae from Rivers Polluted by Heavy Metals," *Freshwater Biol.* 12:17-39 (1982).
20. Say, P. J. and B. A. Whitton. "Changes in Flora Down a Stream Showing a Zinc Gradient," *Hydrobiologia* 76:255-262 (1981).
21. Cairns, J. Jr., G. R. Lanza, and B. C. Parker. "Pollution Related Structural and Functional Changes in Aquatic Communities with Emphasis on Freshwater Algae and Protozoa," *Proc. Acad. Nat. Sci. U.S.* 124:79-127 (1972).
22. Winner, R. W., J. S. Van Dyke, N. Caris, and M. Farrel. "Response of the Macroinvertebrate Fauna to a Copper Gradient in an Experimentally-Polluted Stream," *Verh. Int. Ver. Limnol.* 19:2121-2127 (1975).

23. Ramusino, M. C., G. Pacchetti, and A. Lucchese. "Influence of Chromium (VI) Upon Stream Ephemeroptera in the Pre-Alps," *Bull. Environ. Contam. Toxicol.* 26:228-232 (1981).

24. La Point, T. W., S. M. Melancon, and M. K. Morris. "Relationships Among Observed Metal Concentrations, Criteria, and Benthic Community Structural Responses in 15 Streams," *J. Water Pollut. Contol. Fed.* 56:1030-1038 (1984).

25. Yasuno, M., S. Hatakeyama, and Y. Sugaya. "Characteristic Distribution of Chironomids in the Rivers Polluted with Heavy Metals," *Verh. Int. Ver. Limnol.* 22:2371-2377 (1985).

26. Clements, W. H., J. L. Farris, D. S. Cherry, and J. Cairns, Jr. "The Influence of Water Quality on Macroinvertebrate Community Responses to Copper in Outdoor Experimental Streams," *Aquat. Toxicol.* 14:249-262 (1989).

27. Clements, W. H., J. H. Van Hassel, D. S. Cherry, and J. Cairns, Jr. "Colonization, Variability, and the Application of Substratum-Filled Trays for Biomonitoring Benthic Communities," *Hydrobiologia* 173:45-53 (1989).

28. Karr, J. R. "Assessment of Biotic Integrity Using Fish Communities," *Fisheries* 6:21-27 (1981).

29. Fausch, K. D., J. Lyons, J. R. Karr, and P. L. Angermeier. "Fish Communities as Indicators of Environmental Degredation," *Am. Fish. Soc. Symp.* 8:123-144 (1990).

30. Prosi, F. "Heavy Metals in Aquatic Organisms," in *Metal Pollution in the Aquatic Environment,* U. Forstner and G. T. W. Whittman, Eds. (Berlin:Springer-Verlag, 1979), pp. 271-323.

31. Moriarty, F. *Ecotoxicology, the Study of Pollutants in Ecosystems* (San Diego:Academic Press, 1983), p. 233.

32. Green, R. H. *Sampling Design and Statistical Methods for Environmental Biologists* (New York:John Wiley & Sons, 1979), p. 257.

33. Vannote, R. L., G. W. Minshall, K. W. Cummins, J. R. Sedell, and C. E. Cushing. "The River Continuum Concept," *Can. J. Fish. Aquat. Sci.* 37:130-137 (1980).

34. Crossey, M. J. and T. W. La Point. "A Comparison of Community Structural and Functional Responses to Heavy Metals," *Hydrobiologia* 162:109-121 (1988).

35. Hurlbert, S. H. "Pseudoreplication and the Design of Ecological Field Experiments," *Ecol. Monogr.* 54:187-211 (1984).

36. Stewart-Oaten, A., W. W. Murdoch, and K. R. Parker. "Environmental Impact Assessment: Pseudoreplication in Time?," *Ecology* 67:929-940 (1986).

37. Whitton, B. A. "Toxicity of Heavy Metals to Freshwater Algae: A Review," *Phykos* 9:116-125 (1970).

38. Armitage, P. D. "The Effects of Mine Drainage and Organic Enrichment on Benthos in the River Nent System, Northern Pennies," *Hydrobiologia* 74:119-128 (1980).

39. Rai, L. C., J. P. Gaur, and H. D. Kumar. "Phycology and Metal Pollution," *Biol. Ref.* 56:99-151 (1981).

40. Say, P. J., B. M. Diaz, and B. A. Whitton. "Influence of Zinc on Lotic Plants. I. Tolerance of *Hormidium* Species to Zinc," *Freshwater Biol.* 7:357-376 (1977).

41. Patrick, R. "Effects of Trace Metals in the Aquatic Ecosystem," *Am. Sci.* 66:185-191 (1978).

42. Shehata, F. H. A. and B. A. Whitton. "Field and Laboratory Studies on Blue-Green Algae from Aquatic Sites with High Levels of Zinc," *Verh. Int. Ver. Limnol.* 21:1466-1477 (1981).

43. Cairns, J. Jr. and J. R. Pratt. "On the Relation Between Structural and Functional Analyses of Ecosystems," *Environ. Toxicol. Chem.* 5:785-786 (1986).

44. Schindler, D. W. "Detecting Ecosystem Responses to Anthropogenic Stress," *Can. J. Fish. Aquat. Sci. Suppl.* 44:6-25 (1987).

45. Kelly, M. G. and B. A. Whitton. "Interspecific Differences in Zn, Cd, and Pb Accumulation by Freshwater Algae and Bryophytes," *Hydrobiologia* 175:1-11 (1989).

46. Newman, M. C. and A. W. McIntosh. "Appropriateness of *Aufwuchs* as a Monitor of Bioaccumulation," *Environ. Pollut.* 60:83-100 (1989).

47. Leland, H. V. and J. L. Carter. "Effects of Copper on Species Composition of Periphyton in a Sierra Nevada, Califorrnia, Stream," *Freshwater Biol.* 14:281-296 (1984).

48. Kaufman, L. H. "Stream *Aufwuchs* Accumulation: Disturbance Frequency and Stress Resistance and Resilience," *Oecologia (Berl.)* 52:57-63 (1982).

49. Genter, R. B., D. S. Cherry, E. P. Smith, and J. Cairns, Jr. "Algal-Periphyton Population and Community Changes from Zinc Stress in Stream Mesocosms," *Hydrobiologia* 153:261-275 (1987).

50. Genter, R. B., F. S. Colwell, J. R. Pratt, D. S. Cherry, and J. Cairns, Jr. "Changes in Epilithic Communities Due to Individual and Combined Treatments of Zinc and Snail Grazing in Stream Mesocosms," *Toxicol. Ind. Health* 4:185-201 (1988).

51. Sigmon, C. F., H. J. Kania, and R. J. Beyers. "Reductions in Biomass and Diversity Resulting from Exposure to Mercury in Artificial Streams," *J. Fish. Res. Bd. Can.* 34:493-500 (1977).

52. Selby, D. A., J. M. Ihnat, and J. J. Messer. "Effects of Subacute Cadmium Exposure on a Hardwater Mountain Stream Microcosm," *Water Res.* 19:645-655 (1985).

53. Leland, H. V. and J. L. Carter. "Effects of Copper on Production, Nitrogen Fixation and Processing of Leaf Litter in a Sierra Nevada, California, Stream," *Freshwater Biol.* 15:155-173 (1985).

54. Hynes, H. B. N. *The Biology of Polluted Waters* (Liverpool:Liverpool Univ. Press, 1960).

55. Hart, C. W., Jr. and S. L. H. Fuller. *Pollution Ecology of Freshwater Invertebrates* (San Diego:Academic Press, 1974).

56. Wiederholm, T. "Responses of Aquatic Insects to Environmental Pollution," in *The Ecology of Aquatic Insects,* V. H. Resh and D. M. Rosenberg, Eds. (New York:Praeger, 1984), pp. 508-557.

57. Voshell, J. R., R. J. Layton, and S. W. Hiner. "Field Techniques for Determining the Effects of Toxic Substances on Benthic Macroinvertebrates in Rocky-Bottomed Streams," in *Aquatic Toxicology and Hazard Assessment: Vol. 12,* U. M. Cowgill and L. R. Williams, Eds., ASTM S.T.P. 1027, (Philadelphia, PA:American Society for Testing and Materials, 1989), pp. 134-155.

58. Winner, R. W., M. W. Boesel, and M. P. Farrell. "Insect Community Structure as an Index of Heavy-Metal Pollution in Lotic Ecosystems," *Can. J. Fish. Aquat. Sci.* 37:647-655 (1980).

59. Chadwick, J. W., S. P. Canton, and R. L. Dent. "Recovery of Benthic Invertebrate Communities in Silver Bow Creek, Montana, Following Improved Metal Mine Wastewater Treatment," *Water Air Soil Pollut.* 28:427-438 (1986).

60. Petersen, L. B. M. and R. C. Petersen, Jr. "Anomalies in Hydropsychid Capture Nets from Polluted Streams," *Freshwater Biol.* 13:185-191 (1983).

61. Rolin, R. A. "The Effects of Heavy Metal Pollution of the Upper Arkansas River on the Distribution of Aquatic Macroinvertebrates," *Hydrobiologia* 160:3-8 (1988).

62. Specht, W. L., D. S. Cherry, R. A. Lechleitner, and J. Cairns, Jr. "Structural, Functional, and Recovery Responses of Stream Invertebrates to Fly Ash Effluent," *Can. J. Fish. Aquat. Sci.* 41:884-896 (1984).

63. Clements, W. H., D. S. Cherry, and J. Cairns, Jr. "Structural Alterations in Aquatic Insect Communities Exposed to Copper in Laboratory Streams," *Environ. Toxicol. Chem.* 7:715-722 (1988).

64. Van Hassel, J. II. and A. E. Gaulke. "Site-Specific Water Quality Criteria from In-Stream Monitoring Data," *Environ. Toxicol. Chem.* 5:417-426 (1986).

65. Clements, W. H., D. S. Cherry, and J. Cairns, Jr. "Macroinvertebrate Community Responses to Copper in Laboratory and Field Experimental Streams," *Arch. Environ. Contam. Toxicol.* 19:361-365 (1990).

66. Leland, H. V., S. V. Fend, T. L. Dudley, and J. L. Carter. "Effects of Copper on Species Composition of Benthic Insects in a Sierra Nevada, California, Stream," *Freshwater Biol.* 21:163-179 (1989).

67. Smock, L. A. "The Influence of Feeding Habits on Whole-Body Metal Concentrations in Aquatic Insects," *Freshwater Biol.* 13:301-311 (1983).

68. Krantzberg, G. and P. M. Stokes. "Metal Regulation, Tolerance, and Body Burdens in the Larvae of the Genus *Chironomus*," *Can. J. Fish. Aquat. Sci.* 46:389-398 (1989).

69. Hodson, P. V., U. Borgmann, and H. Shear. "Toxicity of Copper to Aquatic Biota," in *Copper in the Environment, II. Health Effects*, J. O. Nriagu, Ed. (New York:John Wiley & Sons, 1979), pp. 308-372.

70. Surber, E. W. *"Cricotopus bicinctus,* A Midgefly Resistant to Electroplating Wastes," *Trans. Am. Fish. Soc.* 89:111-116 (1959).

71. Waterhouse, J. C. and M. P. Farrell. "Identifying Pollution Related Changes in Chironomid Communities as a Function of Taxonomic Rank," *Can. J. Fish. Aquat. Sci.* 42:406-413 (1985).

72. Anderson, R. L., C. T. Walbridge, and T. J. Fiandt. "Survival and Growth of *Tanytarsus dissimilis* (Chironomidae) Exposed to Copper, Cadmium, Zinc, and Lead," *Arch. Environ. Contam. Toxicol.* 9:329-335 (1980).

73. Klerks, P. L. and J. S. Levinton. "Effects of Heavy Metals in a Polluted Aquatic Ecosystem," in *Ecotoxicology: Problems and Approaches,* S. A. Levin, M. A. Harwell, J. R. Kelly, and K. D. Kimball, Eds. (New York:Springer-Verlag, 1989), pp. 41-67.

74. Leland, H. V. "Drift Response of Aquatic Insects to Copper," *Verh. Int. Ver. Limnol.* 22:2413-2419 (1985).

75. Leland, H. V. and J. L. Carter, and S. V. Fend. "Use of Detrended Correspondence Analysis to Evaluate Factors Controlling Spatial Distribution of Benthic Insects," *Hydrobiologia* 132:113-123 (1986).

76. Perry, J. A. and N. H. Troelstrup, Jr. "Whole Ecosystem Manipulation: A Productive for Test System Research?," *Environ. Toxicol. Chem.* 7:941-951.

77. Clements, W. H., D. S. Cherry, and J. Cairns, Jr. "The Influence of Copper Exposure on Predator-Prey Interactions in Aquatic Insect Communities," *Freshwater Biol.* 21:483-488 (1989).

78. Pontasch, K. W., B. R. Neiderlehner, and J. Cairns, Jr. "Comparisons of Single-Species, Microcosm and Field Responses to a Complex Effluent," *Environ. Toxicol. Chem.* 8:521-532 (1989).

79. Hellawell, J. M. "Change in Natural and Managed Ecosytems: Detection, Measurement, and Assessment," *Proc. R. Soc. London* 197:31-56 (1977).

80. Burrows, I. G. and B. A. Whitton. "Heavy Metals in Water, Sediments and Macroinvertebrates from a Metal-Contaminated River Free of Organic Pollution," *Hydrobiologia* 106:263-273 (1983).

81. Gower, A. M. and S. T. Darlington. "Relationship Between Copper Concentrations in Larvae of *Plectrocnemia conspersa* (Curtis) (Trichoptera) and in Mine Drainage Streams," *Environ. Pollut.* 65:155-168.

82. Brown, B. E. "Effects of Mine Drainage on The River Hayle, Cornwall. (A) Factors Affecting Concentrations of Copper, Zinc and Iron in Water, Sediments and Dominant Invertebrate Fauna," *Hydrobiologia* 52:221-233 (1977).

83. Brown, B. E. "Uptake of Copper and Lead by a Metal-Tolerant Isopod *Asellus meridianus*," *Freshwater Biol.* 7:235-244 (1978).

84. Duzzin, B., B. Pavoni and R. Donazzolo. "Macroinvertebrate Communities and Sediments as Pollution Indicators for Heavy Metals in the River Adige (Italy)," *Water Res.* 22:1353-1363 (1988).

85. Smock, L. A. "Relationships Between Metal Concentrations and Organism Size in Aquatic Insects," *Freshwater Biol.* 13:313-321 (1983).

86. Lynch, T. R., C. J. Popp, and G. Z. Jacobi. "Aquatic Insects as Environmental Monitors of Trace Metal Contamination: Red River, New Mexico," *Water Air Soil Pollut.* 42:19-31 (1988).

87. Bryan, G. W. and L. G. Hummerstone. "Adaptation of The Polychaete *Nereis diversicolor* to Estuarine Sediments Containing High Concentrations of Heavy Metals. I. General Observations and Adaptation to Copper," *J. Mar. Biol. Assoc. U.K.* 51:845-863 (1971).

88. Hare, L., P. G. C. Campbell, A. Tessier, and N. Belzile. "Gut Sediments in a Burrowing Mayfly (Ephemeroptera: *Hexagenia limbata*): Their Contribution to Animal Trace Element Burdens, Their Removal, and the Efficacy of a Correction for Their Presence," *Can. J. Fish. Aquat. Sci.* 46:451-456.

89. Swartz, R. C., G. R. Ditsworth, D. W. Schults, and J. O. Lamberson. "Sediment Toxicity to a Marine Infaunal Amphipod: Cadmium and Its Interaction with Sewage Sludge," *Mar. Environ. Res.* 18:133-153 (1985).

90. Reynoldson, T. B. "Interactions Between Sediment Contaminants and Benthic Organisms," *Hydrobiologia* 149:53-66 (1987).

91. Kay, S. H. "Cadmium in Aquatic Food Webs," *Residue Rev.* 96:13-43.

92. Martin, P. A., D. C. Lasenby, and R. D. Evans. "Fate of Dietary Cadmium at Two Intake Levels in the Odonate Nymph, *Aeshna canadensis*," *Bull. Environ. Toxicol. Chem.* 44:54-58.

93. Krantzberg, G. and P. M. Stokes. "The Importance of Surface Adsorption and pH in Metal Accumulation by Chironomids," *Environ. Toxicol. Chem.* 7:653-670.

94. Van Hassel, J. H., J. J. Ney, and D. L. Garling. "Heavy Metals in a Stream Ecosystem at Sites Near Highways," *Trans. Am. Fish. Soc.* 109:636-643 (1980).

95. Williams, D. R. and J. P. Giesy, J. "Relative Importance of Food and Water Soruces to Cadmium Uptake by *Gambusia affinis* (Poeciliidae)," *Environ. Res.* 16:326-332 (1978).

96. Dallinger, R. and H. Kautzky. "The Importance of Contaminated Food Uptake for the Heavy Metals by Rainbow Trout *(Salmo gairdneri):* A Field Study," *Oecologia (Berl.)* 67:82-89 (1985).

97. Dallinger, R., F. Prosi, and H. Back. "Contaminated Food and Uptake of Heavy Metals by Fish: A Review and a Proposal for Further Research," *Oecologia (Berl.)* 73:91-98 (1987).

98. Douben, P. E. T. "Metabolic Rate and Uptake and Loss of Cadmium from Food by the Fish *Noemacheilus barbatulus* L. (Stone Loach)," *Environ. Pollut.* 59:177-202 (1989).

99. Harrison, S. E. and J. F. Klaverkamp. "Uptake, Elimination, and Tissue Distribution of Aqueous Cadmium by Rainbow Trout *(Salmo gairdneri* Richardson) and Lake Whitefish *(Coregonus clupeaformis* Mitchill)," *Environ. Toxicol. Chem.* 8:87-97 (1989).

100. Hatakeyama, S. and M. Yasuno. "Chronic Effects of Cd on the Reproduction of the Guppy *(Poecillia reticulata)* Through Cd-Accumulated Midge Larvae *(Chironomus yoshimatisui),*" *Ecotoxicol. Environ. Safety* 14:191-207 (1987).

101. Ney, J. J. and J. H. Van Hassel. "Sources of Variability in Accumulation of Heavy Metals by Fishes in a Roadside Stream," *Arch. Environ. Contam. Toxicol.* 12:701-706 (1983).

102. Jefree, R. A. and N. J. Williams. "Mining Pollution and the Diet of the Purple-Striped Gudgeon *Mogurnda mogurnda* Richardson (Eleotridae) in the The Finniss River, Northern Territory, Australia," *Ecol. Monogr.* 50:457-485 (1980).

103. Clements, W. H. and R. J. Livingston. "Overlap and Pollution-Induced Variability in Feeding Habits of Filefish (Pisces: Monacanthidae) from Apalachee Bay, Florida," *Copeia* 1983:331-338 (1983).

104. Livingston, R. J. "Trophic Responses of Fishes to Habitat Variability in Coastal Seagrass Systems," *Ecology* 65:1258-1275 (1984).

Index